An Ecology of Communication

An Ecology of Communication

Response and Responsibility in an Age of Ecocrisis

William Homestead

LEXINGTON BOOKS
Lanham • Boulder • New York • London

Published by Lexington Books
An imprint of The Rowman & Littlefield Publishing Group, Inc.
4501 Forbes Boulevard, Suite 200, Lanham, Maryland 20706
www.rowman.com

6 Tinworth Street, London SE11 5AL, United Kingdom

British Library Cataloguing in Publication Information Available

Library of Congress Cataloging-in-Publication Data Available

ISBN 978-1-7936-1814-6 (cloth)
ISBN 978-1-7936-1815-3 (electronic)

Dedication

This book is dedicated to my children, Solan, Liam, and Ursula, and all the children who will inherit the aftermath of fitting and unfitting responses. Also, to the memory of Rufus, my family's devoted and beloved companion for twelve years, and all nonhuman species who will also inherit the consequences of human responses in an age of ecocrisis.

Contents

Acknowledgments

Many thanks are due to Diane Gruber and Maura MacNeil, who provided feedback on the whole manuscript, as well as much support and encouragement. I most certainly needed all three. Thanks are also due to Meghan Houlihan and Rachael Groner, who provided feedback on early parts of the manuscript, helping me to get on the right track in regard to content and style. Thanks also goes to Steve Schwarze, who commented on an early version of this project at the University of Montana back in 2005.

I would like to further thank the good work done by the leadership and members of the International Environmental Communication Association (IECA). I was fortunate to be able to present an essay version of this work at the 2019 IECA conference at the University of British Columbia. And finally, thank you to Nicolette Amstutz, my editor at Lexington Books, for her continual support.

Introduction

Ecocrisis as a Crisis of Communication

Anyone paying attention, even if lightly, knows the litany: climate crisis, species extirpation and the loss of biodiversity, overconsumption and overpopulation, rainforest destruction, ocean acidification, failing fisheries, agribusiness abuse of land, animals, and people, polluted water and air, and on and on. It would be nice if there were someone to blame, someone to whom we could not-so-politely request to just *cut it out*! Instead we have techno-industrial capitalist structures; anthropocentric attitudes; entrenched patriarchal practices of dominance and control; systemic racism, with historical links between disposable people and a disposable planet; and most prominently, the myths of unlimited growth (bigger), technology as savior (better), and linear progress (more of the same), myths that support these structures, attitudes, and practices, myths that we live, habitually and daily, at work, school, church, and within the government. Thus, the twentieth and twenty-first centuries are clearly marked by ecological crisis, as well as a crisis in narrative. Or, as the eco-theologian Thomas Berry wrote: "We are in trouble just now because we don't have a good story."[1]

Henry David Thoreau was prescient, recognizing in the mid-nineteenth century that our allegiance to growth, technology, and progress was leading us along a problematic path. His cabin at Walden Pond was less than two miles from town, and Concord itself was already a mix of farms and burgeoning industry, a mere sixteen miles from the expanding city of Boston. Two-thirds of New England, excluding Maine, was cleared of land by 1837. Industrialization was creeping forward, displacing nature's web with an interconnected web of railroads. Concord was a pleasant place to explore, filled with pastures, meadows, swamps, and woodlands of six to ten acres each, but Thoreau wrote in his journal that it was impossible to walk in the woods during the day without hearing the chopping of an axe. And he could see and hear trains from his cabin, providing a constant reminder of land transformed by human dictates.[2] In *Walden*, he wrote: "We do not ride on the railroad; it rides upon us."[3] His response in *Civil Disobedience*, directed toward the government of his time, but timeless in regard to injustice, was a bit of advice: "Let your life be a counter-friction to stop the machine."[4]

Other prescient voices followed. In the early 1900s, John Muir fought against the Hetch-Hetchy dam in Yosemite National Park. He lost, and we lost, and the wild salmon that cannot return to their spawning grounds due to numerous similar dams most certainly have lost. For Muir, damming Hetch-Hetchy was an act of sacrilege because it flooded grand gardens and mountain temples. The myths of growth, technology, and progress, when not deconstructed and redefined within an eco-spiritual context, are marked by such desecration. In 1949, Aldo Leopold responded to this arrogant disrespect by extolling humans to "think like a mountain." Such thinking demands a non-anthropocentric perspective, practicing ecological humility instead of dominance, leading to a land ethic devoted to stability, integrity, and beauty.[5]

Rachel Carson followed in 1962 with *Silent Spring*. Carson's scientific warning against the indiscriminate use of pesticides was met with catcalls from the chemical industry, including attacks that she was an overemotional woman. But Carson's thorough account of the dangers of pesticides—or biocides, as she called them, since they kill forms of life—won over the scientific community and much of the public. Her warning, however, extended beyond biocides to include a much greater danger: humans now possessed the power to alter the planet. In the late 1980s, Bill McKibben furthered this argument in *The End of Nature*, documenting the planet-altering problem of global warming and the end of nature as a force independent from human influence. In other words, warming signifies a further shift in our relations with the natural world: there is nothing humans cannot reach, effect, and destroy.

Thoreau, Muir, Leopold, Carson, and McKibben are heroic figures who may be used to track marker moments, but there is also a long history of working-class folks and people of color fighting for eco-social change and to protect their communities. The historian Chad Montrie calls their significant contribution the people's history of environmentalism, which includes farmers, mill workers, autoworkers, coal miners, and union leaders, among many others. Thoreau and his brother John, for example, bypassed Lowell, Massachusetts during their famous boat trip down the Concord and Merrimack rivers, taking a nature-lined canal to avoid an industrial landscape filled with mills and dams, which blocked fish-runs, flooded fields, and polluted waterways. Fisherman and farmers protested: some taking crowbars to the dams, some pushing petitions and filing lawsuits. Inside the factories, the Lowell "mill girls" fought against prison-like working conditions, lamenting separation from and seeking solace within the balm of nature. The good fight, as always, comes from the grassroots.[6]

And there is also a long history of data-crunching scientists who have tried to wake us up. The fundamental understanding of carbon dioxide as a greenhouse gas dates back to 1824 musings of French mathematician Joseph Fourier and Irish scientist John Tyndall's 1850s experiments on the properties of gases. In 1895, the Swedish scientist Svante Arrhenius studied the effect that doubling CO_2 would have on global climate. In 1958, atmospheric CO_2 was first measured at the Mauna Loa Observatory in Hawaii by chemist Charles David Keeling, leading to the "Keeling curve" mapping the yearly increase in CO_2.[7] In Jule Charney's 1979 U.S. National Academy of Sciences report, observatory data showed that if CO_2 levels were to double, global warming would likely amount to 3° Celsius, matching current estimates.[8] And in 1985, atmospheric CO_2 was at 346.12 parts per million, getting closer to 350 ppm, the threshold that climatologists argue it would be wise to not cross if we hope to have a habitable planet.[9]

In 1992, the Union of Concerned Scientists, including over 1,600 of the world's scientists and the majority of living Nobel Laureates, wrote a "Warning to Humanity." The first line reads: "Human beings and the natural world are on a collision course." The warning goes on to document environmental stresses on the atmosphere, water resources, oceans, soil, forests, and living species. Their main message? Limits. The earth is finite, and if we continued on our current course, it cannot indefinitely absorb our waste and pollution and provide our energy and food. They further warned that uncertainty over the extent of the crisis must not delay a precautionary response.[10]

And yet the myths of growth, technology, and progress marched on.

The 2005 United Nation's Millennium Ecosystem Assessment, a study by 1,360 researchers in 95 nations, documented that in the past 50 years rising populations have polluted or overexploited two-thirds of the ecological systems upon which life depends. As a result, there is a rising risk of abrupt environmental collapses that could spur disease and dead zones.[11] The 2019 IPBES Global Assessment Report on Biodiversity and Ecosystem Services, building on the landmark 2005 report and the most comprehensive ever completed, assesses changes over the past five decades and where we are heading, including the extinction of a million species.[12] Also in 2019, over 11,000 scientists from 153 countries signed the "World Scientists' Warning of a Climate Emergency," stating that scientists have a moral obligation to "tell it like it is."[13] Not surprisingly, both reports find the global response woefully insufficient.

And as of July 2020, data from the Mauna Loa Observatory indicates that atmospheric CO_2 is at 416.39 ppm, increasing every year and far above the 350 ppm level that renowned climatologist James Hansen, former head of

NASA's Goddard Institute for Space Studies, argues is optimal for preserving planetary conditions within which civilization developed and biological life adapted, and this measure of our folly does not include other greenhouse gases like methane.[14] The earth, of course, will survive, as it has five earlier periods of species extinction due to asteroid hits and volcanic eruption, but the sixth "great" period of species extirpation, a marker of the Anthropocene, may well include us.

It is difficult to listen to warnings. No one likes to hear bad news, and some early books on ecocrisis, like Paul Ehrlich's 1968 *The Population Bomb*, made predictions for the future that did not happen, such as the claim that widespread famine and social upheaval would occur by the end of the 1970s despite increased food production. The dates came and went, without the predicted eco-social tragedies, and so has some respect for warnings. However, famine did become a major worldwide problem, with millions starving, and population tripled to nearly seven billion by the end of the twentieth century and is expected to grow to nearly ten billion by 2050.[15] Calling this unprecedented increase a "bomb" seems an apt metaphor. Ehrlich successfully alerted the public to dangerous trends—overpopulation and famine—but failed to respect uncertainty and complexity in his attempts to make precise predictions.

We don't need to be prescient to know that dominant institutions beholden to dominant myths are conducting a grand ecological experiment. The ecosystem is too complex for us to know exactly what will happen, but we know the litany of eco-social ills that have already happened, and we know the numerous signs of what is happening and will likely happen. A constant stream of scientific reports, and reports on the reports, warn of potential collapse multiplying as emissions rise, yet denial and inaction, or insufficient action, continue to rule. McKibben's 1989 warning was countered by ExxonMobil despite their own scientists telling top management that warming was real as early at 1978, and criticized by Rush Limbaugh, as well as the many Limbaughs of the world, for not having all the facts.[16] But, in fact, since the publication of *The End of Nature* the scientific consensus on anthropogenic warming has grown to be overwhelming.[17] Taking a precautionary, and rational, stance demands that we act based upon a willingness to listen and respond to warnings; but if the science seems too abstract, too distant, we may also turn to our own experiences, stories, and senses.

When I was a child growing up in Staten Island, New York, my mother warned me, over and over, never to go near a winding brook behind our home. I do not recall specifics—I was too young to understand the realities of sewage and pollution—I only knew that it was bad, off-limits, not to be messed with under any circumstances. Of course I disobeyed. I found myself in it one day, wading in the gurgling water in white sneakers. Everything else near our home was innocuous—the overarching oak trees in the backyard, our neighbor's pool where kids swam in the summertime, the red dirt and shale behind the garage where we dug a giant hole within which to nestle and hide—and it didn't make sense for the brook to be outside this orbit of innocence. But innocence was wiped away when I got home with sopping shoes. I was in trouble with Mom, but that was nothing compared to when she took off my sneakers. My feet were covered with tiny white worms, no doubt necessary to a sewage ecosystem, yet causing me to cry, wheeze, and hyperventilate as my mother took me to the bathroom and picked them off one by one. I had explored the inviting burbling brook, but felt invaded within my safe world.

When I was seven years old, the oak trees surrounding the neighborhood, some as high as sixty to seventy feet, were plowed down by bulldozers so that developers could squeeze duplex homes and condominiums onto 40 x 100 feet lots. My dad began the story with him sipping coffee on the back porch and hearing a rumbling noise. He watched, heartbroken, as a machine army razed the land, including his favorite tree, not an oak but a seventy foot tulip tree that had a circumference of ten feet at the base and was the straightest tree he had ever seen, not a crook or bend. But there was nothing he could do. It seems that more housing was needed for the growing population of another island, Manhattan. Staten Island was a "logical" choice for this development, since it's only a half-hour ferry commute to the big city and a much cheaper place to live. Along with commuters, Manhattan also sent us another product intrinsic to its daily operations: tons and tons of trash, every day. My parents surveyed the worsening situation and promptly moved our family to a small property in New Jersey, which, we were happy to declare, contained twenty-six trees.

Our new neighborhood had woods and small farms nearby, as well as a lake across the street from our house. It was a beautiful place to grow up, even though we weren't allowed to swim in the lake, because it was filled with brown, polluted water. But that didn't dissuade my friends and me from spending summers searching under lily pads for bullfrogs and painted turtles. And winters were paradise—the polluted lake froze, leaving a glistening surface for ice-skating. When it snowed, our neighbors quickly grabbed shovels and created an interconnected web of ice rinks. Some were used for intense hockey games, others for general skating. Post-holiday time, we would collect discarded Christmas trees and build bonfires on the thick, frozen lake. The fire thrashed upward into the starry night in bursts of flame, crackling and popping like tiny firecrackers; the scent of pine filled the air, invigorating nostrils and permeating every inch of our clothes. I would stay out until my scrunched, frozen toes became too painful to ignore, forcing me to head home for the comfort of warm blankets.

Our family lived in New Jersey for over two decades, but after thirty-five years of work—dad as factory worker and mom as school secretary—they retired and moved to a small lake-house in the Adirondack Mountain region of upstate New York. My sister bought the Jersey home, and we sometimes gather there to celebrate holidays. Things have changed, however. The small farms are gone, turned into mega-supermarkets, shopping malls, and deceptively named housing developments like "Willow Estates." If you can avoid it, it's best not to go anywhere in your car between four and six o'clock—traffic jams, commuter hell. Still, relatively speaking, it remains a pleasant place to grow up. Some woods are standing, and the lake, while remaining polluted, still has turtles and frogs. But something has happened to the winters. When I was a teenager, we always had at least a month of solid, skate-worthy ice. My sister's kids were lucky if they got one or two days.

My story is one of countless accounts of cherished childhood places that have been degraded or destroyed—the "railroad" riding over us—and with each removal of a cherished place, we become another step removed from natural processes and ecstasies and another step removed from ourselves. Our physical health is hitched to the health of ecological systems, but as many ecopsychologists have shown, our psychological well-being depends on experiences of wonder, connection, and beauty. Thus, while not wanting to hear bad news is understandable, the refusal to respond is a sign of immaturity, of stunted individual and collective development, played out in a failure to listen to the scientific consensus on warming, the wisdom of mothers, and our own experience. Today, I am more mature and know better than to wade in a polluted brook, but I also know that it is wisest to clean up the brook by going to the roots of eco-social problems.

Gregory Bateson, writing in 1972, simply stated: "The processes of ecology are not mocked."[18] In other words, what we do to global and local ecosystems is eventually communicated back to us, and in the age of climate disruption we probably won't like what we hear. And yet, despite hearing the warnings and experiencing the desecration and loss of beloved places, we still don't seem to be getting the message. To Berry, we have become emotionally and spiritually withdrawn, unable or unwilling to participate in the "great conversation," with the natural world no longer flowing into our physical and psychological being.[19] This argument raises serious concerns, especially when we consider Bateson's claim that the biosphere is best understood as a communicative system of information exchange and aesthetic responsiveness.

We face an ecological and communicative dilemma: the universe, earth, and socio-cultural lifeworld are resoundingly dialogic—everything exists in reciprocal relationships of nonverbal presence—and yet we have created modern and postmodern cultures that are largely governed by monologue.

This is not to suggest that instrumental, or monologic, attitudes and forms of knowledge have no function—the earth supplies resources—but it does mean that it is unresponsive and irresponsible to forget that nature is the source and sustenance of our very being. Thoreau wrote: "The highest we can attain to is not Knowledge, but Sympathy with Intelligence."[20] Such sympathy demands that our penchant for monologue be contextualized within a dialogic attitude, or better yet, dialogic wisdom, recognizing the primacy of relationship and reciprocity within a more-than-human world, and thus a wider range of possible responses in particular situations. But the myths of bigger, better, and more march on, governed by a rigid duality between mind/culture and nature that is inculcated in institutions and then hammered into us. It is one thing to perceive nature as resources, quite another to be expected to find our place within an economic system that only perceives nature as resources.

Our potential for dialogue is often socially and structurally denied, and unconsciously and habitually repressed, and no amount of greenwashing or supposedly green capitalism is going to change that. But that makes deep listening a radical act: the more we practice it, the more we will fight for structural change. And deep structural change will more fully allow us to practice deep listening. The upshot: our illusory separation from and subsequent domination of the natural world, as well as our potential to live more harmonious and enlivening myths, are best understood via communicative metaphors, insights, and experiences.

We remain immature because we have been socialized out of our communicative souls. Nature was an animating and articulating presence for earlier storytelling cultures; it is mute in modern industrial times. At the beginning of the twenty-first century, we desperately need to live different, non-dogmatic stories that vivify our links with the natural world, redefining our notions of progress and growth and the role of technology. And we need creative storytellers and thoughtful listeners who are rationally and radically critical and yet open to other-than-human voice, utterance, and presence. We need to pay attention, deeply, responding to what we learn with responsible action.

The ecological crisis, broadly construed, is a crisis of communication.

To consider ecocrisis as a crisis of communication forces us to confront assumptions about the nature and meaning of communication. An ecology of communication redefines competence by exploring a range of responses, including rational, spiritual, mythic-animistic, and aesthetic forms of communication. The term "ecology," of course, refers to scientific study of the interactions among living beings and their environment, especially as a whole system. I retain this sense of the term, while incorporating differing communicative experiences among these interactions, including bodily/sensuous, psychological, and spiritual connections with the earth and nonhuman beings.

I argue for a redefinition of communication that places a premium on dialogic process expressed as openness to the possibility of being addressed by the other-than-human world. An ecology of communication, then, is characterized by listening and responsiveness within a larger system of reciprocally interacting human and nonhuman others. Human and nonhuman others respond to each other differently based on their unique qualities, but a relational sending and receiving is experienced, whether through nonverbal presence, exchange of energy, or the oxygen, carbon-dioxide cycle. However, I make this argument while doing my best to be critical and self-reflexive concerning my own assumptions, and also respectful of the criticisms of others.

In *Speaking into the Air: A History of the Idea of Communication*, John Durham Peters states that communication is the project of reconciling self and other, or "the natural history of our talkative species," not the perfect dialogic meeting of differing minds.[21] Peters calls the desire for instantaneous mind connection the "dream of communication," arguing that this "dream" is deeply problematic because it leaves us continually lamenting our ineptitude or our existential condition as miscommunicators rather than recognizing otherness and celebrating our potential for shared decision-making and community.

Peters blames the "dream" on a vision of communication begun with Socrates in the *Phaedrus*, which critiques the new media form of writing, privileging instead direct contact and a coupling of desires described as eros. Eros is the longing to recover a lost wholeness or unity, and for Peters this longing has influenced, or infected, our ideas of communication ever since. Communication becomes the soul-to-soul sharing of presence, consciousness, and love; anything less leads us to be pathologically prone to lying and deceit. For Peters, privileging eros errs because it neglects logos, or human reliance on "word, argument, discourse, speech, story, book, and reason."[22] It also errs because it suggests a definition of communication as telepathy—a definition that is tyrannical in its impossible expectations, forcing us to pendulum-swing between fusion and failure. Peters writes: "In renouncing the dream of communication I am not saying that the desire to connect is bad; rather, I mean that the dream itself inhibits the hard work of connection."[23]

Peters' focus on rational discourse, and the existential difficulties that accompany communicative praxis, is necessary and compelling. However, in "Dialogical Relations with Nature," Scott Friskics argues for a more encompassing view of logos. Logos bespeaks "the way of things," and the Word made flesh as primal speech, song, and voice "in and through the beings and things of creation." Everything speaks via their presence, and thus everything participates in logos. Friskics defends his position via Martin Buber's I-Thou ontology, in which others are not merely objective "its" but sacred subjects, yet his views are ultimately rooted in his experience. He begins his essay with a description of his rustic cabin home near the east slope of the Montana Rockies, and then proceeds to describe daily walks to Crown Mountain through forest, hillside, and grassy knoll, which he does for typical reasons—to stretch his legs, get some sun, check the weather—but also because he feels called by the mountain. Crown Mountain "speaks" to Friskics via its "silent, hulking presence," and he feels compelled to respond.[24]

So, we are left with a dilemma: Is communication best understood as a melding of minds and presence, like Spock's silent communing with alien life forms in *Star Trek* or does it reflect Spock's logical side, characterized by word and rationality? Or perhaps we must be men and women of the future before we can combine some of his finer qualities. Spock is more than a playful metaphor; he bespeaks the popularity of this dilemma and the ways it is handed down to us in popular culture. Peters quotes Emerson to argue against telepathic mind melding: "I hold our actual knowledge very cheap. Hear the rats in the wall, see the lizard on the fence, the fungus underfoot, the lichen of the log. What do I know sympathetically, morally, of either of these worlds of life?"[25] In other words, there seems to be a non-communicable gulf between humans and other beings, not to mention between humans. But Emerson also proposed the experiential existence of a unifying Oversoul or larger Mind within which we, and everything, move, breathe, and have our being, and, by logical extension, communicate. Thus, Peters' appeal to Emerson continues the dilemma of defining communication.

Peters' history of communication privileges the impossibility of dialogue, countering my argument that practicing an ecology of communication is both possible and essential. Peters is concerned with reaching too high, raising expectations beyond what is reasonable for everyday communicative praxis, arguing that "we need ways to respect the limits of each other's souls and the demands we can place on each other rather than yearning for bridges and lamenting walls."[26] For him, communicative ideals are dangerous, because they invite despair rather than the work of fruitful cooperation. Miscommunication is a blessing, reminding of the inevitable gaps between self and other and the difficulty of fully knowing ourselves and communicating interior thoughts and feelings to others.

The history of communication also shows that sharing our inner life is not necessarily a good thing, as it has resulted in much fundamentalist fervor and destruction. After all, despite *Star Trek*'s celebration of mind-melding and alien-language translating computers, it portrays a future of warring Romulans, Klingons, and humans who are unable to quell destructive thoughts and emotions. Today, among many physical and ideological conflicts, we have the seemingly insurmountable gulf between climate "deniers" and "alarmists," pandemic pro and anti-maskers, Black Lives Matter protestors and "all lives matter" status quo-ers, or the culture wars generally, which infects all our relations but especially online discourse. For Peters, we remain alien to each other, and this hard look at our existential situation provides us with considerably more wisdom for decision-making and community-building than does a mistaken ideal of dialogic wisdom.

But Peters' emphasis on difference, miscommunication, and alien-otherness also invites despair, most typically expressed as alienation in a world governed by the monologic separation of mind from nature and the myths of growth, technology, and progress. His stance dismisses too quickly the fact that the dialogic interplay between unity and diversity is existentially and experientially real, leading to compassionate identities, identifications, and responses. Friskics, for example, does not expect or ask for perfection, but immersion and attentiveness, making possible the sense of being called. Following Buber, he highlights reciprocity and the claim that living means being addressed.

Peters emphasizes difference and disconnection, yet admits that communication is the ultimate border-crossing concept. Key questions for communication theory are "philosophical, moral, and political," exploring "how wide and deep our empathy for otherness can reach."[27] He also states: "Communication has become the very field upon which to sort out the place of the human in the great network of being."[28] This is weighty stuff, and I could not agree more, as it suggests a weakening of long-standing metaphysical walls between humans and the so-called brute beasts.

Aristotle characterized humans as the "speaking animal," and rigid hierarchical divisions between species have been with us ever since. He is right, of course; human difference is marked by language use and rationality, as well as tool-making and technology. And this difference in complexity makes a difference in regard to human roles and responsibilities within the context of evolution. But if unfolding human identities were constructed from understanding ourselves as a "communicating animal"—a designation that includes speech and rationality but also nonverbal experience within the larger context of communication systems—then our lofty, overly special placement among fellow beings would be brought down to earth. We would become a species governed by unique communicative abilities that may disconnect us from our earthly abode and other-than-human beings but which also provide opportunities for connections—connections that are necessary if we are to thoughtfully respond to ecocrisis. There are always calls from others, whether human, animal, or plant, and thus we are continually confronted with the ethical question of how we should respond. Or rather, we have "response-ability," but our monologic muting of the natural world has caused us to miss out on way too much.

The emerging project of ecopsychology explores this widened response-ability by using communicative metaphors to describe the crisis and cure. In

The Voice of the Earth, Theodore Roszak posits an ecological unconscious, described as Jung's collective unconscious grounded in evolutionary process. Mind evolved from, and is embedded within, nature, and thus the unconscious is not only a storehouse of cultural symbols but connects us to the history of matter. The "ecological intelligence of our species," or the voice of the earth, exists within us, most often blocked yet accessible through deep listening. Such listening awakens healthy selves within healthy environs, while making us more aware of destruction accrued from working our will against the planet, projecting unconscious neuroses that result in toxicity, depleted resources, and species extirpation. The good and the ill speak to us, "if we would hear, of our deep self."[29]

Ecopsychologists are particularly fond of pre-Neolithic to present-day indigenous cultures, arguing that such cultures experience permeable boundaries between mind and nature, resulting in a high degree of ecological wisdom. The work of Paul Shepard is often highlighted, especially his seminal *Nature and Madness*, in which he claims that the long history of hunter-gatherers is the ideal from which we have strayed but can never actually leave because it is so much a part of us. Thus, many ecopsychologists argue that our individual and collective sanity depends upon what I call mythic-animistic communication, where we recover the permeable boundaries and perceptual insights of oral cultures. In other words, the active, interconnecting energy of eros has a history prior to Socrates, maintained by myths that emphasize reciprocity and listening via the songs and signs of animate others.

This communicative openness sounds ideal, but lurking thoughtfully and persistently are Peters' warnings against falling prey to the dream of communication and the loss of logos. Do not such idyllic visions of dialogue merely set us up for failure, or worse, the illusion of success? Some New Age adherents quickly come to mind, with claims of channeling dolphins and dead yogis, communing with crystals, and becoming urban shamans. New Age is a broad category, encompassing those who respect diversity and have a genuine desire for spiritual connection, as well as those who celebrate spiritual oneness over diversity, getting lost in what the spiritual philosopher Ken Wilber once called an amorphous "Divine Goo," making boundaries look pretty good. But what of the permeable boundaries celebrated by ecopsychologists? Are hunter-gatherer cultures really all they're cracked up to be or did the evolution of rationality demand we break free from the dialogic reciprocity displayed by pre-Neolithic peoples? In other words, do the mythic-animistic communicative experiences of hunter-gatherers reflect superstitious beliefs and projections or do they offer potential insights into a widened and more complete response to ecocrisis?

Such inquiry leads to broader philosophical questions. Where should we draw lines of difference between self and other? How subtly connected are the interdependent communicative systems within which we dwell? What is logos, and does our understanding of rationality, mind, and identity need to be redefined via the lens of communication and the reality of our experience? What are the relations between narrative and lived ethics, and how can we place mythos, eros, and logos in dialogue? And, going further, what role does a transcendent dimension, or theos, play in our spiritual, and communicative, lives?

These questions, as they pertain to response and responsibility in an age of ecocrisis, drive this project. More specifically, this project will consider these questions by analyzing the role that communicative metaphors, insights, and experiences play in various discourses, including ecopsychological and New Age, assessing how they promote, or restrict, what the philosopher Calvin O. Schrag calls the "fitting response." These discourses often act as counter-narratives that respond to and offer alternatives to the modern myths of unlimited growth, technology as savior, and linear progress. As such, they merit attention, both for practical reasons—are they effective forms of resistance or do they merely reproduce the status quo?—and for theoretical reasons, encouraging us to reconsider our assumptions about the nature and meaning of communication.

An ecology of communication also participates in and is informed by ongoing scholarly discussion within the field of environmental communication. Donal Carbaugh, for example, studied the Blackfeet practice of "just listening," in which tribal members willfully, and humbly, become co-present and co-participant with the spirits that inhabit places, conversing with ancestors and nonhumans, keeping traditions alive, expanding self and awareness, and learning what they need to learn.[30] Scholarship like Carbaugh's is open to Native expressive systems and is critical of the academic habit of focusing on rhetoric at the expense of the actual world, as the environment is doubly quoted, and both "the word and the world speak."[31]

Emily Plec's anthology, *Perspectives on Human-Animal Communication: Internatural Communication*, provides an overview of rhetoricians who have explored differing expressive systems, with essays on the possibility of crossing the human-animal divide. The subtitle, "Internatural Communication," is instructive, going beyond interspecies communication by emphasizing that all species are nature, and humans are also animals. And, interestingly, in John Durham Peters' more recent 2015 book, *Marvelous Clouds: Toward a Philosophy of Elemental Media*, he redefines media as both natural and cultural; in other words, there are messages everywhere.

Communicative metaphors, insights, and experiences have been present in environmental literature for decades, although largely unnoticed. I will continue to reference them in this project, but it is impossible not to notice these days, with more and more books published, some bestsellers, which focus on listening and voice within larger communicative systems. To name a few: Peter Wohlleben's 2015 *The Hidden Life of Trees: What They Feel, and How They Communicate*, Robin Wall Kimmerer's 2015 *Braiding Sweetgrass: Indigenous Wisdom, Scientific Knowledge and the Teachings of Plants*, and David George Haskell's 2017 *The Song of Trees: Stories from Nature's Great Connectors*. All these books integrate scientific knowledge with other modes of knowing. Kimmerer, for example, who is a scholar, botanist, and Native American, utilizes the grammar of animism and aesthetics, rather than only I-It scientific language, to explore listening beyond the human.

Such deep listening calls forth a range of ethical responses and responsibilities, but despite Peters' turn to a widened understanding of media, his earlier cautionary insights disclose potential pitfalls. For example, while ecopsychology seems to elucidate well our dialogic reciprocity with and responsiveness to the natural world by gathering insights from pre-Neolithic and indigenous cultures that linear notions of progress deny, some New Age texts that focus on oneness may promote a communicative perspective that ignores diversity, shuns rationality, and dismisses collective direct action in favor of the power of individual intention and creating our own reality. Thus, a thorough and critical exploration of what constitutes an ecology of communication must be articulated.

In chapter 1, "The Fitting Response: Calvin O. Schrag and Rational Communication," I begin in Maine, telling my own hiking story juxtaposed with Thoreau's three hiking trips in the Maine woods. My story and insights from Thoreau's trips, especially his third one which was guided by Joe Polis, a Penobscot leader, sets the stage for the communicative theory that is to come, grounding an ecology of communication in experience. I then proceed with the theory, which helps us make sense of experience so that we may more fully understand it, embrace it, and learn from it, as well as have more of it. The theory begins with the practice of transversal rationality, which cuts across claims for modernist absolute Truth and postmodern cultural relativism, exploring the "in-between" where we may participate with or distance ourselves from various traditions. More specifically, Schrag's postmetaphysical philosophy recovers logos and the possibility of enacting fitting responses by embracing agape love and the gift event, whether in the Maine woods or in our daily cultural decision-making or in agriculture and eating, which I use as a frequent exemplar to discern what is unfit. I also turn to Ramsey Eric Ramsey, who adds to Schrag's communicative insights by more fully considering sociocultural blocks that inhibit fitting responses, while arguing for a fitting responsiveness characterized by a wider awareness of structural constraints and our capacity to act. Ramsey also writes of the necessity of eros, or the desire for connection, as a prerequisite for responsiveness, allowing an integration of logos and eros and a "coming through" of what is most fitting in particular situations.

In chapter 2, "Integral Meta-Theory: Ken Wilber and Spiritual Communication," I turn to Wilber's developmental holarchy, or various levels of growth that we transcend and include, the higher enfolding and transforming the lower, with higher levels reflecting a wider and deeper awareness. His model also explores differing lines of development that proceed through the differing levels—we may have a high cognitive line but low spiritual intelligence, to give one example—and various spiritual "state" experiences, or what I am calling spiritual communication. Wilber's work does double duty, helping me to explicate state spiritual communication as well as the dialogic interplay among differing modes of communication at differing levels. Wilber, while positing a transcendent dimension and unfolding development as inherent qualities of the Kosmos, claims that his meta-theory is postmetaphysical because it provides partial truths, or perspectives, contingent on intersubjective agreement that are woven into a larger, evolving yet integral perspective. But while Wilber and Schrag both champion the necessity of logos in postmodern times, there is a tension between their perspectives, with Wilber exploring transrational experiences or spiritual communication, and Schrag recovering religious feeling in relation to communicative ethics while attempting to avoid anything that seems to ring the chimes of metaphysics. I unpack this tension, which is key to a wider awareness and thus a fitting responsiveness, using American transcendentalism as an exemplar.

In chapter 3, "To Learn but Not Return: Paul Shepard and Mythic-Animistic Communication," I explore Shepard's traversing of our pre-Neolithic past, asking what we may learn from a communicative lifeworld filled with respect for other-than-human voices. Like Wilber, Shepard focuses on human development, and between the two of them we are given plenty of insights into why humans too often remain immature. However, another tension arises, with Wilber relegating mythic-animism to lower developmental levels and Shepard championing hunter-gatherer attentiveness to place and nature's diverse voices in relation to sane growth. I also unpack this key tension, exploring mythic-animism expressed by ecopsyschology, including David Abram's arguments for pre-verbal bodily based converse, and the ecophilosophy of deep ecology along with the "Council of All Beings" ritual, to gather possible fitting responses for today's world while being mindful of the unfit.

In chapter 4, "The Pattern that Connects: Gregory Bateson and Aesthetic Communication," I explore Bateson's core argument that ecological crisis reflects a crisis of perception. We are living a mistaken epistemology, attending to isolated parts of our world rather than realizing we are immersed in an evolving communicative system of information exchange, which he called Mind or the pattern that connects. For Bateson, responding to this larger pattern is an aesthetic challenge, in which the apprehension of beauty guides communicative skill. Bateson championed imagination and rational rigor, integrating mythic-animistic sensibility and systems science expressed as aesthetic responsiveness. Such steps to an ecology of mind allowed him to approach a transrational spiritual awareness of immanent "sacred unity" without affirming a transcendent dimension. I turn to Robin Wall Kimmerer's work as an exemplar here; ironically, her more poetic language reflecting a grammar of animism, integrated with science, allows her to better capture his eco-aesthetic vision.

Schrag, Wilber, Shepard, and Bateson provide a rich set of conceptual resources that lay the groundwork for a rational yet eco-spiritual vision of communication, although the terms rational communication, spiritual communication, mythic-animistic communication, and aesthetic communication can be a bit clunky, especially when repeated throughout this project. However, they are also evocative and necessary for demarcating differing and yet interwoven communicative experiences, which should not be construed as predetermined or rigid criteria but the flowering of communicative praxis in time and place. More specifically, Schrag's theoretical arguments for transversal rationality expressed as rational communication, Wilber's experiential inquiry into a transcendent dimension expressed as spiritual communication, Shepard's psycho-historical interpretations of hunter-gatherer life expressed as mythic-animistic communication, and Bateson's ecological/systems science vision expressed as aesthetic communication must inform each other if we are to enact a fitting responsiveness to ecocrisis. The dialogic interplay among these forms of communication also alerts us to the unfit.

In chapter 5, "Discerning the Unfit: New Ageism to Ascension," I introduce New Ageism from a communicative perspective and then critically analyze this nebulous spiritual movement, which has morphed into Ascension literature and practice with visions of a "new earth" consciousness. Spiritual communication, which may be difficult for some readers to embrace as a form of communication, perhaps because they conflate it with New Ageism, is the focus of this chapter, along with Rupert Sheldrake's work on morphogenetic fields and the possibilities of telepathy, which, of course, makes Peters cringe, as well as messages received via other psi phenomenon. The wheat must be separated from the chaff, especially since New Age and Ascension thought and practice assure that a higher awareness and better world are coming. Evaluating spiritual experiences is not easy, and I certainly don't claim to be the final arbiter; nonetheless, this chapter makes the attempt to discern the fitting from the unfitting in relation to our response to ecocrisis.

I do the same for interspecies communication in chapter 6, "Discerning the Unfit: Interspecies Communication," which has grown to include inter-natural and inter-spiritual communication, in which we recognize that all species are nature or express Spirit, and then attempt to communicate based on that understanding. The ability to communicate across species boundaries, and when to cross boundaries and when it is wise not to, is the focus of this chapter. Similar to my analysis of New Ageism and spiritual communication, it is not always easy to demarcate the wheat from the chaff, but I once again make the attempt by exploring several discourses, including J. Allen Boone's *Kinship with All Life* and Michael Roads' *Talking with Nature*, which seem to be infected by New Age and create your own reality solipsism, and those I consider more fitting, and thus less prone to self-deception, Monica Gagliano's *Thus Spoke the Plant* and Stephanie Kaza's *Conversations with Trees*. All these texts move us more toward dialogic relations within a more-than-human world and away from incessant monologue within a solely human one; however, given that interspecies discourses are also often framed as responses to ecocrisis, they merit attention in relation to the four modes of communication and what may be fitting and unfitting.

In chapter 7, "The Call to Responsibility: Thoreau and the Voice(s) of Nature," I turn to the seminal work of Thoreau, and Thoreau himself, as exemplars of the communicative vision and practice argued for in previous chapters. More specifically, I explore the integration of transcendent and immanent dimensions by analyzing Thoreau's experiences of spiritual communication—or the language that all things speak—along with rational, mythic-aesthetic, and aesthetic communication expressed as a grounded transcendentalism. Thoreau practiced differing modes of communication and thus provided many insights into how we may more fully embody response-ability, including in activism, whether against slavery or eco-destruction. Thoreau also provides an opportunity to explore the ways humans can and cannot speak for nature, as well as other likely criticisms of an ecology of communication.

And finally, in the epilogue, "A Fitting Responsiveness: Communicating Our Way into the Future," I return to rational communication debates, focusing on misinformed criticisms of Rachel Carson, as well as misinformation generally in a post-truth age. The systemic distortion of the communicative system explored in earlier chapters is re-emphasized, such that we may better respond to ecocrisis but also a host of other related crises, including a pandemic, systemic racism and sexism, financial instability, and a crisis of leadership, all interwoven with the myths of bigger, better, and more. I also review communicative responses to related eco-social crises—Extinction Rebellion, BLM and environmental justice, Greta Thunberg, the Sunshine movement, biomimicry, ecological design, degrowth and voluntary simplicity, Deep Adaptation, Project Drawdown, new digital "smart" technologies that collect data on nonhumans, etc.—exploring commonalities and differences and the fitting and unfit. And, I once again extend the idea of the fitting response, arguing that an ecology of communication characterized by the dialogic interplay among rational, spiritual, mythic-animistic, and aesthetic communication is best understood as a fitting responsiveness. Such responsiveness focuses less on particular responses and more on the communicative awareness we bring to particular situations and the many ways we are called to act.

Throughout this project, I also turn to the wisdom of my own experience. My inquiry into ecocrisis as a crisis of communication, and then an ecology of communication as a fitting responsiveness, began with communicative experiences within urban, rural, suburban, and wild environments, along with plenty of struggle. As a young Communication Studies undergrad at Rutgers University in the mid-1980s, I did not fit well within dominant narratives, including religious ones, which I considered irrational and foolish. A month prior to graduation, I took a practice commute to New York City and then walked to Madison Avenue among the throng of suits, thinking, just maybe, I would get a job in advertising, but when I reached the massive, angular buildings, I could not bring myself to go inside any doors, having an anxiety attack on the sidewalk instead, breathing heavily, barely able to move. I was not new to the city, going often on weekends for the vibrancy of art shows, independent and old movies, and diverse food, but this was different, a message from body or unconscious or I didn't know what. I did not have wisdom or words.

After graduation, I eurailed in Europe for two months, senses receptive and soul free, with every day different from the next, and then returned penniless and in need of work. My dad called, saying they needed temporary help at the factory, and thus began an hour-long commute past the stench of garbage dumps on my way to a deadening assembly-line. Commuting with my father was an education—I now knew why he came home so tired—but despite lessons, the line was denuding the energy of Europe. During a fifteen-minute mid-morning coffee break, I sat under one of the few trees on the plant property reading an article in the *New York Times* about ozone holes and the clear-cutting of forests. I dropped the paper in disgust, providing a clear view of pollution spewing from my building and my participation in what felt like a cycle of death and self-destruction. The vibrant, palpable presence of life experienced while traveling was gone. All I knew then was that nature, in the form of a nearby park adjacent to campus, was a balm, my only refuge. What I know now is that my college education, while waking me up to world filled with eco-social problems, was thoroughly anthropocentric, and thus did not supply fitting words or wisdom in response.

Today, I am a father of three children and teacher of young adults, which, like being a student, are deeply communicative and challenging. Thus, while struggle never ends—like Peters, I deeply honor the existential, the bumbling and stumbling, the absurd, especially the human folly that creates eco-social problems—I try to put some degree of wisdom into words, for the young will inherit the consequences of our responses to the climate data and warnings but also to the loss of beloved places and denuded landscapes and seascapes, which goes beyond data to the heart of what it means to be whole, communicating human beings.

Reading environmental literature for thirty years also woke me up to the extent of our predicament, and in the process, I noticed all those communicative metaphors, insights, and experiences. For Emerson, reading was a spiritual practice that inspired self-culture, and Thoreau wrote a section of *Walden* on reading, arguing that it was a noble exercise, elevating mind and spirit. Reading, then, was a journey in itself. And so, in honoring my experiences, I hope to call forth similar experiences in others or support and further call them forth.

The good news is that something new, yet with ties to the deep past, is emerging in the Anthropocene, which Carbaugh, in *Voice and Environmental Communication*, another recent anthology that widens our view of communication, calls a listening-based model of communication.[32] I like to think that an ecology of communication is that model. However, from my own experience, I have discovered that differing forms of communication are not always easily separated; in particular, the mythic-animistic may include elements of the aesthetic, which may border on the spiritual.

This difficulty to demarcate may not matter—because what matters is integrating rationality with various expressions of deep listening elicited by mythic-animistic, aesthetic, and spiritual communication—but it may matter, if, as Wilber warns, the pre-rational (go back to nature) and rational (systems science) is mistaken for the transrational. The result, for Wilber, is unfit responses, including mythic-animistic hunter-gatherers romantically celebrated as spiritually advanced and web-of-life systems science taken as a new eco-spiritual paradigm, leading to a flattened, and thus flawed, model of spiritual development. To Wilber, the most extreme negative consequence is doom and gloom environmentalism that tends toward ecofascism. In other words, if the ecological golden age was in the past, the only solution to ecocrisis is to de-industrialize the world back to beginnings of human emergence. There is much to transversally investigate here, and an ecology of communication, when viewed within the context of Wilber's holarchic model of development, allows us to fruitfully do just that.

Treating the ecological crisis as a crisis of communication elicits possibilities and pitfalls. Peters makes thoughtful arguments against the too-easy, telepathic dream of communication, and I will listen and learn from his insights while also embracing Buber's claim that living means being addressed. An ecology of communication is marked by this commitment, respecting difference and difficulty *and* the depth of our relationships experienced via mythic-animistic, aesthetic, and spiritual communication, along with rational communication that differentiates and honors pre-rational and transrational forms of experience.

An ecology of communication, then, informs all experience; or rather, in keeping with the postmetaphysical critique of monologic absolutes explored via Schrag, communication is always present but unpredictable, providing underlying influence without rigid foundation. The communicative experiences critically explored in this project include living within a dialogic cosmos, dialogue with nature and nonhumans, and the dialogic interplay among differing modes of knowing and communication. As such, this project also explores the process of addressing and being addressed by others, and the possibility of speaking as, speaking for, and being spoken through by others.

An ecology of communication navigates the fruitful tensions between mind and nature, unity and diversity, self and other, as well as several other dualities. Dualisms are an aspect of abstract thought and can be valuable thinking tools; or rather, they're valuable when they are dialogically related rather than monologically opposed. Communicative praxis must not be reduced to either pole or totalistic synthesis, but should reflect the creative art of keeping the conversation going. This would be good, very good, for the well-being of the earth, nonhuman species, and ourselves as communicating animals. The litany of ecological ills with which I began this introduction, while reflecting damage that often cannot be undone, does not mean that our responses are set in stone. Despair, while a reasonable response to damage done and damage yet to be done, dualistically opposed to seeking communicative ideals beyond our reach, need not be our default positions. There is lots of room in-between for new narratives, new insights, and new actions.

In sum, an ecology of communication moves us beyond the anthropocentric bias of communication study and moves us toward a listening-based model of communication; moves long in coming, late in arrival, and essential to discerning fitting responses and the call to responsibility in an age of ecocrisis.

NOTES

1. Thomas Berry, *The Dream of the Earth* (California: Sierra Club Books, 1988), 123.
2. Robert D. Richardson, Jr., *Henry Thoreau: A Life of the Mind* (Berkeley, CA: University of California Press, 1986), 139.
3. Henry David Thoreau, *The Portable Thoreau*, edited by Carl Bode (New York: Penguin Books, 1975), 345.
4. Ibid., 120.

5. See Leopold's seminal *Sand County Almanac*, particularly the sections "Thinking Like a Mountain" and "The Land Ethic."

6. Chad Montrie, *The People's History of Environmentalism in the United States* (London and New York: Continuum International Publishing Group, 2011), 13–16.

7. Climatechange12.com, *Disruption*, September 7, 2014, YouTube video, 52:27, https://www.youtube.com/watch?v=uWPj6CxtsGo.

8. Chris Mooney, "30 Years Ago Scientists Warned Congress on Global Warming. What They Said Sounds Eerily Familiar," *Washington Post*, July 11, 2016, https://www.washingtonpost.com/news/energy-environment/wp/2016/06/11/30-years-ago-scientists-warned-congress-on-global-warming-what-they-said-sounds-eerily-familiar/.

9. NOAA, August 2020, https://www.esrl.noaa.gov/gmd/ccgg/trends/data.html.

10. Union of Concerned Scientists, August 2020, https://www.ucsusa.org/resources/1992-world-scientists-warning-humanity.

11. Millennium Assessment Report, https://www.millenniumassessment.org/documents/document.356.aspx.pdf.

12. IPBES, https://ipbes.net/global-assessment.

13. Oxford Academic, November 5, 2019, https://academic.oup.com/bioscience/article/70/1/8/5610806.

14. NOAA, August 2020, https://www.esrl.noaa.gov/gmd/ccgg/trends/data.html. James Hansen's remarks on CO_2 ppm and optimal conditions have become well-known thanks to 350.org.

15. Paul R. Ehrlich and Anne H. Ehrlich, *Betrayal of Science and Reason: How Anti-Environmental Rhetoric Threatens Our Future* (Washington, DC: Island Press, 1996). As of 2019, the UN estimates that population will grow to 9.7 billion by 2050, although they also report that the growth rate is slowing, https://www.un.org/development/desa/en/news/population/world-population-prospects-2019.html.

16. Naomi Oreskes and Geofrey Supran, "What Exxon Mobil Didn't Say About Climate Change," *The New York Times*, August 22, 2017, https://www.nytimes.com/2017/08/22/opinion/exxon-climate-change-.html.

17. NASA, https://climate.nasa.gov/scientific-consensus/.

18. Gregory Bateson, *Steps to an Ecology of Mind* (New York: Ballantine, 1972), 504.

19. Berry, *The Dream of the Earth*.

20. Henry David Thoreau, "Walking," in *Nature, Walking*, edited by John Elder (Boston: Beacon Press, 1991), 113.

21. John Durham Peters, *Speaking into the Air: A History of the Idea of Communication* (Chicago: The University of Chicago Press, 1999), 9.

22. Peters, *Speaking into the Air*, 9.

23. Ibid., 30.

24. Scott Friskics, "Dialogic Relations with Nature," *Environmental Ethics*, 23, Issue 4 (Winter 2001), 392–393.

25. Peters, *Speaking into the Air*, 245.

26. Ibid., 66.

27. Ibid., 230.

28. Ibid., 229.

29. Theodore Roszak, *The Voice of the Earth* (New York: Simon & Schuster, 1992), 5.

30. Donal Carbaugh, "Just Listen: 'Listening' and Landscape Among the Blackfeet," *Cultures in Conversation* (New York: Routledge, 2005).

31. Donal Carbaugh, "Quoting 'The Environment': Touchstones on Earth," *Environmental Communication: A Journal of Nature and Culture, 1(1 (2007))*, 64-73.

32. Donal Carbaugh, "Response Essay: Environmental Voices Including Dialogue with Nature, Within and Beyond Language," *Voice and Environmental Communication,* edited by Jennifer Peeples and Stephen Depoe (England: Palgrave MacMillan, 2014), 249.

Chapter 1

The Fitting Response

Calvin O. Schrag and Rational Communication

INTERLUDE: A RETURN TO EXPERIENCE

In August of 1988, I hiked 350 miles of the Appalachian Trail in New Hampshire and Maine, spending two weeks preparing by taking stamina-increasing swims in a cool, clear lake cradled by pine-filled mountains near my parents' small retirement house in the Adirondacks of upstate New York. After each swim, I waded by the shore and waited for Bluegills to investigate my leg hairs, which swayed like tentacles in the softly rippling water. Then, to complete my training, I daydreamed by the water's edge in a nest of brown pine needles, enveloped within their sweet, earthy scent while the sun reflected off the surface of the lake, shooting beads of light up the base of giant pines.

I also built a rock wall. My parents had put in a septic system—the old one was built for seasonal use and they feared it might be polluting the groundwater and lake—and so the yard was pretty torn up. I took newly exposed rocks and put them to good use in the wall; the click of rock against rock signaling compatibility. If the sound was too harsh, without music, I knew that I was too present, too much in charge. I eventually got into a groove, working attentively and quickly, allowing the different shapes to fall into a jigsaw-style mosaic.

I left my parents' place on a bright, sunny morning, greeted three hours later by rain streaking the windshield like tears on glass in Gorham, New Hampshire. After a stormy night in my tent, I arrived at a trailhead. Backpack attached, I splashed through a puddle. Cold, stinging droplets hit my face.

"It's begun," I said to myself.

The trail was flat but soon inclined with large boulders blocking the path. I climbed and jumped with 50 pounds of gear and Kraft macaroni and cheese jangling on my back, leaving behind dual footprints on the muddy trail. Soaked shorts and t-shirt plastered my skin, but carrying extra weight kept me warm. Adrenaline spiked, and I was hyper-aware of the rhythm of my steps and sensuous new surroundings.

Wet, cool air.

The scent of pine: pungent.

The presence of growth and decay.

The cycles of abundant life and death that civilized culture obsessively attempts to hide.

Ralph Waldo Emerson, in *Nature*, calls for us to listen to Nature, or Spirit; the natural world; and our inner nature.[1] All are teachers; all interwoven, elevating and yet grounding human awareness. The trail beckoned: there were lessons afoot.

After three days of mostly leaving human relationships behind, I made my way into Maine and one of the most mountainous sections of more than 2,100 miles of trail. I hiked one peak after the next, covering little total ground. Although only twenty-eight years old, my knees ached from carrying a weighty pack up and then especially down numerous slopes. It could have been worse. I met a young hiker in those first few days who lay in a trail shelter, huddled in a ball like an old tabby cat, barely able to move a paw.

But the effort it took to climb mountains provided gifts. I focused physical and mental energy on each step. When I reached a summit, the faraway landscape rushed my senses, sling-shotting awareness outward into vast terrain of trees and open sky, the seemingly separate becoming one seamless life force. After two weeks of alternating between intense focus and expansive vistas, my perception became clear, acute, receptive.

At one point, I was entranced by a small piece of moss along the trail. Sunlight danced upon it, illuminating a bright, jewel-like verdant green. I kneeled, as if receiving a message from the Divine itself—wake up, all is holy—and then turned my head in all directions, hoping to share my discovery. I was the only human present but did not feel alone; everything was alive with presence. I remained on my knees with a single thought: "My God, I'm losing it over a piece of moss."

The next day, I sauntered along the trail with head down, making sure not to trip over roots and rocks. I looked up, and there was an immense creature with full antlers standing ten feet ahead. I had never seen a moose before, and was dumbstruck by its long tree-like legs and huge head and body. And those eyes. I never imagined such large and captivating eyes. We stared like lovers and I was entranced again; the danger of the situation never entered my awareness, and I sensed that it never entered the moose's awareness that I was dangerous. I'm not sure how long we stood there—time was not a construct of timecards or workweeks—but I eventually moved off to the side and proceeded on my way, smiling, and then contemplating my encounter with a wild, other-than-human life form with whom I am somehow kin.

After thirty days, I reached the summit of Mount Kahtadin in Baxter State Park, gazed skyward at blackbirds that looked like spirited silhouettes circling in the wind, rested on solid, sun-flooded boulders, and then descended to a parking lot, car, and friends to drive me home, and yet away from home.

Many have gone on lengthier hikes, ascending death-defying peaks in the far wilder American west and numerous beautiful countries. Still, my month-long hike, with four town stops, signaled an escape from the "civilized" world. Or did it? A lesson from the trip was that there was no escaping human dictates, human impositions. The mountain vistas occasionally included clear cuts off in the distance, and during one town-stop, I saw a weather report that included an acid rain index.

I also carried $1,000 worth of camping technology on my back and brought along a Walkman and a few tapes for occasional musical interludes. Shutting out the sounds and silence of nature is un-wild-like behavior, but like my moss and moose experiences, dancing on the top of mountains while listening to the bright sounds of Pat Metheny also induced peak moments of freedom and joy.

Thoreau made three excursions into the Maine woods, without toys like Walkmans, without Gore-Tex boots, Helly-Tech rain jackets, and who-knows-what-tech thermal socks and underwear, and with much more wildness.

On his first trip, he climbed Mount Kahtadin with Maine relatives and a white guide, clouds enveloping the summit. Only a few non-native parties had made the excursion prior to his own, and the rugged terrain altered his vision. Nature in Concord was far more pastoral, with Thoreau writing in his journal about the constant sound of the axe. He initially encountered more of the same in Maine—loggers, mills, camps, trails, some houses, farms, and dams—but signs of human hands eventually gave way to untrammeled wilderness of uncut forests, lakes without cabins, and undammed streams in the northern portion of the state.[2] And then, the rocky Mount Kahtadin; his understanding of nature as harmonious and benevolent was challenged by awe-filled primitive power.[3] His famous response was a euphoric howl: "What is this Titan that has possession of me? Talk of mysteries!—Think of our life in nature,—daily to be shown matter, to come into contact with it—rocks, trees, wind on our cheeks! The solid earth! The actual world! The common sense! Contact! Contact! Who are we? Where are we?"[4]

More lessons were afoot during a second trip with his relative and Maine connection, George Thatcher, this time to hunt moose and led by Joe Aitteon, son of the Penobscot governor. Thoreau, without a gun, was hunting experience and knowledge, especially Indian ways. Aitteon was able to identify sounds like treefall, walk noiselessly and paddle powerfully, and he taught Indian names. When Aitteon shot a moose mom, Thoreau was much less enthused, although impressed by his tracking and movements. But then the butchery. He lamented his participation, especially since they did not fully use the animal; for Thoreau, it was too much like sport hunting, too much like "shooting your neighbor's horses," which were ultimately "God's own horses." They hauled out as much meat as possible, and the hide, on Thoreau's insistence; a lot of weight for a birch bark canoe.[5]

They ended the day at an Indian camp. Thoreau was within their world more fully, not just studying from afar, not just books and arrowheads, witnessing them speak to each other, yet unintelligible, a "distinct and comparatively aboriginal race" speaking a language "spoken in New England who shall say how long." Thoreau considered himself a man of the woods, and rightfully so, but these were the true people of nature, causing him to question the limits of his knowledge.[6]

On his third trip, he once again sought out a Penobscot guide; to his good fortune, he found Joe Polis, who would also expose him to more logged landscape leading to untrammeled wilderness, contacting bedrock truth upon which all things are measured, as well as a more intense course in Native ways.

It was 1857, the hiking party including Edward Hoar from Concord, a hearty friend and Thoreau admirer—who hiked with raw, wet feet after slogging through swamps and got lost, wondering how long he could survive on berries—and Polis, a Penobscot tribal leader familiar with New York City and Philadelphia who represented his people before the governments of Maine and the United States, yet deeply at home in the woods. The Penobscot had lost most of their lands but retained Indian Island, an ancestral home, where they fought to sustain themselves via trade, especially crafts like basket-weaving and canoe-making, market-hunting, and farming, although many had gone, some due to an outbreak of small pox.[7]

Polis was honored by his tribe as a meteoulin, a teacher and shaman, and guiding whites was an opportunity to share the Penobscot way of life; early in the trip, with Thoreau and Polis still feeling each other out, finding equal footing, they agreed to tell each other what they knew. Thoreau, despite unreceptivity to creation stories told with "dumb wonder which he hopes will be contagious," thinking them superstition, was eager to learn.[8] He would have learned more if he had done what he often did, see the mythic-animistic resonance, the poetry and wonder, elicited by oral traditions. But Thoreau was a child of the Enlightenment, which likely included conscious or unconscious bias of superiority, at least at the start of the trip.[9]

Thoreau had plenty to teach Polis from his substantial reading, including botany, and Hoar tried to explain the laws of whites in regard to land, but Polis, who, Thoreau noticed, took a newspaper, had spent his life reading nature.[10]

Thoreau was impressed by Polis' multiple skills, from carrying a 100-pound canoe on his head and only taking a blanket, axe, gun, and the clothes on his back, to starting a fire with wet wood by gathering dry bark from underneath dead, leaning trees, making a different tea from available herbs for each day of the week, and never getting lost. He learned, and learned of his ignorance.[11]

In response to Thoreau and Hoar's chattering questions, Polis responded, "May be your way of talking—may be all right—no Indian way," and he was sometimes coy.[12] When Thoreau asked about not getting lost, Polis was blunt: "Oh I can't tell you—Great difference between me & white man."[13] But his bluntness was not meant to disparage, it was just difficult to explain the many ways he perceived direction, reading trees and rocks. Thoreau saw a "sharpened and educated sense," as Polis did not use memory, recalling routes, but refined instinct and perception in the moment. He did not need the white man's knowing, and thus did not seek it.[14]

More lessons ensued: how to carry a canoe—the trip included many portages, although sometimes Thoreau and Hoar hiked while Polis navigated river rapids—how to call to animals, including snakes and muskrats, how to tell white spruce from black spruce, and how to dig up spruce roots and turn them into tough, flexible string for sewing canoes.[15] Still more: Thoreau woke up at night to find the unburned end of burned wood glowing brilliant white, as well as part of a decayed stump five feet from the fire. Artoosoqu', remarked Polis, the Penobscot had seen it high in trees, moving, making a noise. Thoreau did not need a scientific explanation, making an "empty chamber" of an "inhabited house" of "spirits as good as himself." Science could be left for another time, while he became receptive to wonder for its own sake, ready for more direct experience, for more moving of the gift: "Nature must have made a thousand revelations to them which she still keeps secret to us."[16]

Polis named stars for Thoreau, led them through rough, muddy terrain, which he navigated with ease, going back to get their gear, and was confused by their inability to track. But then he killed a moose, big-eyed and majestic to be sure, and a spirit as good as himself, starting a debate. For Polis, hunting moose was part of tradition, and he needed money to buy back land and feed his family. For Thoreau, however, the latter was the "white man's argument," as if his family had no other options. Thoreau, as always and for good reason, focused on eternal ethical truths, higher laws, but did not seem to recognize cultural privilege.[17]

More lessons: the medicinal properties of plants, how to paddle a canoe properly, how to write on birch bark with black spruce twig, and how to make a candle and pipe out of birch bark.[18]

They wanted to climb Katahdin—Kataadn, for Natives, meaning highest land—but Hoar's feet wouldn't allow it; Polis was saddened, he would have brought home more moose meat if he knew they were going home.[19] Near the end of their journey, Polis playfully challenged Thoreau to a race, Thoreau with their gear and Polis with the canoe. The trip covered 325 miles over nearly two weeks; Thoreau asked Polis if he was glad to be home, and he responded: "It makes no difference to me where I am."[20]

Polis also shared stories of his own life and the tribe's history, and his belief in education as the means for protecting Penobscot sovereignty and property; he claimed his son was the best student at a white school in Old Town. He also told Thoreau a story of visiting the politician and lawyer Daniel Webster, the perpetrator of Manifest Destiny, at his home near Boston; but he was not open to conversation and seemed to threaten him by raising his hand as if to strike.[21]

Thoreau, back in Concord, tried to make sense of Polis: he navigated the white world, calling himself a Protestant, going to church and honoring the Sabbath while living in a spacious house and reading the newspaper, and yet led his tribe, tanned animal hides in the front yard, and was completely at home in the woods, displaying "so much intelligence that the white man does not," increasing his "own capacity, as well as faith."[22]

Polis was not perfect: Thoreau also catalogued mistakes, like leaving matches and his shoes out in the rain, his failure to discern the outlet for a lake—Thoreau was right on that one—and wrote that he was like an excited fifteen-year-old, rather than a seasoned hunter, when initially shooting at and missing the moose.[23] But when Thoreau found an Indian fishing basket filled with fish after the trip, his deep appreciation was stimulated again, stating its maker was "meditating a small poem," aware that for Indians words were objects, and objects like words; everything spoke, and language was a force that influenced the world and thus must be used with care.[24]

From then on, Thoreau defended Indians against prejudice, not just from book learning or talking to them around Concord, but from personal and profound experience with Polis. And he defended their right to land, writing of the need for "natural preserves" in which "the bear and the panther, and even some of the hunter race, may still exist," and not be "civilized off of the face of the earth."[25]

But Maine was changing, increasingly civilized: eleven years earlier, in 1846, only a house or two existed where now stood villages, saw mills, and a store, and they encountered logging and dams that greatly altered the ecosystem.[26]

I did not experience the wilderness and wildness Thoreau encountered. The marked and well-worn trail up Katahdin had iron steps to help hikers maneuver over the rockiest terrain, and during a quick town-stop, I ran into a general store, bought a pint of Ben and Jerry's Coffee Heath Bar Crunch ice cream, and then hiked while shoveling the sugary mixture with a plastic spoon. The trail had its dangers—several memorial signs were posted honoring dead hikers, and I fell several times, due to a shifting pack—but it was marked, and a typical day of survival included a myriad of early evening camping chores: find water, filter water, refill water bottles, set up clothes line, hang up wet socks, blow up Thermarest, set up tent, set up Whisper-Lite stove and cook meal on bare rock surrounded by dense forest, the fake-orange macaroni and cheese tasting like it cost way more than two for a dollar. But when the sun set—after finishing clean-up chores: wash pot with liquid castile soap, unhook fuel bottle, refold stove and tin fire guard—I imagine we shared similar experience, gazing up toward the star-filled night into infinite space, listening, far enough away from home to discover a deeper sense of home.

My hike was compromised by culture and technology, but we were both bit by "no-see-ums" and awoke to the thrilling voices of loons. And I had transformative experiences of expansiveness, timelessness, deep connections, and radical otherness, the currents of universal being circulating through me, finding, like Emerson, a faith that all is well, or that all could be well.

Early in the hike, I received a trail name from fellow hikers, inspired by leaving behind dual footprints: Kangaroo, or Kanga for short, signaling a shift in identity. Late in the hike, peace increasing with each step, I walked early in the morning with a startling but natural clarity—no boundaries between self and other, no filter—and soon after napped on bare rock near a small waterfall, awaking to deep contentment.

I met other transformed souls. I will never forget a conversation with a fifty-year-old vice president of a bank who had been out in the wilderness for over four months (with periodic town stops). He didn't look like a banker to me—I had trouble envisioning a suit covering his grime and grin—so I asked him what he was going to do once the trip was over. With seriousness and without hesitation he responded that he was going to start getting rid of stuff because "you just don't need it." Months of carrying everything he needed on his back cured him of his addiction to overconsumption.

But not all meetings with fellow hikers were so profound.

I stopped hiking just shy of the Mahoosic Notch, a half-mile rocky gorge that many consider the most treacherous section of trail.

"Five o'clock. Looks like I'm going to have the trail shelter to myself," I thought. But then an Outward Bound instructor arrived ahead of a group of twelve urban-dwelling teenagers.

"I need adult conversation," she said, sitting down next to me on the wooden floor of the shelter. One of the kids had stolen their group lunch, and they had to have an hour-long sit-down to talk about trust and the reasons why they were here.

Before we could chat further, two army-fatigued, non-English-speaking Italians strolled into camp and attempted to ask for directions. A dazed, sixty-or-so-year-old gentleman hobbled in from the Notch, proclaiming that he was the oldest of a church group of six, "Two ministers and the rest regular folk." An American soldier entered, informing us that he was on joint maneuvers with the Italians—a kind of military exchange program—and quickly led them off amid smiling goodbyes. And another Outward Bound group appeared filled with teenagers out in the woods for the first time.

The hobbled gentleman struggled to prime a camp stove.

"Do you want some help?" I asked.

"I know how to operate it," he replied, still out of breath, "but I could use a match."

He placed the stove inside the shelter, reeking of fuel, and took the match.

"Could you light the stove on the ground, as far away from me as possible?" I chuckled, imagining the shelter ablaze.

He dutifully placed it on the ground and lit the base. Two-foot flames shot straight up, knocking him on his ass.

While I suppressed another laugh, the entire troop of Italian and American soldiers marched through camp with plugged rifles. Then the new Outward Bound gang complained about the outhouse and an extremely large, long-legged Jackrabbit hopped up to the base of camp, causing a minister to scare it away with a camera attached to a foot-long zoom lens.

"Have I been transported inside the pages of freakish fairy tale?" I wondered.

But the older gentleman brought me back to reality, cooking food with bluish-white flames shooting up the sides of the pot and circling over the lid. The Outward Bound instructor got her burn-kit ready just in case.

Who are we? Where are we?

Thoreau knew that he was not braving deep wilderness at his Walden cabin, but conducting an experiment in simplicity and self-reliance, confronting vital facts of life. He wrote about spiritual transformation, aided by the grace of the natural world, because he knew that the sanity of the burgeoning industrialized world would depend upon it. His experience of wildness near the top of Mount Kahtadin did not undermine his Walden experience, it added a realization: there are limits to man's control. We are part and parcel of nature, but not its lord and master. He now placed the nourishing and connecting aspects of nature within the context of humility. Nature's power must be respected.

But what of the parade of humans who graced the Mahoosic Notch campsite? Did they also deserve respect? What of new-found faith that all is well?

The psychological experience of wildness—expressed as the embrace of life and death and humility before nature's power—would go a long way toward preserving the world; and thus how should we respond when wilderness has been increasingly tamed, and we bring our cultural lives to the woods, reducing potential depth of experience?

A good place to start is realizing that we live within communicative systems prior to speech, to attempts at rational utterance, and that an embrace of a wilder and much wider range of communicative experience provides insight and guidance.

My Appalachian Trail experiences support this affirmation:

Receptivity was honed by discovering the sensuous allure of nature during my pre-hike swims in the Adirondacks of upstate New York, communing with fish and pines.

My parents expressed rational foresight and the wisdom of reciprocity by attending to the needs of the lake, and therefore their own needs, adding a new septic and preventing the water from being polluted.

Building the rock wall took considerable effort, but it was also accomplished via attentive listening and harmonious moments of effortless effort.

Hiking in the rain made me alive, alert, and better able to read surroundings, senses woven with contemplation revealing cycles of life and death and the illusory separation of nature and culture.

Receiving a trail name reflected a shift in identity to a more relational self, revealing the power of language to connect and disconnect.

The myriad set-up and clean-up chores, not to mention aches and pains, grounded the hike in the existential and the body, providing counterpoint to feeling wonder while gazing at the stars, relating to a dialogic cosmos.

Attentiveness to roots and rocks and placing each step with care, lest I fall, provided information, as did falling from shifting weight and wet terrain; lessons were learned from trial, error, and practice.

Signs of human hands, of human imposition, were everywhere, starting with the blazed trail itself but also in gear, music, ice cream, clear cuts, and acid rain indexes, as were signs of human striving, and danger, in the form of memorials.

I conversed with a bright and brilliant jewel-like piece of moss, experiencing beauty and disclosing the divine—messages of sacredness, as I interpreted them—and communed with a big-eyed moose, discovering awe and wonder and radical otherness. Both experiences arrived as grace, as gifts or revelations rewarded to open minds and hearts, which move when embodied and communicated.

A more relational self led to moments where a separate self-sense dissipated when my awareness expanded along with mountain vistas, but also during an early morning with no thoughts clouding my walk and after napping by waterfall, awakening to deep peace and interbeing, each experience a receiving from becoming bodily, aesthetically, and spiritually attuned.

The receiving of spiritual insight led to a message of faith and hope that, in some sense, all is well and thus all might be well, despite human foolishness.

Adult conversation with the banker disclosed the power of rational reflection supported by felt experience, while learning of the teenage sit-down on trust and witnessing the follies of the strange grouping at the campsite—elderly church-goers, Outward Bounders, military trainees—provided support for Peters' concern regarding the dream of communication and expectations of easy connection, as well as a reminder that faith and hope are works in progress.

And the hike led me to question my habitual life, deepening a call to study environmental philosophy and ethics, which later led to a call to teach and the moving of the gift.

Thoreau's hikes in Maine also support the affirmative embrace of wider communicative experience and thus the arguments of this project, especially in regard to Native knowledge and the limits of his own knowledge.

Joe Polis read signs of nature, called to animals, and practiced wilderness skills born of experience within the land. Thoreau listened and learned, although stopped short of embracing mythic-animistic creation stories, failing to see poetry and potential lessons embedded within oral traditions. Thoreau also disagreed with Polis on hunting, displaying cultural difference and different perspective, which supports Peters' arguments that communication and consensus are difficult and ongoing. Still, Thoreau, who championed "the language that all things speak" in *Walden* and attended to nature's voices, their oratorios, on his walks, learned that nature speaks in more ways than imagined, not to mention Native language itself, which embraced relationship to the land and the power of naming. And upon returning and beholding a Native basket, he saw how artifice communicates culture and values and knowledge. Such experiences led him to be the first to advocate for nature preserves, so that wilderness and wildness may teach those with the ability to listen.

Despite vastly different times and experiences, I walked changed yet similar ground as Thoreau, and perhaps uncovered similar insights: Unity and diversity intimately related within subjective experience of an intersubjective, more-than-human world. This world includes human-Divine relations—Nature always speaks of Spirit, Emerson wrote—human-nature relations, including the insight that we are nature and nature is always already communicating; human-human relations, filled with folly and possibility; and relations mediated by artifice, by tools and technology, for good and ill.

TRANSVERSAL RATIONALITY: THE PRACTICE OF LOGOS

My Appalachian Trail hike was invigorated by transrational experiences, or what I am calling spiritual communication, and further confirmed that awareness and attention determine the degree to which we are responsive to others, including nonhuman others. This caused me to question the role of rationality in blocking, eliciting, and making sense of transrational experience and this awareness and attention. Ken Wilber, Paul Shepard, and Gregory Bateson will provide an eco-spiritual context for rethinking and redefining communication via spiritual, mythic-animistic, and aesthetic experiences that support a more responsive awareness and attention, but to fully support this rethinking and redefining a precise articulation of the role of rationality is needed, as well as a theoretical focus and rich concepts. I have found all of these in Calvin O. Schrag's practice of transversal rationality in search of fitting responses.

Significant insights are often learned in good conversations with good friends, and that is how I discovered Schrag's writings. After my hike, which occurred after factory work, print shop work, study with a spiritual teacher, a first try at graduate school at the University of Maine, and more struggle and searching for vocation, I moved to Indiana, home of Purdue University, where a Rutgers friend was pursuing a Ph.D. and taking classes with Schrag. Before reading him myself, I absorbed plenty from the art of talk in cafes, like the salons of Europe, and before the distraction of screens. Emerson wrote that friendship is marked by two qualities, truth and tenderness, and Thoreau argued that friends should be complements, not doubles without difference.[27] These qualities defined our friendship, but they are also a good start toward rational communication.

While the appeal of Schrag's work began in friendship and converse, what hooked me was that phrase: the fitting response. When you are searching, such a phrase resonates, as it does when you read the World Scientists' Warning to Humanity, like I did in 1992 while living in Indiana, and wonder why everyone isn't talking about it and acting in response. During that same year, the first major climate conference, the Earth Summit in Rio de Janeiro, ended with no significant action, no significant response. Then-President George H. W. Bush had remarked that he would combat the greenhouse effect with the White House effect, and then changed his tune, stating the American way of life was not up for negotiation.[28] This was a crucial time for a bold response, as climate science had not yet been politicized, and not yet been spun by merchants of doubt who profit from no action. Taking responsible, and thus rational, steps then would have meant a different climate world now.

Schrag, then, focuses on a key question—what is fitting?—but I also start with his work because transversal rationality describes and authorizes my method of inquiry. Transversal rationality, of course, is not the only theory of rationality in relation to communication, but it is the most fitting for this project and for living thoughtfully and well in postmodern times. An ecology of communication needs transversal rationality, or the critique, articulation, and disclosure of the possibility of fitting responses, to avoid participating in "the dream of communication." In fact, without the kinds of parameters Schrag sets out, spiritual, mythic-animistic, and aesthetic communication are prone to misinterpretation and may supply unfit responses to ecocrisis. Contrarily, I go further than Schrag by utilizing transversal rationality to traverse boundaries and explore communicative experiences that are often ignored because they fall between rigid dichotomies, including environmentally damaging dualities like mind and nature.

In *The Resources of Rationality: A Response to the Postmodern Challenge*, Schrag writes compellingly on the modernist misreading of logos expressed as logocentric metanarratives of rational progress and truth, which ultimately leads to rational man instrumentally aligned against nature. This conception of logos is different from the ancient Greek view, which bound the rational soul to the cosmos; rationality, while a mark of human distinctiveness, is not merely a human product that separates and divides but a property of being itself that connects and enlivens. Thus, the "human mind is rational insofar as it participates in the rational structure of the cosmos."[29]

According to Schrag, the Greek concept of logos encompassed multiple senses and usages, including "utterance, conversation, oration, rendering an account through story, measure, proportion, general principle or rule, argument, and definition" but was replaced by technical rationality, shifting "logos to logic." Such logic is characterized by calculation, predication, and control, displaying a universalizing tendency expressed as scientism. In other words, modernity's differentiation of the culture-spheres of science, morality, and art, each developing unhindered by the others—which brought much genuine progress, unshackling the Galileos of the world from fundamentalist religion—also eventually led to the "uncontested sovereignty" of science.[30]

Schrag's articulation of premodern logos participating with the rational order of the cosmos reminds of Thoreau's call for sympathy with intelligence, as well as his assertion that the true philosopher should "so love wisdom as to live according to its dictates," solving "some of the problems of life, not only theoretically, but practically."[31] Technical rationality, of course, has solved problems, yet without sympathy and love it has also created new ones.

Agribusiness, for example, is a technocrat's dream, abstracting us from the land, as well as other animals, human, and nonhuman, treating all life like "its"; the result has been plenty of inequitably distributed food and plenty of destruction, including poop lagoons that overflow during storms, polluting waterways; wasted energy and increased CO_2; worker injury and death due to pesticide exposure and dangerous machinery; and assembly-line animal abuse, among many other examples of techno-solutions leading to new eco-social problems. Techno-rationality is clearly not in sympathy with the Greek conception of logos, or the pattern that connects revealed by systems science and articulated by Bateson, and thus, when valorized as thinking as such, it is not a fitting response to ecocrisis.

Schrag's articulation of Greek thought is also consistent with Philip Hadot's arguments in *What Is Ancient Philosophy?*, in which he quotes Thoreau's views on the philosopher and then turns to logos to argue for "philosophy as a way of life." He writes: "Attentive people live in the constant presence of the universal Reason which is immanent in the cosmos. They see all things from the perspective of the Reason, and consent joyfully to its will."[32] Hadot further states that the ancient Greek conception of universal Reason, or Logos, expresses itself in nature as physics, in the human community as ethics, and in individual thought as logic.[33] The pre-Socratic philosopher Heraclitus, while emphasizing constant flux, also claimed that some type of intelligence connected and guided all things, and Plato extolled the intelligence of archetypes or transcendental first forms, while Aristotle maintained that moments of wisdom might be attained through contemplating "divine Intellect."[34] All these views extend our conception of mind, or "nous," suggesting that the intelligence of the cosmos is reflected in the human mind. All these views were lost in the modernist revision of logos to logic.

For Hadot, logos was practiced via spiritual exercises like contemplation, philosophical discourse, questioning, and dialogue and served inner transformation. Socrates, while extolling the interconnecting energy of eros, practiced these exercises to make himself and others as rational as possible, as well as respectful of not-knowing, leading us to better know how to live. Thus logos and eros, and the rational and spiritual, as defined by Hadot, dialogically inform each other, with philosophy as a way of life expressed as sympathy with inner and outer intelligence.[35]

Schrag writes that the initial postmodern reactions to premodern Greek logos and its modernist re-conception as technical rationality are incredulity and deconstruction.[36] Both are too totalizing, and thus too marginalizing of multiplicity, difference, and indeterminacy. The presumption that we can know ourselves, others, or anything with certainty via rational inquiry or communication is just that: a culturally constructed presumption, or dream, reflecting the power of dominant narratives. Such narratives, of course, include the myths of unlimited growth, technology as savior, and linear progress, which are rooted in modernist misreadings of rationality interwoven with the discourse of scientism. These myths not only perpetuate ecocrisis but have historically resulted in exclusionary clubs that literally club others into submission. Hunter-gatherer oral cultures, past and present, that do not fit within dominant narratives have been systematically devalued and often destroyed, something Joe Polis knew well, as the Native population in New England was reduced by 90 percent due to Manifest Destiny expansion and exposure to diseases like small pox, with displacement and attempts at Christian re-education for those that survived.[37]

Despite the merits of incredulity and deconstruction, Schrag argues that the tradition of logos remains operative, although in a different way, providing constructive possibilities for postmodern times.[38] In response to Greek logos and the postmodern turn, he offers his post-metaphysical philosophy, asserting that we exist within the "in-between," splitting the difference between "robust consensus and heterogeneous dis-sensus," or "vertically anchored universals and horizontally dispersed particulars," and it is this very condition that makes dialogue and situational understandings, however incomplete, both possible and desirable.[39] While there are no criteria in advance, no essentialized ethics, no monologic foundations for guiding praxis, by dwelling communicatively within the complexity of life and the limits of knowledge we find differing traditions, meanings, and mysteries that inform, but do not rigidly decide, our responses. Thus, postmodern times, and our eco-existential situation, demands that we practice transversal rationality or the ability to traverse the in-between, moving among various times and places, and thus contexts of meaning, privileging some situationally but none finally in our search for what is more fitting.

This gathering of insights from different times and places has three aspects: critique, articulation, and disclosure. Given that there are no rigid moral criteria known in advance of communicative praxis, critique demands that we both align with and distance ourselves from the commitments of our communities and traditions. Thus, critique is composed of two interrelated parts: *participation* and *distanciation*. We think with and against the traditions and meaning-formations within which we are born and are expected to find our place, including, of course, the dominance of techno-rationality, the narratives of bigger, better, and more, and an array of eco-social crises. Schrag acknowledges the inevitability of living within traditions, yet provides strategies for the continual reassessment of those traditions via a dialectics of participation and distanciation.[40]

Articulation and disclosure emerge from this dialectic. Articulation is the process of generating meanings that are *social* and *intersubjective*, and not the product of the internal workings of a modernist, Cartesian self. We don't know things and then share our knowledge with mutual understanding being the result—that's the dream—instead, there is always an *interpretive* component to the articulation of meanings. And, I was happy to discover, Schrag extends such articulations beyond rational discourse to "a time and space of action, mood, desire, bodily and institutional inscriptions—a vast arena of *nondiscursive* dispositions and practices that also exhibit an articulatory function."[41] For Schrag, then, logos is an "event of articulation" that is "as much at work in the texture of our emotions and desires as in the texts of our discourse" and that opens a door, which Schrag does not walk through but I will: logos that includes the practice of nonverbal converse within an other-than-human world, which, of course, discloses more perspectives, more interpretations, not rigid truth.[42]

And finally, disclosure is an expression of *discernment*, contrasted with certainty, which reveals insights via the openness required for critique and the articulation of interpretations. As such, it is critical of rigid traditions that foreclose possibilities for new meanings, actions, and responses. Thus, the "task of transversal rationality is that of scanning the terrain of our lifeworld involvements, enabling a critique, articulation, and disclosure of the manifold forms of life that these involvements produce."[43] In other words, we must traverse differing traditions, including premodern Greek thought and modern scientific investigation, discerning, as best we can, what responses are fitting and unfitting in particular situations. And we are better at it when we widen our awareness by practicing transversal rationality.

Schrag maintains that critique, articulation, and disclosure make rational action possible, but these actions no longer come from an easy appeal to a universal logos or moral code. The modernist self has no universals to provide certainty; instead, postmodern selves have transversal rationality as a means for exploring the possibility of expressing fitting responses. This lack of final criteria or moral injunction is not an absence, but a presence. What is fitting can never be absolutely determined, but what has proven to be unfit is more final—consider ecocrisis—and from this backdrop, we make determinations and consider a range of actions. The past provides lessons, or "repetitions of difference," of what is more or less fit, but these repeated lessons must be rationally revisited if we are to consider fitting responses in the present, as well as see our way forward into the future.[44]

For the deconstructive postmodernist, the decentering of logocentrism, or any form of centrism for that matter, is crucial for liberation. The only response fit for repeating is continual critique expressed as the decentering of meanings. But such decentered meanings often resemble tumbleweeds set adrift in a relativistic landscape in which no values, principles, or responsibilities inform our actions, leaving us bereft of responses and full of despair. Contrarily, modernist certainty, in the form of narratives that have created ecocrisis in the first place, will only dig deeper holes. One need only look at the 2015 "Ecomodernist Manifesto," written by a gaggle of techno-optimists—many respectable, yet the document has no sourcing to back its claims—to see that the logocentric metanarratives of unlimited growth, technology as savior, and linear progress are still dominant.[45]

Once again, Schrag responds with transversal rationality and communicative praxis, splitting the difference between modernist techno-rationality and postmodern continual critique by traversing them, along with premodern thought and practice, in order to gather situational insights. Thus, the deconstruction of universal logos does not demand a dismissal of logos, but its reconfiguration marked by the transversal play of rationality and communication that rejects both "hegemonic totalization" and "anarchic multiplicity." For Schrag, reason remains in force despite not providing metaphysical guarantees, leading him to state that "the universal logos of logocentrism is dead," but the "transversal logos of communicative rationality is alive and well."[46]

Schrag's lexicon of academic language—post-metaphysical, logocentric meta-narrative, transversal rationality, praxial critique, articulation and disclosure, repetitions of difference, and the fitting response—may seem much ado about something simple: we need rational action in response to eco-social problems. But good theory leads to more thoughtful, and therefore more ethical, actions. Also, it is worth noting that he wrote during the deconstructive postmodern heyday of the 1970s to 1990s, which left little in its wake. Deconstructive postmodernism was right, of course, and still is right—the grand metanarratives that govern our institutions and lives must be taken apart—but someone, especially in philosophy, needed to pick up the pieces and construct a way forward toward what is more fitting. I can recall too many café conversations with graduate students at Purdue in the 1990s, in which they argued for deconstruction and cultural relativism, while I affirmed that some positions are better. After all, we can't fight ecocrisis with a tumbleweed. We also can't fight post-truth disinformation, which is a hallmark of Trumpian authoritarian leadership, unless we move away from relativism and toward rational consideration of the fitting response.

Schrag's articulation of the in-between honors the postmodern deconstruction of claims for ultimate Truth—consider the immense damage done by fundamentalist fervor generally—and the insight that all responses are ultimately interpretations, yet he provides a reconfigured logos that helps us to discern better interpretations. However, it goes too far to say that he argues for sympathy with intelligence, despite his willingness to learn from ancient Greek logos; rather, transversal logos reflects an ongoing synthesizing intelligence characterized by adjustment and modification with no final synthesis. This is a step forward from incessantly ping-ponging between modernist absolutes and postmodern relativism. Yet, for Wilber, and for me, something is missing from such transversal play: the universal resonance of unfolding Spirit and the diverse earth, or sympathy with unity-in-diversity, whether disclosed by a variety of eco-religious experiences, like mine on the trail, or in so many ways, at so many times by so many others, or ecological science, for that matter. In other words, a transversal gathering of insights from esoteric religion and systems science reveals various degrees of unity-in-diversity, from "one in many, many in one" to ecosystem interdependency, and we participate in these degrees of unity-in-diversity no matter the time or place.

So, for me, the question becomes: If we lived in sympathy with these experiences, these realities, these expressions of intelligence, while acknowledging that there is no final intelligence and that these experiences, realities, and expressions must be interpreted, would we not be more likely to enact fitting responses?

Schrag deconstructs ahistoric, and thus universal, appeals, valorizing the transhistorical gathering of insights via praxial critique, but utilizing transversal rationality this project argues that unity-in-diversity repeats throughout history, although, of course, expressed in different ways, and living in sympathy with this awareness, as much as possible, leads to what is more fitting. Or, because unity-in-diversity repeats, and thus we live in relationship with others, the question of what is the fitting response also repeats.

We can affirm, then, a larger logos, which is not, of course, a static Creationist intelligent design, but dynamic, unfolding intelligence marked by the dialogic interplay among order and disorder. Such interplay leads to creative emergence, within which all species participate—there are some "traditions," like the evolutionary epic, that we can't distance ourselves from, nor would we want to. Instead, we must widen our communicative abilities to live in sympathy. And this in-sympathy affirmation is not a claim for ahistorical Truth, but a historically grounded description of our eco-spiritual and actual existential situation, informed by ancient Greek logos, but also the desire for wholeness via eros, and religious experience expressed as theos; all are gathered via transversal rationality, disclosing what is worth bringing forward via participation and disregarding via distanciation.

Ramsey Eric Ramsey, building on Schrag's work in *The Long Path to Nearness*, also focuses on our actual existential situation, but in a different and much-needed manner. He applies transversal rationality to the communicative constraints placed upon us by social structures and debilitating habits and the possibilities for liberation, adding a unique contribution to discerning fitting responses. In other words, while I utilize but also attempt to further Schrag's work by articulating forms of communication that make us responsive to and responsible for repeating eco-spiritual conditions, Ramsey articulates the power of conditioning to repeatedly constrain us, making us unresponsive to institutional powers-that-be, as well as eco-spiritual realities.

Socialization leads us to participate in far more than we are aware, and thus distanciation is imperative to individual and collective action.

Ramsey argues that communication, especially modernist monologue, is the means by which oppressive forces position us, placing limits on our ability to imagine new possibilities for engagement. He turns to Wilhelm Reich's argument that the tendency to unthinkingly participate in our oppression is closely related to "character structure" or the ways in which identities are formed by social structures and ideology. Economic institutions, government, family, work, church, school, and the narratives of bigger, better, and more, and so on, all operate to structure our identities until our character traits "are functionally identical with certain factors of our social structure and ideology."[47]

Ramsey also turns to critical theorist Louis Althusser, who argues in "Ideology and Ideological State Apparatuses" that we are born into ideologies that shape perceptions, a process he calls "interpellation," but Reich takes interpellation further by stating that "armoring" causes us to incorporate our oppression into our body's habitual tendencies, or our "habit bodies." Ramsey describes armoring as "bodily rigidity" that carries "many inhibitions" and "prohibitions of actualizing many of the possibilities" of our world.[48] Habits are intrinsic to agency—without them we would not be able to function in the world, like driving a car in our community while thinking about places far away—but because we are often unconscious of our habits we tend to continually reproduce societal formations via our habit bodies. Thus, we are socialized into a world of language and meaning systems that by their very nature attempt to speak for us.

Agribusiness once again provides a telling example. Wendell Berry argues that the good farmer is like an artist, in sympathy with patterns of intelligence. However, in "Whose Head is the Farmer Using? Whose Head is Using the Farmer," he laments the fact that conventional farmers tend to rely solely on techno-scientific knowledge while disregarding thought generated by their own experience, including mimicking nature's wisdom via no-till, no pesticide plant arrangements. Berry advises farmers to listen to experience and nature, or inner genius and the genius of place, but for conventional farmers daily decisions concerning farming practices are made by listening to agribusiness corporations, techno-scientific experts, and, of course, the lure of profit. Their heads, and habit bodies, have been thoroughly inculcated by societal structures: big business, big profit, and bigger, better, and more generally.[49]

It would seem that we are rather screwed, having little to no freedom to actualize fitting responses, especially since our communicative habits are interwoven with the oppressive features of dualisms like mind and nature, but also between self and other, in which care and concern are lacking, whether from racism, sexism, class disparities, or religious differences, etc. However, the fact that we are interpellated via communication is the very reason that transversal rationality and communicative praxis can liberate us from reproducing societal structures.

Communication is culprit and cure.

Citing Ernst Bloch, Ramsey argues that while our lifeworld is greatly determined by societal forces, it is not fully determined. Our freedom is compromised and degraded, but freedom and possibility are still regulative features of daily life. New possibilities for fitting responses arise out of the conjunction between the inherent potential embedded within objective circumstances and the capacity for subjects to recognize this potential. Our hope for liberatory possibilities and praxis depends on our being aware of this conjunction between potentiality and capacity.[50]

Ramsey states that our awareness of this conjunction between objective and subjective forces is best understood as "a-whereness," as it represents a thorough awareness of our existential situation. "A-whereness" includes the post-metaphysical awareness that there are no universal Moral Codes, as we are on shifting ground and thus critique, articulation, and disclosure must include an attitude of humility; an awareness that habits are produced in consort with societal structures and embedded within the unconscious and our bodies; and an awareness of the needs of particular, concrete social circumstances. A-whereness of these factors increases our possibilities for liberatory praxis; however, our humility causes us to recognize that there is no complete a-whereness: "a-whereness, rather than being true or false, is taken as being either wider or narrower with respect to the disclosure of possibility."[51]

How, then, might we embody a wider awareness of our eco-existential predicament, including material constraints? How might we increase freedom and our capacity to act? Schrag's reconfigured logos is a good place to start, as it provides a means for evaluating past practices for addressing new encounters, and Ramsey is a good next step, because we can't fight ecocrisis unless we free ourselves from interpellation and bad habits.

Agribusiness, to further the example, is most certainly not the final word, as regenerative agriculturalists, permaculturalists, local farmers, and backyard and urban gardeners critically question its structural power and the eating habits they produce, while also drawing insights from the past and rethinking and re-enacting the present. But what can we learn from pre-Neolithic, pre-agricultural lifeways? Critique, articulation, and disclosure also provides the means for considering mythic-animistic traditions, or mythos, prior to the premodern Greek, including Shepard's insights into the communicative practices of hunter-gatherer life. Yet, as we will see, transversal rationality also demands that we highlight authors who take a different view of hunter-gatherer lifeways, including Wilber, who argues an inevitable evolution of scientific rationality thesis with a more critical interpretation of animistic orientations. The point, however, is that the practice of transversal rationality, in contrast to monologic expression of logos as scientism and instrumental rationality, dialogically traverses various arguments, acknowledging possibilities for learning from hunter-gatherer traditions, along with many other traditions, while remaining critical via distanciation.

Thus, the dialogic interplay between participation and distanciation is key to articulating an ecology of communication, or what is fitting and unfitting when gathering insights from hunter-gatherer life that informs mythic-animistic communication, ancient Greek logos that informs rational communication, systems science that informs aesthetic communication, and religious or mystical traditions, or theos, that informs spiritual communication.

The past may be embraced as a recollection and reconsideration of tradition, the present may be embraced as a critical opportunity for interpretive decision, and the future may be embraced as the anticipation of reinterpretation and new possibilities. Such is the essence of transversal rationality and rational communication.

THE GIFT MUST MOVE: THEOS AND EROS

Friendship provides many gifts, like converse that spirals into new and surprising terrain that aids our growth and being introduced to new friends and more conversations. After I had graduated from Rutgers University, my friend introduced me to Rafael Catala, a poet, scholar, and president of the nonprofit Ometeca Institute. Ometeca, a word taken from Aztec language, means two into one, and the institute was devoted to dialogue between the sciences and the humanities, which too often monologue as if two separate cultures. Rafael was also a spiritual teacher, and to my great non-religious surprise, I started meeting with him once, and then twice a week to study mysticism and practice meditation. There is the old spiritual cliché that when the student is ready, the teacher appears. Cliché or not, Rafael was that teacher for me, as I was more than ready; as much as I honored factory work and work at a print shop—or rather, as much as I honor the workers who must pay their bills and put food on the table—I was searching for more, coming to the point where Monty Python absurdity, which I loved, was not enough of a fitting response to the inanities of the world.

Rafael began our studies by focusing on Christian mysticism and met lots of resistance. My Lutheran church experience growing up was less than stellar—the typical boredom during services, sure, but also all that incomprehensible God talk. As a teenager, I watched two boxers about to enter the ring spewing confidence because God was on their side. Well, one of them had to be wrong, but of course both were wrong if they thought God, whatever God was, gave a hoot about two guys pummeling each other. No wonder I was drawn to Python absurdity. But then Rafael also began using the dreaded word, and that stopped the conversation. At that time, in my mid-twenties, I could not fathom how anyone, especially someone with a Ph.D., could believe in God; but, of course, his interpretation was quite different than the boxers, but also my pastor. God was omnipresent for Rafael, and thus God, or whatever word you want to use to designate the source of all things but also the life force within all things, was transcendent and immanent. I was still resistant, as if scarred by too much Sunday School or fearful of joining a cult, but, bit by cognitive dissonance bit, I was able to use the word "G-o-d."

Ometeca Institute conferences in Mexico, Costa Rica, and Venezuela ended up being where I cut my academic teeth—Ometeca was also devoted to dialogue between scholars in the United States and Latin America—but as important as that was, what influenced me the most was the mystical God-talk, and reading, thinking, and meditating together within a supportive, yet discern the fitting and unfitting environment. My friend was insightful enough to introduce me to Rafael, and in a telling twist, I introduced him to a former professor who guided him to the doctoral program at Purdue that he needed to find. And that move also ended up greatly influencing me. So it goes with the gifts of friendship when supporting the developing intellect, but also more. The more, when studying mysticism with Rafael, was the active and irreducible soul, which, when freed from interpellation and bad habits, continually desires more: more experience, more learning, more growth. And if I didn't like the word 'God," he suggested I read Emerson or Whitman or eastern sages, who don't use the word much but took one to the same place: unity-in-diversity and the call to responsibility. And, it was Rafael who introduced me to the work of Bateson and Buber, grounding my spiritual searching within patterns of larger mind and relationships.

Schrag, to my surprise, uses the word "God," but, unsurprisingly, has still another interpretation. He argues that the history of mysticism reflects "strong" or "metaphysical" transcendence, preferring a this-worldly self-in-transcendence, "moving beyond that which it has become and going over to that which is not yet"—or the desire for more experience, learning, and growth.[52] Such self-transcendence engenders other selves: a self-in-community, self-in-discourse, and self-in-action. Thus, transcending limitations and responding to others fosters dynamic and unfolding self-conceptions, decentering the Cartesian notion of an isolated, sovereign self. In doing so, Schrag writes that we deconstruct Descartes' dictum, "I think therefore I am," re-inscribing it as "We interact, therefore we are."[53] This sounds like a universalizing claim for ongoing relationship—or, said from the perspective of Buber's I-Thou ontology: "We exist in relationship, therefore we are"—and thus perhaps Schrag and I are on the same page in regard to principles that continually repeat, although such principles are not the same as rigid criteria.

Despite differences in emphasis, the call for such relational selves are many: Wilber posits an eco-noetic self, Shepard a mythic or animistic self, and Bateson a cybernetic self, and then there is ecopsychology's embodied self, deep ecology's ecological self, and ecofeminism's relational self, etc. All are responses to the hyper-individualized modernist self influenced by Descartes dualistic separation of mind and body, as well as mind from nature. We embody many decentered selves, containing multitudes as Whitman howled, and yet "ever atom belonging to me as good belongs to you."[54]

In *God as Otherwise than Being: Toward a Semantics of the Gift*, Schrag adds still another self: the responding self. After the deconstruction of the modernist self, postmodern responsibility remains; there is no "I" separate from a "you," instead, there is the presence of embodied others and ethical demand. But Schrag's articulation of the responding self and the call to responsibility goes beyond common notions of ethics, such as rights, duties, and justice. In fact, while Schrag argues against transcendental, set-in-stone ethics, he does posit a kind of transcendent dimension, described via the semantics of the gift. Others call forth the responding self and thus the gift event in which we give without expectations of a return reward or benefit; otherwise, reciprocal relations degrade into mere exchange. Such responsive giving reflects agape love—which sounds like it also has universal resonance—expressed as unconditional care and charity.[55] From the perspective of mystical traditions, such giving for the sake of giving mirrors the overflowing of transcendent Spirit unfolding in and as the world, but for Schrag, the gift event is transcendent because it cannot be fully realized in the world, and yet continually calls us to be responsible for the world we create.

For Schrag, the gift event must be transcendent to preordained ethics. Again, rigid criteria, even claims for rights, duties, and justice, may be misinterpreted with communicative praxis degrading into exchange relations, which, due to interpellation and habit bodies, are often economic or master-slave in which oppressors elicit obligation from the oppressed. The gift event transcends forced reciprocity, as the giving of the gift should not obligate in ways made habitual by consumerist and hierarchical social relations. Agape love and the responding self come first, calling forth the responding self in others; and thus after ethics, rights, duties, and justice are transcended, they reappear with a difference as the gift event. Schrag calls this the logos of the gift, in which the gift event is both transcendent and immanent: "The gift, although otherwise than being, transcendent and wholly other, suffers disclosure in the time and space of human interactions."[56]

The gift event furthers the reconfiguration of logos in postmodern, and crisis, times, and its purview, or structure of meaning, is to keep the gift moving rather than merely returning.[57] To illustrate, Schrag provides the example of a mentor who gives without expectation, with the student receiving the gift later becoming a mentor who gives to others. And this, of course, perfectly describes my relationship with Rafael, and the relations between Thoreau and Emerson for that matter, or any genuine student-mentor encounter. The gift event expressed as agape love does not reflect conditional giving, or socially conditioned giving expressed via our habit bodies—consider Christmas materialism and stampedes in shopping malls, the "great" plastic garbage patch in the ocean, and overconsumption generally—but the liberation of new possibilities within the world. But the logos of the gift event also requires Kairos, or the "right moment for a gift to be given and received."[58] The responding self must be mindful of this situational integration of logos and Kairos for ethics to re-emerge and the gift to move.[59]

The gift event is an impossible ideal—an admission that Peters would applaud—but the ideal remains in effect and effectual. It is not too present nor too distant; instead, the gift expressed as agape love allows for fitting, or rather, more fitting responses to come through us. The semantics of the gift, as an expression of what we may call God, is transcendent—Schrag is clearly salvaging a religious dimension to human existence in postmodern times—yet we can pre-enact our hopes for the future. He writes: "This preenactment of what in its fullness of time resides in the future takes shape in the form of the fitting response."[60] In other words, the transcendent ideal of the gift event not only liberates the possibility for fitting responses within the world, it calls forth future worlds. We must be the change, as the saying goes, if we want change to come. Still, as Ramsey showed, we are inevitably complicit in social structures, and infrastructure—driving and polluting, buying and wasting, in various degrees, and so on—at the same time a widened, gift-giving awareness increases our capacity to act, fighting to radically change social structures and infrastructure.

Schrag focuses on fitting responses within the context of the polis but is persuaded by Maurice Merleau-Ponty's claim that the "body as lived is at once nature and culture, at once experiencing and being experienced." His agreement with Merleau-Ponty on embodiment "provides an avenue for overcoming an unacceptable dichotomy and dualism that has been with us all too long," that is, the nature-culture dualism.[61] While not an eco-philosopher, Schrag gives the earth its due by questioning the construct of ownership, stating that while we can privatize nature's services, personal ownership and property rights are subordinate to dwelling within a shared-by-all-species planet. In other words, we must practice transversal rationality when considering the tradition of ownership, which is late and short in relation to our long history as hunter-gatherers, by participating but also distancing ourselves. When we do so, Schrag asserts: "Rather than the earth belonging to us, parceled out as our property, it might advance an understanding of our relation to the earth by viewing it as that to which we belong."[62] This turn to the earth is not fully developed in Schrag's work, but is most welcome. From the perspective of an ecology of communication, this belonging is a larger context for gift giving expressed via the call to responsibility.

While Schrag explores the transcendent, or theos, infused in the world as the gift event, Ramsey again adds to his articulation of the fitting response, this time by focusing on immanent eroticism. In "Procreation in a Beautiful Medium," he makes a distinction between response and responsiveness, arguing that while fitting responses are impossible—in the sense that there are no final criteria to make such a determination—we can embody a fitting responsiveness marked by the wider awareness we bring to encounters with others. Like Peters, Ramsey makes his argument concerning communication by traversing the ancient Greek conception of eros. Unlike Peters, Ramsey affirms the erotic attachment of self and other.

In Greek myth, Eros is conceived in Zeus' garden, born of a union between Plenty and Poverty. Due to this mixed parentage, Eros is neither god nor mortal, and thus becomes an intermediary between the divine and human, which remain other and unable to directly meet. Eros exists in-between having and not having, causing him to wander in search of what he desires. Ramsey argues that communicative praxis is like Eros, neither wholly idealist and divine nor wholly materialist and mortal. Dwelling within the in-between, or a space of procreation in which we continually make ourselves and our world, we encounter both abundance and lack, and this "places demands on us and piques our desires, in a word it eroticizes us." This panerotic in-between is "divinely hued," and yet a grounded, provocative otherness calls forth our responses. When eroticized as such—creatively connected and attuned to others yet existentially different and desiring—we are energized to express a fitting responsiveness.[63]

Ramsey turns to Thoreau's "Walking" to further elucidate a fitting responsiveness, especially the art of sauntering, described etymologically as "Sainte-Terrer" or Holy-Lander, or as "sans terre," meaning without land or home but at home everywhere. One who saunters is not a mere vagabond or idler but walks in search of the holy like a meandering river returning to the sea. Thoreau plays out the Holy-Lander metaphor, stating we must "go forth and reconquer this Holy Land from the Infidels." Ramsey explores both senses of sauntering, stating that "infidel" refers to "those who are thankless" and thus while sauntering we should "be thankful for the erotic attachment to otherness in all its manifestations." Such thankfulness is a conquering in the sense that we overcome bad habits that de-eroticize us. And, not surprisingly, such bad habits emerge when we lack thankfulness: "Walking is being thankful, being thankful is being grateful for the erotic attachment to the radical in-between of existence. All of this is a way towards a fitting responsiveness."[64]

Along with thankfulness Ramsey also emphasizes humility, care, goodness, and justice, all emerging because we exist in relationship or because there is no eroticized "I" without a "you," and this responsiveness creates the capacity for what is most fitting in particular situations to come through us. Such "coming through" is not solely an individual act; instead, responses emerge from reciprocal engagement. Thus, instead of completely losing ourselves in the other, we continually constitute ourselves via an erotic attending to others. As with the gift event, we are called forth, and this in turn calls forth responses from others. For Ramsey, however, eros more than agape makes us responsible for responding. He writes: "So walking humbly with the near-god-like-Eros would be a sauntering, a homelessness that is equally at home in every situation because within each situation it would be erotically attached and responsive enough to care about justice and goodness."[65]

Peters blames the Socratic conception of eros for the "dream of communication," in which subjective interiorities are supposedly revealed, interwoven, and understood; yet this dream is merely another block to the hard work of communication and community building. Ramsey would not disagree, because living within a panerotic in-between respects our existential impoverishment. We must work and wait: "Fitting responsiveness would have to know how both to take matters into its hard working hands and how to practice letting go and waiting if it were to meet the requirement to do justice and love goodness."[66] Thus, while we are impoverished, it is this very condition that supplies us with plenty of possibilities, including the possibility of future acts of justice and goodness. Consider Schrag's example of mentoring and the moving of the gift, in which timing matters and fitting responses take time to bear fruit; or, in the context of ecocrisis, we may not get an immediate response to various eco-engagements, but we do all we can—working and waiting—with the hope that intersubjective learning leads to collective actions and collective change to a more just and sustainable world.

And so, while possibility does not include complete understanding, as otherness always remains, it does include a divinely hued reciprocity and "coming through"—for Schrag, this coming though is the fitting response as gift event, inspired by agape love; for Ramsey, it is a fitting responsiveness characterized by thankfulness, goodness, and justice, called forth by eros or the panerotic in-between.

Ramsey's views on eros are more encompassing than Peters'. His turn to Thoreau is especially provocative, citing "Walking" but also "Life without Principle" to argue that while bad habits "cannot obliterate the erotic," they "remain an ever-present danger." Thoreau makes a distinction between divine and profane relations, with the latter provoking conversation that "degenerates into mere gossip."[67] And yet, the choice is up to us: "The same ear is fitted to receive both communications. Only the character of the hearer determines to which it shall be open, and to which closed. I believe that the mind can be permanently profaned by the habit of attending to trivial things, so that all our thoughts shall be tinged with triviality."[68] Thoreau is concerned with where we habitually focus our attention, especially when we turn a deaf ear to nature. In fact, he states that when he reads more than one newspaper a week, he does not dwell in his "native region," and the "sun, the clouds, the snow, the trees say not so much to me."[69] While concern about newspapers may seem quaint today given the internet, it is also quantifiably more true in our social media, nature-deficit disorder age.

Ramsey, then, adds another self to our list, a sauntering self that overcomes a rigid self-concept and other societal habits and blocks that inhibit responsiveness. However, like Schrag, and despite his turn to Thoreau, he does not explore a fitting responsiveness within the wider "a-whereness" of eco-spiritual unity-in-diversity, and thus an ecology of communication. But he does argue that eros begets logos and responsibility, which work in consort in the pursuit of justice and goodness. And he makes an important distinction between mundane and world-historical responses. Passing the salt when requested is "close to inconsequential," and yet it is "on the continuum of fitting responses possible for mortals." Everyday responses still matter, or as Thoreau put it, why shouldn't we "congratulate ourselves on the ever-glorious morning?"[70] Yet on the other end of the continuum are world-historical responses that come through persons embedded in trying and telling circumstances.

Mahatma Gandhi and Martin Luther King and Rosa Parks and many others are great because of their responses to injustice—as is Thoreau, who inspired Gandhi and King with his response of civil disobedience and virulently spoke out against slavery—but the circumstances within which they found themselves were also great. Thus, inspired by Ramsey we can say that ecocrisis is the "great" circumstance of our time addressing us via the call to responsibility and deserving of great responses; although, it is not the only one: the 1 percent versus the 99 percent, BLM, the Me Too movement, a pandemic, and, unfortunately, plenty of others. If we dig, however, freeing ourselves from interpellation by dominant structures and narratives, we will find that they are all hitched together by dominant structures and narratives, whether bigger, better, or more, or white supremacy and systemic racism, or patriarchal inequalities, etc. Climate crisis, of course, will negatively affect everyone, especially the poor and people of color, while the 1 percent will be better able to insulate themselves from climate, but not their failures in response.

To respond to one crisis is to find ourselves called to respond to all. Thus, we must speak out while being spoken through via thankfulness, humility, care, justice, and goodness, giving gifts by embodying a fitting responsiveness that creates the possibility for fitting responses both small and great, or mundane and world-historical.

We may be existentially and communicatively impoverished, but potentiality and capacity remain, and thus there is plenty of dialogic play to be experienced, borders to be crossed, and great work to be done. However, we are still left with the question of the subtlety, degree, and range of our intersubjective communications and connections with others, like the many gifts of my Appalachian Trial hike. Wilber, for example, argues for the possibility of nonduality, in which the transcendent and immanent are not-two and moments of union between self and other are experienced. I approached such experience with the verdant green piece of moss and awe-inspiring moose, but found myself in it more, experiencing no-self, when on Maine mountaintops viewing the vast terrain after intense focus and after I woke up one morning with an exceedingly quiet mind feeling part and parcel of everything. Such experiences provide insights, and turning to Wilber we will explore them as forms of spiritual communication. For Schrag, however, such experiences are likely too totalizing, running counter to his post-metaphysical philosophy; yet, as we will see, Wilber also claims his spiritual philosophy is post-metaphysical.

Ramsey's articulation of eros would seem to argue somewhat for but mostly against nondual experience. The argument for is given in his description of a divinely hued "coming through" in which what needs to be said or done is said or done in a particular situation: there would seem to be a transcendent coming together in such experience, a situational centering or sympathy, or, from the Taoist tradition, wu-wei or effortless effort. However, while he states that our responses are "not strictly ours" when they come through us from the panerotic in-between, we bear responsibility for what we say and do. We must account for our responses, and "they are more ours when they fail," or when they are unfitting.[71]

The argument against is given when he emphasizes the distance between self and other, which is necessary if there is to be desire for erotic attachment. This would seem to run counter to Wilber's emphasis on transcendence as nonattachment, culminating in the nondual embrace of all things. For Ramsey, nonattachment is the letting go of bad habits, liberating possibilities for goodness and justice, not union. There may not be much difference between Wilber and Ramsey on nonattachment—Wilber also argues for the letting go of bad habits, or addictions—but Ramsey calls his position "areligious religiosity," arguing that a divinely hued eroticism does not lead to pantheism, or by extension, nondualism, as "G-d" represents an "unbridgeable distance." He turns to the Jewish transcendent tradition here, in which "G-d" is so radically other that the divine cannot be named. Ramsey embraces radical otherness as a prerequisite to desiring, and thus communication, leading to the panerotic claim that "responsiveness must be everywhere."[72]

Ramsey, like Schrag, practices transversal rationality to reconsider traditions, including religious interpretations of transcendent and immanent, gathering insights in search of fitting responses, or so we may embody a fitting responsiveness. For Ramsey, because we are not one, we desire connection, but religious traditions include and are informed by nondual experiences, which reveal that because we are one, or one-in-many, many-in-one, we desire connection. Such experiences do not lead to complete understanding—I am not arguing in support of supposedly enlightened gurus—but may act as catalyst to a fuller understanding. Exploring such distinctions, rather than being problematic, allows us to better practice rational communication, learning from differing traditions and being more fully responsible.

While not embracing nondual experience expressed as spiritual communication, including Thoreau's pan*en*theism and divine revelations in *Walden*, Ramsey argues for world-historical responsibility and an increased range of responses, resulting in an "ethics of relief."[73] Such ethics provides "relief" in a double sense: we have a-whereness of our circumstance that resembles a relief map, raised out of the doldrums of one-dimensionality to a three-dimensional world of potential and capacity, and thus we may have relief from our oppression. Such relief would liberate interpersonal relations with other humans; the dominant culture, when seen in three dimensions, yields fissures for marginalized voices to speak. To me, however, we must go farther (or wider and deeper) to a fourth dimension of cosmos, and a fifth dimension of transcendent and immanent mystery, which would liberate all our relations, and thus our capacity to converse within a more-than-human world.

It is, of course, often a struggle to be responsive, or as Schrag puts it, a "loving struggle" that is also our deep desire: at worst, from the perspective of ecocrisis, species die or struggle to survive; at best, we flourish by developing in sympathy with intelligence. Such intelligence is expressed as unfolding, and repeating, relationships that call forth dialogic relations, whether with human or nonhuman others. When such reciprocity is fully embraced, we realize the fruits of our struggles, discovering that agape love and an erotically charged "coming through" are playful, pleasurable, and empowering.

And so, Schrag's articulation of transversal rationality and the fitting response, aided by Ramsey's articulation of a fitting responsiveness, provides a means for critically considering insights from ancient Greek philosophy, and, as we shall more fully see, religious or mystical traditions and hunter-gatherer lifeways, as well as systems science and aesthetics via Bateson's vision of the pattern that connects. If we are to gather insights from differing traditions and embody a fitting responsiveness, we must invite the interplay among rational, spiritual, mythic-animistic, and aesthetic communication.

Transversal rationality is open to traversing these differing modes of communication, each informing the other, while remaining critical via distanciation. No mode of communication, or historical age that represents a particular mode, is final. Instead, we must discern the positives and negatives that each mode and age produces, or what has proven to be fitting and unfitting, applying what we learn to a litany of eco-social crises.

A fitting responsiveness, then, reflects the widened awareness, rather than rigid criteria, we bring to communicative interactions, as well as what comes through us in the form of responses and actions. After all, my range of responses on the Appalachian Trail included nondual moments on mountaintops and during early morning sauntering, a sharing of I-Thou presence and reciprocity with a verdant green piece of moss, the radical otherness of and yet kinship with a fortunately docile Moose, and a human comedy that demanded the hard work of community. All these encounters occurred without the benefit Schrag's and Ramsey's theories, or much theory generally—the reading came later. And those lost in theory, especially deconstructive postmodernism, should take a hike, getting out of their head and into their bodies, emotions, and the physical world.

But we may take a hike, have profound experiences, and interpret them through the lens of restrictive dominant narratives rather than the fecund in-between, or we may take a hike and feel liberated from dominant narratives and be inspired to create our own theory. Perhaps I have interpreted poorly, but I am convinced that I have been able to interpret better, more fittingly, after encountering Schrag and Ramsey on the page, because they have added to my range of experience and my range of thought.

The hike influenced the reading and reading influenced my reflections on the hike. It is clear to me that my communicative experiences with degrees of wildness and fellow human animals out on the trail reflected the reciprocal giving of agape love and were erotically charged, and I have felt called ever since.

NOTES

1. Ralph Waldo Emerson, "Nature," in *Nature, Walking*, edited by John Elder (Boston: Beacon Press, 1991). This book also includes Thoreau's essay "Walking."
2. Laura Dassow Walls, *Henry David Thoreau: A Life* (Chicago: The University of Chicago Press, 2017), 240–242.
3. Richardson, Jr., *Henry Thoreau*, 181.
4. Henry David Thoreau, *The Maine Woods*, edited and annotated by Jeffrey S. Cramer (New Haven: Yale University Press, 2009), 64.
5. Walls, *Henry David Thoreau*, 334–337.
6. Ibid., 339.

7. Ibid., 407–408.
8. Thoreau, *The Maine Woods*, 160.
9. Walls, *Henry David Thoreau*, 408–409.
10. Ibid., 419.
11. Ibid., 409–412.
12. Thoreau, *The Maine Woods*, 269.
13. Walls, *Henry David Thoreau*, 409.
14. Ibid., 411.
15. Ibid., 412.
16. Thoreau, *The Maine Woods*, 168–169.
17. Walls, *Henry David Thoreau*, 416.
18. Ibid., 412.
19. Thoreau, *The Maine Woods*, 251.
20. Ibid., 276.
21. Walls, *Henry David Thoreau*, 414 and Thoreau, *The Maine Woods*, 235.
22. Walls, *Henry David Thoreau*, 419.
23. Thoreau, *The Maine Woods*, 246, 269
24. Walls, *Henry David Thoreau*, 428.
25. Ibid., 422.
26. Thoreau, *The Maine Woods*, 267.
27. Harmon Smith, *My Friend, My Friend: The Story of Thoreau's Relationship with Emerson* (Amherst, MA: University of Massachusetts Press, 2001), 122–125.
28. Scott Waldman and Benjamin Hulac, "This Is When the GOP Turned Away From Climate Policy," *E&E News*, December 5, 2018, https://www.eenews.net/stories/1060108785.
29. Calvin O. Schrag, *The Resources of Rationality: A Response to the Postmodern Challenge* (Bloomington: Indiana University Press, 1992), 18.
30. Ibid., 18.
31. Thoreau, *The Portable Thoreau*, 270.
32. Philip Hadot, *What Is Ancient Philosophy?* (MA: Harvard University Press, 2004), 135.
33. Ibid., 137.
34. Ibid., 80.
35. Ibid., 22–38.
36. Schrag, *The Resources of Rationality*, 23–29.
37. Walls, *Henry David Thoreau*, 408.
38. Schrag, *The Resources of Rationality*, 48.
39. Calvin O. Schrag interviewed in *Experiences Between Philosophy and Communication,* edited by Ramsey Eric Ramsey & David James Miller (New York: SUNY Press, 2003), 24.
40. Schrag, *The Resources of Rationality*, 65.
41. Ibid., my italics, 82.
42. Ibid., 82–83.
43. Ibid., 171–172.
44. Ramsey, *Experiences*, 26–31.

45. Ecomodernism.org, August 2020, http://www.ecomodernism.org/.

46. Schrag, *The Resources of Rationality*, 164.

47. Ramsey Eric Ramsey, *The Long Path to Nearness: A Contribution to the Corporeal Philosophy of Communication and the Groundwork of an Ethics of Relief* (New York: Humanity Books, 1998), 119.

48. Ibid., 25.

49. Wendell Berry, "Whose Head is the Farmer Using, Whose Head Is Using the Farmer," in *Meeting the Expectations of the Land*, edited by Wes Jackson, Wendell Berry, and Bruce Colman (San Francisco: North Point Press, 1984).

50. Ramsey, *The Long Path to Nearness*, 17.

51. Ibid., 76.

52. Calvin O. Schrag, *The Self After Postmodernity* (New Haven: Yale University Press, 1997), 111.

53. Ibid., 78.

54. This line from Whitman is from his famous poem "Song of Myself" in *Leaves of Grass*.

55. Calvin O. Schrag, *God as Otherwise Than Being: Toward a Semantics of the Gift* (Evanston, IL: Northwestern University Press, 2002), 111.

56. Ibid., 115.

57. Ibid., 119.

58. Ibid., 123.

59. Ibid., 128.

60. Ibid., 137

61. Ibid., 100.

62. Ibid., 108-109.

63. Ramsey Eric Ramsey, "Procreation in a Beautiful Medium: Eroticizing the Vectors of Communicative Praxis," in *Experiences Between Philosophy and Communication,* edited by Ramsey Eric Ramsey & David James Miller (New York: SUNY Press, 2003), 194–195.

64. Ibid., 196–197.

65. Ibid., 198.

66. Ibid., 199.

67. Henry David Thoreau, "Life Without Principle," *The Portable Thoreau*, edited by Carl Bode (New York: Penguin Books, 1975), 645.

68. Ibid., 649.

69. Ibid., 645.

70. Ibid., 655.

71. Ramsey, "Procreation in a Beautiful Medium," 194.

72. Ibid., 195

73. Ramsey, *The Long Path to Nearness*, 110–115.

Chapter 2

Integral Meta-Theory

Ken Wilber and Spiritual Communication

LEVELS/STAGES, LINES, AND STATES

Ken Wilber provides a model of human development that may be used to interpret the relations among rational, spiritual, mythic-animistic, and aesthetic communication. In doing so, his writings differ from those of Schrag, Shepard, and Bateson in a key respect: he thoroughly explicates the descent and ascent of Spirit.[1] Some, including Shepard, argue that theorizing a transcendent dimension reflects a faulty use of imagination, or the aloofness of abstract thought, creating a world that does not exist rather than living humbly within the wild and diverse created world. Most certainly mistaken, or unfit, interpretations of the transcendent abound, but such criticism neglects the transversal gathering of insights from religions east and west, especially esoteric mystical traditions and spiritual practices, in which transcendent and immanent dimensions, while ultimately ineffable, are experientially explored and articulated. Even Buddhism, which many consider to be without a transcendent dimension, discloses that the world of form emerges from and expresses emptiness. This criticism also rejects insights from seminal philosophers, including Plato, Plotinus, and Hegel, and transcendentalists like Emerson and Thoreau.

My attraction to Wilber is the same as my attraction to Rafael: they embrace rationality and spirituality. I did not just read mystical texts and meditate with Rafael; we also explored the intersections among science, aesthetics, and spirituality. In fact, Rafael wrote science-inspired poems based on his theory of sciencepoetry, in which the separation among forms of knowledge was overcome. Science and religious feeling, while quite different, are not rigidly opposed to each other, like I thought they were after going to church but also after undergraduate education. My college classes rightly focused on critical thinking, but wrongly at the expense of other modes of knowing. It was only when I met Rafael after graduation that I fully considered the possibility of something more, even though I had transrational experiences in childhood that suggested that rationality, as wonderful as it is, does not tell the full story. Of course, we never get the full story, but we get more of it when we embrace multiple modes of knowing and communication and consider the possibility of a transcendent dimension that is also fully immanent.

After being primed by conversation, reading, and meditating with Rafael, Wilber became a logical next step. I came across one of his early books, *No Boundary: Eastern and Western Approaches to Personal Growth*, and that started a bit of an obsession in which I read most of his two dozen or so titles. And while no one knows the full story, he takes a stab at it in *A Theory of Everything* and *A Brief History of Everything*, and in his meta-theory generally. Or rather, he provides an integral framework that helps us to interpret everything, while remaining open to new knowledge.

Wilber is not an academic, having dropped out of a graduate program in biochemistry to read eastern philosophy, world religious texts, and most everything else on the way to writing all those books. Not surprisingly, however, when you attempt to take on everything not everyone is a fan. There is a website filled with articles devoted to Wilber criticisms and debates, but also support, which is complicated by the fact that Wilber's interpretations of what we can know of reality have evolved as he has evolved, and thus criticisms of early work, including that he fails to respond to the postmodern turn, are addressed in later work; he has even gone so far as to label stages of his writing as Wilber 1 through Wilber 5.[2] However, he has some major supporters, including Robert Kegan, a Harvard professor at the Graduate School of Education, who calls him a "national treasure," and his integral meta-theory, which includes interpretations of agape as the involution of Spirit and eros as the evolution of Spirit, is quite influential, spawning books on integral business, integral ecology, integral education, and pretty much, well, everything.[3]

Given that Wilber has so many books and evolved positions—which is what you would expect—addressing his work as a whole is a challenge. This is further complicated by the fact that a recent book, *The Religion of Tomorrow*, is a dense 800-page tome that is not the easiest to wade through even if you have read earlier work. Thus, since his writings are impossible to encapsulate, I won't try, but I will attempt to hit some highlights in relation to communication generally, the interplay among differing forms of communication, and spiritual communication defined as insights emerging from spiritual experiences.

And so, here goes something but not everything.

To me, a transcendent dimension is inevitably a part of our imaginative life—out of nothing came something—but Wilber's exploration of the transcendent, rather than a religious conceit or only an appeal to tradition, is key to his formulation of a developmental path of increasing spiritual depth consisting of eight basic levels or stages: Archaic, Magic, Magic-Mythic, Mythic, Rational, Pluralistic, Integral, and Super Integral. Development begins at pre-rational stages (Archaic, Magic, Magic-Mythic, Mythic) with identification with the immediate physical world, eventually moves to a more mature identity as rational world citizen (Rational, Pluralistic), and then proceeds to the transrational (rationality-plus: Integral, Super Integral) realization that we are Spirit aware of Itself. The stages prior to Integral are considered first tier, while the Integral stage marks the beginning of second tier consciousness, or the first stage in which the partial truths of previous stages are integrated into a wider awareness.[4] From the perspective of this project, it is important to keep in mind that higher levels or stages reflect a different, and higher, level of communicative competence, with second tier integral awareness being far more competent, and thus more fitting, because such widened awareness is better able to transversally gather insights as needed from earlier levels.

Wilber has mapped the "Great Nest of Being" comprised of the evolutionary ontology of matter, life, mind, and Spirit, or the ancient Greek "Kosmos," onto human development, and thus there is an intimate relation among all levels. He uses the phrase "Great Nest" instead of "Great Chain" to be clear that it is not a rigid hierarchy but levels within levels. Development, then, is a holarchic process of transcending and including, or differentiation and integration, because each lower level is transcended and then included into the higher level.[5] Interestingly, while our evolutionary potential is actualized via decreasing ego-centeredness (once an ego is formed) and increasing identification with others as we take stage-steps, ego is included in higher levels but is not the center of our experience. In fact, Wilber argues that we have healthy egos when we are more often centered in unfolding Spirit and thus have transcended, or decentered, the isolated ego's attachments and lack of clarity. However, so-called enlightened adepts, if they exist, have supposedly transcended a separate self-sense.[6]

But, again, this is a simplified overview of Wilber's model, or meta-theory, as he gathers insights from numerous theories, which includes four quadrants of development: interior-individual (introspection, intention; subjective "I"), exterior-individual (biology and behavior, including brain states; objective "It"), interior-collective (cultural values and ethics; intersubjective "We"), and exterior-collective (social structures from tribes to nation-states to planetary; interobjective "Its"). The stages/levels briefly mentioned focus on interior individual and cultural development, but all interrelated quadrants are crucial to a fuller understanding of development and the possibility of enacting fitting responses.[7] Ramsey, for example, disclosed blocks to interior development due to oppressive exterior social structures that we unconsciously carry in our habit bodies. Thus, it is much easier to develop individually and culturally when we have social structures that invite rather than restrict growth. However, Ramsey also disclosed our capacity to resist, and so it is possible to develop individually up the holarchy despite social constraints.

I also do not have space to fully explicate each stage of interior development, which are deeply influenced by Jean Gebser's seminal work, and confirmed by over 100 theorists over the past 100 years, but I will provide brief sketches, as mapping out developmental levels also maps out differing communicative styles, goals, and ethics as pre-rational, rational, and transrational/integral subjects navigate their worlds.[8] However, stages/levels of development are also interwoven with numerous developmental *lines*, including multiple intelligences that proceed through the stages, as well as *states* of consciousness or spiritual experiences that may occur at any level and be interpreted at any level, but which may also act as catalysts to growth. This interweaving of stages/levels, lines, and states discloses a variety of responses to ecocrisis, enhancing the discernment of what is fitting and unfitting.

The Archaic is the infant fusion stage, and thus of not much relevance to this project. This primary stage does not transcend subject and object because there is not a functioning ego that can tell the difference between them. However, it is interesting to note that Freud critiqued the mystic for residing in this state of non-differentiation, which he called the oceanic feeling. But Freud mistakenly confused pre-rational with transrational realms. Mystics do not regress to a pre-egoic state; rather, they transcend the isolated ego or the rigid distinction between subject and object while also acknowledging differentiation as an expression of Spirit.[9]

Generally speaking, the next three pre-rational levels are marked by lots of participation and little to no distanciation. At the Magic level, self and other are poorly differentiated, leading to human characteristics being projected onto nature—"It's flooding because our tribe offended spirit elements" or "The volcano is erupting because it's mad at me" —which Wilber equates with animism.[10] In other words, humans superstitiously attempt to control the outer world or ward off evil spirits, but, taking a next step to Magic-Mythic, this controlling power is given to some great other, like various gods and goddesses who have magical powers, or eventually, God and his miracles, if one truly believes. At the next stage, Mythic, humans more fully conform to family or tribe and ethnocentric mythologies. Today, this would extend to political party, nation, and literal interpretations of religion. Such mythologies are a step forward in regard to lessening magical thinking—Mythic transcends and includes Magic—but, overall, the Magic, Magic-Mythic, and Mythic stages are marked by little critical questioning, as that demands another step to the Rational level, and robust participation in tradition. At the Mythic level, this includes a strong sense of right and wrong, rules and roles, law and order; all positive steps but deeply dangerous when bound up with an "us versus them" authoritarian attitude based on rigid belief.[11]

The tension between Wilber's work and Shepard's championing of hunter-gatherer life and mythic-animistic communication immediately becomes apparent here. Wilber obviously delegates insights from our mythic-animistic past to a lower, Magic or Magic-Mythic level of development, in which projections onto nature are common. Insights emerging from this practice are not dismissed—all levels are necessary to development—but, as we will see, Shepard argues that hunter-gatherer development is marked by valued forms of thinking, including forethought expressed in the hunt, rituals, and storytelling, and given that growth is stimulated by immersion within specific places and interactions with animal others, such thinking is often superior to the dissociated "rational" consciousness of the modern period, in which mind is utterly separate from nature. Projecting animism, or aliveness, onto others may reflect irrational superstition, but may also reveal needed information for survival and stimulate further awareness.

Wilber does recognize the dysfunctional aspects of modernity, including a dissociated rationality, but states that regressing to pre-rational levels will not solve this problem.[12] Instead, he champions the many dignities of a *healthy* rationality, including differentiation among self, culture, and nature leading to ideals of universal human rights and freedom, like the 1948 UN Universal Declaration of Human Rights, and world-centric dialogue required to understand and begin to solve eco-social crises, like the warnings of world scientists and rational discourse in search of fitting responses.[13] Such dignities are much needed given that magic-mythic thinking is still very much with us, whether via New Age insistence that there are no limits to our ability to create our own reality or claims for our God-given right to have dominion over the planet with no limits to consumption.

The Rational level, then, marks the much-needed birth of the scientific mind; yet, ironically, rationality, when reduced to scientism, results in mythic faith in unlimited growth, technology as savior, and linear progress, and thus the source of much damage disclosed by scientific warnings.[14] While these modern myths have delivered the material goods via capitalist structures considered rational (at the Rational level), they are often dissociated from the earth and spiritually bankrupt. Linear progress, for example, suggests a straight line of development in which insights from earlier traditions, like pre-rational listening to the land and transrational religious experience, are not included, or distanciation without participation. As a result, economic growth is considered unlimited and technology by itself will save us, denying the depth and breadth of ecocrisis.

To Wilber, each stage reveals dignities and dysfunctions, or light and shadows, yet each higher stage also reveals further growth, especially if the dysfunctions are transcended and the dignities included at the next level. Moving from Rational to the Pluralistic stage, then, furthers rational development; however, it must be noted that it takes many years to become fully rational, assuming we ever do. After all, humans begin at the Archaic level fused with their mothers and the surrounding emotional environment before moving to Magic and Mythic levels; but at the Rational stage, we think about thinking and discover the potential of introspection, which blooms at the Pluralistic stage such that we more fully identify with differing groups and an ethic of care emerges. Such blooming includes the environmental, civil rights, and feminist movements of the 1960s, and multiculturalism generally.[15]

As we saw with Schrag, this Pluralistic postmodern stage is also marked by the deconstruction of universal truth-claims, whether religious or scientific, due to the partiality and interpretive nature of all truth-claims. This comes in handy when challenging domination and oppression, but, again, a potential dysfunction is the relativizing of value: all cultures are valued, which is clearly a progressive step, yet no distinction is made between dominating hierarchies and actualizing holarchies like stages of development, and thus no cultural values are considered to be better, including, more irony, pluralistic values. This performative contradiction—those at Pluralistic inevitably argue that a multicultural attitude and global perspective of universal care is superior to Magic-Mythic ethnocentric views that don't take the other into account—leaves us bereft when confronting eco-social crises, as no response is more fitting.[16]

Despite this downside, the upside of Pluralistic is that we may begin to practice transversal rationality, and thus rational communication expressed as a synthesizing intelligence able to consider multiple perspectives in search of situational fitting responses. The result is communal feelings and a deeper sense of responsibility, which are central to this stage, in contrast to the rise of individualism, a dignity, and hyper-individualism, a dysfunction, at Rational.[17] The Pluralistic level also marks the beginning of a more complete ecological awareness, in which we integrate whole and part, body and mind, but we may still have a bias toward identifying with the soul of humanity rather than the soul of the world (*anima mundi*, world soul). Pluralistic, when its dignities rather than dysfunctions are embraced, brings us to the brink of the Integral level.

Wilber sometimes calls the Integral level vision-logic, a phrase I prefer because this stage does indeed marry rationality with a larger vision of wholeness. Following Gebser, he also refers to this stage as integral-aperspectival, because while no perspective is final we participate in the continual process of creating higher unities that integrate seemingly divergent perspectives. We willingly tolerate ambiguity and paradox while making an inner commitment to growth and increased meaning, both in our life and in the lives of others. Or, to use Schrag's transversal terminology, we reject "vertically anchored universals and horizontally dispersed particulars," rationally traversing the in-between gathering insights via praxial critique by participating in traditions in search of fitting responses—like our hunter-gatherer past and its communicative practices and the history of mysticism and its communicative practices—while also critically distancing ourselves from unfit interpretations of these traditions. Thus, vision-logic, at its best, represents a *constructive* postmodern and dialogic process integrating the premodern with the modern, or logos, mythos, and theos.[18]

The Integral stage, as the first second tier stage, is the first level in which the partial truths of previous stages are integrated, and this provides relief in the dual sense articulated by Ramsey—relief map with more dimensions, ethical relief from oppression—as well as relief from the various dysfunctions and ironies of these stages. Vision-logic at the integral stage is also similar to Bateson's articulation of aesthetic communication, in which mind and nature are both differentiated and integrated. Transrational state experiences are not required for second tier consciousness, but contemplation that synthesizes various perspectives in search of unity-in-diversity, if taken far enough, has the potential to open us to the *experience* of unity-in-diversity. If this experience is powerful enough, we more fully transcend and include identification with humankind, becoming centered within the sacredness of the earth community as a whole, or in Bateson's terminology, sacred unity or the pattern that connects, or in the language of spiritual traditions, the immanent Divine or world soul.[19]

Growing up to the Integral stage is no small feat: despite each previous stage being a necessary step within a developmental holarchy, each previous stage tends to view its View as the one and only, and thus there is often transcending but very little including of insights, or all distanciation, no participation. This goes a long way toward explaining differing communicative styles, goals, and ethics, as well as the culture wars; much miscommunication, and anger, occurs due to communicative subjects having their "center of gravity" at differing stages with their differing perspectives: generally speaking, premodern, modern, and postmodern. This also goes a long way toward explaining differing responses to ecocrisis, or the ability to even recognize ecocrisis; a Mythic-level religious fundamentalist, for example, who takes scripture literally may deny science because they have not progressed to the Rational level of development, claiming that God would not allow bad things like climate crisis to happen, or when it does happen that it must be God's will. On the other hand, those at the Rational level, perceiving this foolishness, tend to equate spirituality with fundamentalist religion, throwing out the proverbial baby with the bathwater. This knee-jerk response fails to recognize that while ecocrisis is indeed a crisis of rationality—we don't listen to the scientific consensus on climate; it is not rational to destroy the ecosystem that gives us life, and so on—it is also a crisis of spirituality, in which the earth and nonhumans are viewed as a collection of objects rather than a communion of subjects.[20]

Such analysis further discloses the dignities and dysfunctions of the Rational level, but also makes one wonder what might be worth including from premodern Magic-Mythic stages. As we will see with Shepard, there is much to include from our hunter-gatherer past, but theistic religion at these stages also displays values worth keeping, like participation in tradition, law, and morality, or simply that values matter, which deconstructive postmodern pluralism tends to relativize and discard; of course, problems ensue when these values are interpreted from a pre-rational literalistic level. The key, therefore, is to transcend, include, and *transform* these values from a higher view, a higher, more inclusive perspective, which is exactly what is fully embraced at integral levels. God's will, after all, is interpreted differently at different levels: consider Martin Luther King only wanting to do God's will when fighting for civil rights and against oppression, or liberation theologian Archbishop Oscar Romero attempting to see with God's eyes—a good form of projecting—and fighting to liberate the poor in El Slavador or God's will mystically interpreted as Thoreauvian sympathy with intelligence.

The Integral stage is also no small historical occurrence: based on various studies, Wilber estimates that 5 percent of worldwide population is at this stage, which has only appeared in the last few decades, while up to 60–70 percent have their center of gravity at pre-rational, ethnocentric Mythic levels and the rest at Rational and Pluralistic, although with some overlap. We need to let that sink in: 60 to 70 percent are at pre-rational levels. This explains a lot in regard to eco-social crises and also explains why the interplay among education, growth, and communication is so essential. However, moving to new stages is no easy task; citing Kegan, Wilber states that while stage-steps may be accelerated, typically from profound experiences, including confronting crisis and spiritual experiences, they cannot be skipped and typically take five years, if it happens at all.[21] He further argues that the dysfunctional performative contradiction at Pluralistic, which he calls "aperspective madness" marked by the relativizing of value, has become a major roadblock to developing to second tier Integral, while also thoroughly pissing off those at Mythic and Rational—again, culture wars—which, irony of ironies, provided a major roadway for the election of climate crisis denying Donald Trump.[22]

Part of me does not want to get into Trump, or any phenomenon that will fade, in a largely theoretical book that I hope will have universal resonance for a long time to come; but, like eco-social crises, and the coronavirus pandemic, and Black Lives Matter, and Me Too—which are easily linked to his shortcomings—he represents a marker moment in the United States but also world history. Also, given the 60–70 percent at pre-rational levels, Trumpism will be around for some time despite Trump losing the 2020 election. Plus, Wilber takes him on, providing a provocative take from the perspective of the developmental holarchy.

In *Trump and a Post-Truth World*, Wilber argues that Pluralistic aperspective madness degraded into contradictory no-truth nihilism and only-my-truth-is-truth narcissism, leading to various regressions that Trump unknowingly rode to victory. In other words, if leading-edge-of-evolution Pluralistic, represented by 20–25 percent of the population, is governed by its dysfunctions more than its dignities, those at earlier stages will not be clamoring to take next steps; instead, they will attack based on their own dysfunctions, with differing levels becoming narcissistic silos amplified by social media more than stage-steps to wider awareness and fuller responsibility. Ironically (so many ironies when evolution stalls), such silos have become rampant at the Pluralistic level as identity politics, a step forward when fighting oppression yet too often regressing to ethnocentric-like us-versus-them politics and communication. With no guiding universal principles—they have been deconstructed—there is nothing to unify us, nothing to call forth next steps. And without the pull of truth (scientific investigation), goodness (ethical discernment), and beauty (aesthetic response to pattern and spiritual experience), and thus the pull to listen to and integrate various perspectives, we have no way to distinguish what is more true and the result is more ugly divisiveness.[23]

Trump's victory may be attributed to many factors, but Wilber focuses on an ignored and overarching one: a self-correcting response of evolution. When the leading edge is ruled by its dysfunctions, expect regressions, and much pain and conflict, in order to relearn lessons, or fully learn them, so that consciousness can evolve via transcending and including rather than halted by transcending and rejecting.[24] This interpretation is unkind to Trump voters who voted for him for non-Mythic-level reasons (belief, which turned out to be false, that Trump would bring back manufacturing jobs, anti-Clinton politics-as-usual monkey wrench into-the-system sentiment, etc.) and critical of anti-Trump Pluralistic voters who fail to realize that all levels have value and everyone has to take earlier stage-steps. However, it also suggests that there is an unconscious wisdom at work: the developmental holarchy itself, or Spirit in action, or eros defined as the drive for wholeness, or order out of chaos, or self-organization, or Kosmic grooves, or whatever terms one wants to use.

To Wilber, Pluralistic thinking and responding is too often unresponsive to this wisdom, or rather, it is not in sympathy with intelligence due to confusing dominating hierarchies with actualizing holarchies. Dominating hierarchies exist, of course, and must be deconstructed, but this confusion has acted as a block to growth that ushered Trump into office. In other words, just like Rational-level science, when reduced to scientism, throws out all forms of religion and spirituality without distinguishing between pre-rational and transrational expressions, Pluralistic has too often thrown out all forms of hierarchy, including developmental holarchies from which pluralism has emerged.[25]

The good news: the potential for growth provides hope for progressive change via more fitting responses; the blunt bad news: this analysis once again suggests that we are rather screwed, especially since multidimensional wicked problems in need of multidimensional integral solutions, like climate crisis, are accelerating at an unprecedented and alarming rate. Wilber acknowledges this wickedness, stating that our collective evolutionary future has become a race with our survival at stake: the detailed description of stages of development, which began to be articulated around 100 years ago, disclose that we have potential for glory and yet the dissociation between mind and nature may lead to collapse, and "at this time it is too close to call."[26]

Wilber holds out hope, *if* those who have fallen prey to Pluralistic dysfunction respond to their complicity in creating the culture wars by embracing all levels and the dignities of healthy universal pluralism, and *if* the emergence of the Integral stage becomes the new leading edge of evolution, influencing everything.[27] The 5 percent at this stage tend to be movers and shakers, creating new ideas, policies, and technologies, which are also accelerating at an unprecedented rate; yet much will depend on whether new economic and social structures and design infrastructures emerge, and thus new ways of living, guided by postmodern eco-spiritual narratives rather than modern myths of progress, technology, and growth. Much will also depend on communicative competence expressed as the ability to speak the language of differing stage-views. Targeted messaging is nothing new, certainly not for environmental communication practitioners who frame messages to particular publics, but an integral approach provides plenty of insights for crafting those messages, including to Mythic-level religious fundamentalists who may be called upon to love and save God's creation.[28]

Still, as Peters' reminds, communication is hard and takes hard work, especially given that communication and identity are deeply interwoven, and we don't like to change either. When our ideas are attacked, we feel attacked. Cognitive dissonance is necessary for taking a stage-step, but we will resist, searching for reasons not to change, or feel backed into a corner and refuse to change. I have always been fascinated by stories of big perspectival change, like the son of a prominent white supremacist rejecting his father's views and fighting for racial justice, or a daughter who grew up within the Westboro Baptist Church learning hate and then rejecting it. In their previous unenlightened lives, they encountered a lot of anger in response, and rightly so, but what I find most fascinating is that, in both cases, what led to change was others being willing to listen and dialogue with them.

I am not saying that listening and dialogue is the only fitting response, but I am saying that it is a necessary element; it is likely that the combination of being confronted by anger and protest and then finding a sympathetic ear willing to listen are key, or cognitive dissonance followed by a way out. There is also the timing question raised by Schrag: we may give a gift that others are not ready to receive, which is a basic reality of teaching. I have been teaching college students for over twenty years, and have numerous stories of former students thanking me several years after graduating, as if they finally really get it. Of course, I often don't know exactly what they get. There is a time and place for what is most fitting to come through, but also a time and place to receive the moving of the gift. And there may be times and places when it is unwise to engage at all, like against online trolling, or if a Mythic-level militia member with guns is shouting and salivating on courthouse steps.

To complicate things while providing further clarity, we must also consider distinct and yet related *lines* of development, including multiple intelligences like cognitive, interpersonal, intrapersonal, and emotional, as well as values like love and concern, creativity, specific talents, and so on, which proceed through the various stages.[29] There is also a spiritual line, or spiritual intelligence, in which subjects ask religious questions of meaning at differing levels, receiving quite different answers.[30] Religion, while too often residing at the Mythic, fundamentalist level, runs all the way through development. Unfortunately, religious expression has been more exoteric than esoteric, focused on church hierarchy, organization, rules, roles, orthodoxy and tradition—again, Mythic—making it easy to rationally attack, "throwing out the baby" and undermining significant experiences and a developmental line. And that, of course, perfectly describes my view on religion prior to meeting Rafael. But the cognitive dissonance arising from our many debates eventually gave way to much listening and learning. Gifts may take time to be received but when they are received they give and give, both to the receiver and to others we encounter.

My religious education was furthered by Wilber, but also when I stumbled upon James Fowler's work on stages of faith. Fowler has articulated stages that mirror Wilber's, with each stage widening one's ultimate concern, moving from Magic-Mythic egoic selves appealing to God for miracles to Rational and Pluralistic deconstruction of literalism and construction of metaphoric understandings in search of meaning and purpose, to a more universalizing, second tier Integral view embracing insights from wisdom traditions east and west, finding diversity but also much coherence in regard to spiritual experiences and values, especially concern for others, like the "Golden Rule."[31] The greening of religion also discloses much potential for the faithful to practice an eco-spiritual ethic. Witness Pope Francis' 2015 Encyclical Letter, running 180 pages, on "Care for our Common Home" in which he declares our "Sister, Mother Earth, who sustains and governs us . . . now cries out to us" because of our "irresponsible use and abuse of the good with which God has endowed her."[32] While the Pope maintains a theistic bias embracing the Trinity and Church language and commitments generally, he also invokes his namesake Saint Francis, once called the patron-saint of ecology, stating that God speaks of goodness and beauty via nature, if one is willing to listen.[33] Surveys show that the Encyclical, or "Francis Effect," has nudged Catholics to be more concerned about climate crisis.[34]

An eco-activist line of development also discloses a range of responses at differing levels. In *Integral Ecology*, Sean Esbjorn-Hargens and Michael Zimmerman unpack basic eco-attitudes or selves: Eco-Guardians and Eco-Warriors marked by pre-rational animistic identification with nature and aggressive, defensive tendencies; rational-level Eco-Managers and Eco-Strategists who work on passing laws and conserving resources while often beholden to status quo systems; pluralistic Eco-Radicals who deconstruct the dysfunctions and destruction of modernity, often without seeing dignities, while fighting for eco-social justice; pluralistic-integral Eco-Holists who practice vision-logic to map out ecosystem complexity and wicked problems but may focus too much on exterior change at the expense of interior transformation; and Eco-Integralists who see value in all perspectives, hopefully without over-including conflicting views that offer little value. Again, we express multiple selves and thus fitting situational responses may come forth from these differing eco-activist attitudes or in some combination.[35]

While we have a stage center of gravity exploring lines of development reveals that some lines may be at lower or higher stages, and thus we are often somewhat enlightened in some lines while needing much work in others. Problems are difficult to solve when one line, like cognitive, is highly developed but other lines, like emotional or interpersonal or spiritual, reside at lower levels. This is easy to see in personal relationships—good at arguing, bad at relating—but also in organizations responding to various crises, as the lack of coherence among lines may create crisis, making communicating and working together difficult.[36] It is also easy to see when rational-level scientists try to communicate to a broader public without the line-tools to do so.

While community building is difficult, Peters argues that this is not something to lament, as argumentation is part and parcel of rational communication and democracy. But when the hard work of connection becomes toxic among communities at differing levels and lines, like in Congress, dissonance degrades into a big mess of us-versus-them anger. Connection, of course, is easier when communicative subjects easily agree, like at a Trump rally, to use a much too easy example, where attendees welcome Mythic-level authoritarian rhetoric. Trump's stage level of gravity often resides at Mythic, evidenced by his stoking of ethnocentric fears, especially in regard to immigration, and not only denying but mocking Rational-level science and pulling the United States out of the Pluralistic, world-centric dialogue Paris Climate Accord. But while Trump rode the reaction against dysfunctional Pluralistic into the White House—"make America great again" is another way to say "keep America at the Mythic level"—he also embraces the dysfunctional relativizing of values: the needed postmodern deconstruction of authoritarian Truth has been manipulated into authoritarian post-truth "alternate facts" and "fake news," in which no criticism is tolerated.[37]

Not surprisingly, stages of development reveal plenty of confirmation biases, in which "facts" match particular levels of awareness and thus what is taken as true at one level bounces off of other levels, like trying to convince a Mythic-level religious fundamentalist to embrace scientific evolution. Or, to put it another way, belief trumps fact. Yet too easy agreement may also occur at Rational and Pluralistic levels, even though they are higher levels displaying more care and concern, including, at Rational, uncritical belief in modernist myths, or, at Pluralism, quick agreement on the nastiness of all hierarchies in favor of feelings and equality, or, at both levels, an overemphasize on distanciation and progress over participating in traditions worth conserving, all of which a Trump-supporter at a Mythic-Rational level with a high cognitive line may reject with good reasons, as well as poor ones.

The mixing of levels and lines discloses the complexity of human agency but also the potential for growth; as climate crisis deepens and the laws of physics cannot be denied, as Bill McKibben likes to say, the call to transcend our worst selves and communicate across levels and lines becomes imperative. When Pluralistic-level groups, who have previously been divided by identity politics, transversally communicate based on intersectional concerns and focus on old-fashioned coalition politics and grassroots organizing, much may be accomplished. The dignity of deconstructing claims for universal Truth becomes apparent here, as embracing uncertainty is central to the practice of rational communication and making connections with others: to avoid confirmation bias and learn from cognitive dissonance, there must be openness that allows for dialogic exchange. And if one also strategically communicates across levels, at the right time and place, the Integral level may become the new leading edge.

At second tier Integral such communication would mean honoring all stages and displaying empathy by acknowledging that we all proceed through pre-rational levels, although at differing ages; honoring the mixing of levels and lines and thus the complexity of agency and response; gathering insights from all stages but also transforming them from the higher perspectives of higher stages; messaging so others can better listen and understand appropriate to particular levels while also acknowledging that most are not changing anytime soon, if at all; and working on our own growth while planting seeds of change by creating new ideas and agreements and structures such that the gift must move.[38] Still, wrathful reminding from the vantage point of the Integral level may also be fitting, like Chris Hedges' diatribe "Onward, Christian Fascists," in which he calls out the bastardization of Jesus' teachings by the Christian right, including concern for the poor twisted into the magical showering of wealth upon church leaders, the lust for social and political power and thus support for Trump despite obvious moral failings, and Magic-Mythic thinking generally, like the claim that only they know God's will. Hedges, an award-winning journalist and minister who graduated from the Harvard Divinity School, further calls out the failure of the progressive church, in the name of tolerance and dialogue, to denounce this hypocrisy.[39]

All this suggests that the developmental holarchy is working even when we don't see it at work. And yet, Integral is not the highest stage. Wilber also explicates third tier Super Integral, or the "growing tip" of evolution, which he further breaks down into four levels: Para-Mind, Meta-Mind, Overmind, and Supermind, this time influenced by the thought and language of Sri Aurobindo.[40] Given that less than 1 percent of the population is estimated to be at these levels one might wonder about their practical worth. Wilber is not oblivious to this critique; given the approximate stage percentages, the world would be infinitely saner if more of us reached the Rational level. Vision-logic would be heaven. Still, it is important to map out the highest levels, especially since many have experiential glimpses that inform their sense of self and the direction of their lives, although I have not met any enlightened mystics lately, nor do I expect to.

Wilber describes these levels in much depth (and they are marked by depth and breadth), but a very quick overview is sufficient. In contrast to second tier integral, in which unity-in-diversity is conceptually understood or perhaps glimpsed, third tier levels are characterized by a more fully lived awareness, often catalyzed by experiences of unity-in-diversity, moving from immanent communion to transcendent union, to immanent and transcendent nondual Supreme Identity. Thus the highest stage, Supermind, while not final, might be best described as a simultaneous centering and decentering, or identification with Spirit unfolding whose center is everywhere and circumference nowhere. Or, in Buddhist language, the nondual embrace of emptiness and form and thus the gift of each moment.[41]

Wilber, then, goes further than Fowler by focusing on religious experiences and deeper, esoteric meanings and identity in regard to developing and expressing spiritual intelligence. For me, when working with Rafael and then after, I tried to at least approximate third tier awareness via a variety of practices, including meditating on the "I" that is all things, placing my personal sense of "I" within a larger context by meditating on God or the Divine or whatever in all things, and then moving on with my day by consciously recognizing God within people—basically, saying Namaste inwardly—as well as with trees and plants and animals. Of course, these practices did not make me anywhere near enlightened, a goal I youthfully thought was possible but in time came to happily assume is impossible, but some degree of spiritual intelligence is worth bringing to our communicative encounters within a more-than-human world. And, like any skill, you need to practice.

Esbjorn-Hargens and Zimmerman call the self-orientation at third tier Super Integral Eco-Sage, in which one experiences diverse nature as an expression of Spirit with a requisite call to serve the good of all species. However, given that all stages have potential dysfunctions, the Eco-Sage may be too focused on transcendent contemplation at the expense of immanent action. But when transcendent and immanent are experienced as not-two, action in the world mirrors the Bodhisattva Vow of deep commitment to all sentient beings. It is important to note, however, that the Bodhisattva Vow, like God's will and all forms of thinking, experience, and action, may be interpreted differently at differing stages. The Eco-Sage, at its highest expression, gathers insights from other eco-selves while buoyed by rational discernment and a variety of spiritual experiences.[42]

Along with levels and lines of "growing up," then, Wilber also explores five states of consciousness or "waking up," including gross, subtle, causal, empty witnessing, and nondual unity spiritual experiences. As mentioned, the discovery of stages of development, or hidden maps that cannot be seen by looking within, goes back approximately 100 years, but states of awareness, which can be explored by looking within, often via meditation traditions, go back 50,000 years to the first shamans exploring altered states of consciousness.[43] Evelyn Underhill's seminal work on the history and practice of mysticism reveal that states are typically experienced in three ways, communion, union, and identity, while William James wrote that religious experiences are commonly marked by four characteristics, ineffability, transience, being out-of-doors, and unity.[44] Wilber includes and goes beyond both, thoroughly exploring state experiences that disclose insights, and, once again, such insights can be understood as forms of spiritual communication; however, it is wise to consider that spiritual communication must be balanced by mythic-animistic, rational, and aesthetic communication to keep us grounded and guard against unfit interpretations.

The five states can be equated with expressions of mysticism: gross nature mysticism, subtle deity mysticism, causal formless mysticism, mirror-mind witnessing mysticism, and nondual mysticism, reflecting a progressive focus on the earth or immanent Spirit, deities or spirits or first forms pointing toward transcendent Spirit, transcendent Spirit, and nondual immanent and transcendent Spirit, with communion, union, and identity potentially occurring within each.[45] Wilber makes clear that these states are naturally occurring, this-worldly experiences with corresponding brain-wave patterns, yet they should not be reduced to brain waves. In other words, spiritual states and brain states are intertwined and arise mutually, and, more generally, the unfolding complexity of material evolution and unfolding complexity of levels, lines, and states, or spiritual evolution, arise mutually.[46]

Wilber has used the writings of Ralph Waldo Emerson as his exemplar for the experience of nature mysticism or the world soul, or Over-Soul as Emerson referred to it: "We see the world piece by piece, as the sun, the moon, the animal, the tree; but the whole, of which these are the shining parts, is the soul [the Over-Soul, the World Soul]."[47] The focus here is on diverse nature, but the seemingly insurmountable distance between subject and object is transcended within the Over-Soul. Again, this transcendence often proceeds out of vision-logic, or contemplation, in which imagination guides the interplay between thought and feeling toward a relationship of oneness with the rest of nature. In these intuitive moments, the unity of all things is revealed as an obvious pre-given of existence, but only through contemplating the sensual aspects of these same things in all their diversity.

Ironically, Peters turned to Emerson to emphasize separation rather than connectedness, yet he is right (once again) to point out that Emerson also had instructive experiences of the distance between subject and object, in which he questioned the limits of our knowledge of others, and we spend most of our communicative time negotiating this distance. After all, both Emerson and Thoreau wrote of double consciousness, of ecstatic, mystical moments inspiring devotion to higher laws and more frequent mundaneness when we are not quite feeling it.[48] Or, as Emerson put it, "our faith comes in moments, our vice is habitual." We may be God made manifest, expressing "likeness to God" and inner light, but, too often, we are "a God in ruins," Emerson's way of saying that we tend toward lives of quiet desperation. Yet by practicing self-reliance, we may keep the "sweetness of solitude" amid the crowd, and Thoreau wrote that "moments of inspiration are not lost though we have no particular poem to show for them; for those experiences have left an indelible impression, and we are ever and anon reminded of them."[49]

Despite wisely admitting limits to transcendence, Emerson's transcendentalist worldview is based on moments of transcendence that reveal our higher nature and higher laws like unity-in-diversity. Positing higher laws would likely make Schrag uncomfortable, as they sound too much like rigid criteria; yet, as I have shown, the in-between is filled with evolving relationships. Lived principles like unity-in-diversity and relationships that resonate and repeat are law-like but non-dogmatic; they inform but do not rigidly decide.

We are always left with postmodern interpretation in differing situations—that also repeats—but, as Thoreau wrote, life without principle produces no value to society, no contribution to the common good, and thus no gift event.

And so, while smiling knowingly at our all-too-human propensity for being a God in ruins, I will further articulate our potential for experiencing differing forms of state spiritual communication, knowing their god-like power from my time on the Appalachian Trail. However, despite such experiences disclosing glimpses of human potential, which often leave an indelible impression that enhances communicative competence, I am not arguing for continuous enlightenment or perfect communication.

Wilber uses Emerson to describe experiences of nature mysticism, rather than thinkers like Shepard or Bateson, because he wants to be clear that the world soul should not be mistaken for Mythic-level biospheric immersion, which reveals animistic reciprocity but resides at pre-rational levels, or ecologically inspired systems thinking, which reveals interdependency but is still based in science and thus resides at the Rational level of development. Bateson's vision, however, adds aesthetic responsiveness to systems science, or aesthetic communication that leads to experiences of sacred unity, although without appeal to a transcendent dimension. Regardless, Wilber argues that we must transcend and include Mythic and Rational—and thus differing expressions of relationship at those levels—by traversing mystical religious traditions and articulating the experiential insight of a transcendent dimension that is also fully immanent, or Spirit in and as the world.

Subtle deity mysticism, with its focus on state experiences of deities, spirits, or first forms, is harder to rationally make sense of and accept, as many spiritual traditions consider these beings as ontologically real, including devas in Hinduism, angels in Christianity, and guiding spirits generally. Reading about angels and devas always give me pause because I can't relate—I have never had such experiences, whether on the trail or elsewhere—but subtle state experiences also include "interior luminosities and sounds . . . bliss currents and cognitions, and expansive affective states of love and compassion," guiding us to a higher, more integrative self. These states are likely more relatable, especially blissful moments of love and compassion, which I have often experienced.[50] But, of course, whether I have had any spiritual experiences is not a condition of their existence. To make such a claim would not only be the height of ego and arrogance, but irrational and laughable. The point, then, is that people have intense subtle state experiences of nonphysical entities or luminous forms and energies, often in visions or inner impartations arising out of meditative or trance states, and they will be interpreted based on one's culture and stage level of gravity. To Wilber, at Magic or Mythic levels they are more likely to be interpreted as real while at higher levels they are likely interpreted as involutionary first forms or patterns emerging from a storehouse of consciousness or larger shared mind.[51] Either way, an insight is received, although, as always, the quality of the insight communicated depends on the quality of the interpretation.

Causal and witnessing mysticism are steps to more profoundly experiencing that we are Spirit in and as the world, but with focus on the unmanifest Source of all form. In other words, we fully embrace a transcendent dimension before incorporating it into our sense of self. This shift often requires a temporary psychological break from the manifest world, leading to a witnessing awareness in which we detach from limited notions of self, dis-identifying with body, emotions, desires, and thoughts in order to identify with what remains: the unmoved Witness of body, emotions, desires, and thoughts. We are "not this, not that" because we are the Seer of "this and that," aware of awareness and witnessing all form with equanimity.[52] We dwell in the fullness of Emptiness on our way to feeling liberated from the things of this world, discovering a healthy non-attachment that lessens the influence of the isolated ego's tendency to be weighed down by illusory desires. However, this liberation then allows full embrace of the world as the glorious manifestation of Spirit.

Nondual mysticism is marked by this embrace, in which form is Emptiness and Emptiness is form and "Consciousness and its display are not-two." Or, as Wilber further writes, "you will not have experience, you will suddenly become all experience," and "you are not looking at the mountain, you are the mountain."[53] Thus the word "experience" is problematized, as it suggests a separate subject experiencing an object rather than nondual awareness or consciousness. This does not mean that distinctions are not recognized within this state "experience"; rather, they are seen with clarity, and thus simultaneously experienced as aspects of Spirit. Wilber advises: "abide as Emptiness, embrace all form." Put simply, or in utter simplicity on the far side of complexity, we discover nondual awareness in which transcendent and immanent are experienced as One Taste.[54]

Nondual awareness is closely related to causal and witnessing mysticism, since by detaching from the world of form we are finally able to fully embrace the world for what it is: the perfect manifestation of Spirit. Wilber is fond of quoting Sri Ramana Maharshi's seemingly paradoxical syllogism in order to convey the relation between the causal/witnessing and nondual: The world is illusory; Brahman alone is Real; Brahman is the world. The first two lines reflect the beginning of causal state awareness, while the third completes the ultimate realization of the nondual.[55] Thus, I emphasize once again, while the world is transcended it is also fully embraced.

Wilber writes that answers to our deepest questions seem obvious within this state when paired with Super Integral stages: Who am I? Why am I here? What is the purpose of life? What value do I have? Where did this world come from? Why is there something rather than nothing?[56] Such communicative certainty sounds dangerous, and it is, since all levels have their dysfunctions. Again, all states may be experienced at all levels, yet may be interpreted with more awareness at higher Integral levels. At lower levels, a transrational peak experience, or peeks into wider awareness, may act as a catalyst to growth, or, if interpreted dysfunctionally, may result in ego aggrandizement. At Super Integral, healthy nondual awareness is expressed as Eco-Sage highly effective effortless effort, in which what needs to be done is done. Wilber explains: "Resting in simple, clear, ever-present [non-dual] awareness, you will arise with the qualities and virtues of your own highest potentials." This may take many forms: compassion, discriminating wisdom, healing presence, artistic accomplishment, athletic skill, great educator, or "perhaps something utterly simple, maybe being the best flower gardener on the block."[57] In other words, liberated from separation, our highest potentials are freed to flow forth as fitting responses, or rather, as responses that are more fitting.

While spiritual states are rare and often transient, or problematic when interpreted at pre-rational levels, or denied at rational levels, or not experienced at all, Wilber argues that we may take a scientific approach to inducing them, as material and spiritual methodologies are similar: if you want to know this, do this (formulate a hypothesis, or learn about a meditation practice or contemplative prayer or plan a long hike); do the experiment (with controls) or spiritual practice (consistently); and then communally confirm results by comparing findings with other competent experimenters or practitioners. And, like peer-reviewed science, confirmation is rigorous among a demanding spiritual community, with false claims and misinterpreted experiences challenged by the teacher and advanced student-practitioners, backed by a long history of trial and error that may be transversally revisited.[58]

Such methodologies, whether scientific or spiritual, engage the process of knowing more, but transversal rationality demands we ask questions about the limits of knowledge and whether one may fully possess wisdom, despite Wilber's claim that nondual awareness is the most certain of all human experiences.[59] Ultimately, however, nondual insights cannot be rationally demonstrated or proven via the eye of flesh (science and senses) or the eye of mind (rationality, logic), but is revealed via eye of contemplation (spiritual practice and experiences).[60] Thus, communal confirmation among fellow contemplators is necessary for determining the unfit, especially self-deception, which is as possible as the possibility of spiritual experiences. And so, whether I have had an experience does not matter, but if I have engaged in a practice and had experiences of spiritual communication, then I must compare, contrast, and confirm with history, literature, and fellow practitioners, and once confirmed, I may be part of the community that compares, contrasts, and confirms.

Sounds a bit like a secret society, and that is off-putting, but it is actually a community of critical interpreters.

Given that states may be experienced via communion, union, or identity, interpretations may also vary from personal to impersonal, and thus one may experience what feels like personal conversations with God or the great Thou, or impersonal dialogue with Spirit.[61] Wilber is receptive to personal converse, which surprised me a bit given his embrace of nondualism. But he honors differing traditions and argues that when such converse is practiced in right spirit, it becomes a sincere form of worship, in which we are receptive to God's will or sympathy with intelligence; but, again, we must participate in communal confirmation, and not only with those from our community. And that, of course, is asking a lot, as it opens us up to the possibility of dissonance.

In the progressive Quaker tradition, for example, practitioners sit together in collective silence, no doubt battling inner noise like thinking about hard chairs on butt cheeks or thinking about thinking, but if a degree of inner silence emerges, aided by the peaceful energy of other souls, God may speak from within—after all, that's where the kingdom can be found—and thus out of collective silence comes individual impartation that may be shared with the group. At the least, each is able to speak their piece coming from inner peace.[62] But, again, what is received depends on stage-interpretation, and dignity or dysfunction, lest we forget that many morally bankrupt evangelicals support Trump because God told them to, with one selling Trump prayer coins for $40 so that the faithful are reminded to do collective prayer on his behalf. As always, experiences are far-ranging, from the profound to the absurd, but personal relationship with Godhead or great Thou or transcendent Spirit, which we ultimately are, may bear fruit, especially at higher stages.

Before continuing, I must re-emphasize that this is a simplified overview of Wilber's meta-theory, which includes four quadrants of development: interior-individual (introspection, intention; subjective "I") and interior-collective (cultural values and ethics; intersubjective "We"), and exterior-individual (biology and behavior, including brain states; objective "It") and exterior-collective (social structures from tribes to nation-states to planetary; interobjective "Its").[63] The stages/levels, lines, and states briefly articulated thus far mostly focus on interior individual and cultural development in relation to communication, but all interrelated quadrants are crucial to a fuller understanding of development and the possibility of enacting fitting responses. Along with Ramsey's disclosure of how exterior social structures bound with dominant narratives may block much communicative possibility, we also need healthy exterior individual development, or healthy bodies and brains, which, sadly, is often bound up with social structures: people of color and the poor, for example, are most at risk of being infected with and dying from coronavirus, and often live in unhealthy environs that lead to asthma. We cannot develop if we can't breathe.

And so, all quadrants, and levels and lines within them, and states interpreted from differing levels, must be taken into account to more fully understand and respond to the variety of wicked problems we face, including, of course, the most wicked: climate crisis. Otherwise, our responses are partial yet often put forward as complete and fitting. Gathering insights from all quadrants exposes the old saw question, what do you change first, individual consciousness or collective societal structures, as a false dichotomy.

Wilber's writings also continually make it clear that we do not easily waltz through stages of development. Numerous pathologies may keep us stuck at a certain level or cause us to regress, and thus he recommends shadow work, or what he calls "cleaning up," especially for those expressing Pluralistic dysfunctions that block growth to second tier Integral.[64] And while Wilber considers the potential for creativity and growth to be universal aspects of the manifest world, or Spirit in action, he argues that stages/levels, lines, and states are not predetermined; rather, they are involutionary archetypes or first forms—Spirit unfolding comes with orienting structures—which evolve as patterns of self-organization. Wilber refers to such structures as emergent Kosmic grooves or habits (in the good sense), revealed by phenomenological and empirical inquiry.[65] Again, such habits are similar to repetitions of difference, but for Schrag such repetitions emerge only from cultural history. Regardless, like Schrag, he argues that his meta-theory is post-metaphysical, as he puts together ongoing available evidence in the form of orienting generalizations, while acknowledging that new evidence and patterns may emerge. However, unlike Schrag, he fully embraces religious experience as a valid knowledge-claim or form of communication, which includes revelations of evolving involutionary first forms.

Wilber's meta-theory is central to this project because it includes insights into mythic-animistic communication at pre-rational levels; insights into rational communication at Rational, Pluralistic, and Integral levels, and, as we will see, insights into aesthetic communication at the brink of Integral and then at Integral. However, the debate must continue, via participation and distanciation, as to whether mythic-animistic communication resides solely at pre-rational levels or if Wilber has failed to fully include the merits of mythic-animism. For Wilber, the debate centers on the pre-trans fallacy, or the tendency to mistakenly equate pre-rational animistic experience with transrational aesthetic and spiritual experience simply because they are both non-rational.

While Wilber's meta-theory is highly instructive, this debate shows the limits of schemas of this type; we do not always fall neatly into differing levels and lines. In fact, Wilber's integral approach may be a bit problematic due to his choice of metaphors. Integration, especially from a post-metaphysical perspective, suggests totality, even though Wilber consistently reminds that while emergent structures display universal resonance, no perspective is final, including nondual awareness. Also, we must remember that while situational coherence among perspectives may be desirable—the possibility of fitting responses—fruitful contradictions that resist synthesis will remain, which is why I prefer the phrase "dialogic interplay" in support of fitting responsiveness and responses that are more fit.

Again, Wilber has his critics, usually focused on specific claims related to a particular quadrant rather than his vision as a whole, which makes sense given that his meta-theory puts together over 100 congruent theories on development. Such overlap, or communal confirmation, suggests enduring structures. And, as his supporters Zimmerman and Esbjorn-Hargens show, his work may be used to address ecocrisis. They argue for bringing into dialogue all partial and pertinent perspectives that may provide responses to eco-social problems, while acknowledging that some perspectives are incommensurate; the Greenpeace activist, for example, may find solutions from eco-economists, like carbon trading, as more of the capitalist game rather than challenging destructive modernist myths. Still, the Greenpeace activist and eco-economist may have similar concerns and worthwhile perspectives to share, which, if possible, may be woven into a wider tapestry of partial truths. Following Wilber, Zimmerman and Esbjorn-Hargens state that all we have in a postmodern, post-metaphysical world are perspectives, but rather than leading to post-truth with no perspective being better, the weaving of perspectives from various levels, lines, and states provides a more comprehensive and therefore more evolved and superior response, or, again, a fitting responsiveness.[66]

Wilber's meta-theory brings depth and clarity to our lives, and thus to an ecology of communication. In other words, Wilber's work does double duty: his articulation of stages/levels and lines of development provides insight into the dialogic interplay among differing modes and moments of communication, and he articulates the potential for communication via spiritual states, the highest being nondual state experience, in which we perceive one in many, many in one. Such communication does not mean that the many, or the diverse earth, disappears; rather, depth of relationship has been discovered with clarity: the other is other but also us, thou art that. Or rather, psychological distance dissolves, material distance remains. Again, such experience is communicative because an insight is received, and the depth of the insight communicated cannot be discovered in any other way.

Wilber takes religious experience seriously, and such experience, while not sufficient to climbing the ladder of development—we must do much rational work—may point the way. Again, a glimpse of unity, of *state* spiritual communication, while initially interpreted from one's stage center of gravity, may catalyze transformation; after all, the Sufi poet Kabir remarked that perceiving the unity of the universe "for fifteen seconds" made him a "servant for life."[67] Peters would obviously point out the dangers of such lofty ideals, which are far loftier than Schrag's appeal to agape love, as they suggest that communication that does not reach such heights reflects failure. But the transformative potential of spiritual communication provides context for other valued and necessary modes of communication. Still, stage center of gravity, along with differing developmental lines, thankfully reveals that state experiences and talking a good game does not mean one is enlightened, as many gurus and their many misdeeds have shown, including authoritarian sexual and emotional abuse. What marks our degree of enlightenment in differing situations is response and responsibility, but what we sometimes get is a strange brew of spiritual insight and mythic thinking. After all, even the Dalai Lama, following Tibetan cultural traditions that often serve him and others well, once condemned homosexuality as "sexual misconduct" before changing his tune.

Ultimately, Wilber's meta-theory connects development, identification, and communicative ethics: "If you identify only with you, you will treat others narcissistically. If you identify with your friends and family, you will treat them with care. If you identify with your nation, you will treat your countrymen as compatriots. If you identify with all human beings, you will strive to treat all people fairly and compassionately, regardless of race, sex, color, or creed. If your identity expands to embrace the Kosmos, you will treat all sentient beings with respect and kindness."[68]

Thus, in its simplest expression, we must take progressive steps from egocentrism to ethnocentrism to worldcentrism, or world-centric dialogue where we can begin to address global eco-social problems, and beyond into second tier, having our center of gravity within the world soul, and still further into third tier Spirit unfolding. Or, the more we are aware of Spirit manifested in and as all four quadrants, the more adequately we may "interpret the intuition of Spirit," and "the more that Spirit can speak," and "the more the channels of communication are open, leading from communication to communion to union to identity."[69]

For Wilber, our most enlightened responses reflect our always already Supreme Identity and the call to responsibility such identification entails, or, in his terminology, "showing up." Such presence is way beyond most of us, or all of us, but, like Schrag's gift event, informs our responses as possibility and ideal.

SHADOWS AND LIGHT, UNFITTING AND FITTING: TRANSCENDENT TRADITIONS

It is not difficult to find criticisms of mysticism. However, I did not expect Buber to supply one of them. I have always read *I and Thou* as a mystic-friendly text, although, similar to reading differing translations of the *Tao Te Ching*, comparing the earlier Smith translation with the later Kaufman one says a lot about interpretation, as Smith uses the traditional "thou" but Kaufman uses the more modern "you." "Thou" suggests sacred connection to the other while "you" suggests a more existential respect for the otherness of the other, which would make connection more difficult. To me, these are interrelated truths, although I know what interpretation Peters would prefer; regardless, whether thou or you, we exist within and honor relationship when we treat others like thous or yous instead of its.

Despite differing translations of *I and Thou*, Buber makes it clear in a foreword to *Pointing the Way* that he passed through a mystical stage on his way to relationship and dialogue. He does not deny "ecstatic" religious experiences of unity, but he's critical of the escapist attitude that may issue from this experience, returning to the world but regarding "everyday life as an obscuring of the true life." Instead of bringing unity to whole existence, including hardship and sickness, there may be a fleeing into "detached feelings of unity of being, elevated above life." For Buber, in such "higher hours," there is no other, and thus "the great dialogue between I and thou is silent; nothing else exists than one's self, which one experiences as *the* self."[70]

In an essay, "With a Monist," Buber further criticizes this escapist attitude in his response to the question of whether he is a mystic: "No, I answered, and looked at him in a friendly way, for I still grant to reason a claim that the mystic must deny it. Beyond this, I lack the mystic's negation . . . I am enormously concerned with just this world, this painful and precious fullness of all that I see, hear, taste. I cannot wish away any part of its reality. I can only wish that I might heighten this reality."[71] Wilber, of course, would agree with the desire to heighten our experience of this world, while stating that reason is not negated but transcended and included. Still, Buber wisely criticizes a regressive, anti-rational mysticism; in fact, this is Wilber's critique of New Ageism, which he considers solipsistic, mistaking pre-rational for transrational awareness.

Buber wisely criticizes potential shadow elements associated with causal and witnessing states, in which one identifies with the transcendent without embracing the immanent. Wilber uses mystical language of escape, but always follows with embrace: "flee the Many, find the One; having found the One, embrace the Many as the One."[72] Or: "If you are *only* after the things of this sensory world, then you will not discover higher or deeper realities. But if you go overboard and deny or repress this world, you will never find the Nondual, the radical estate that includes *both* the One and the Many, otherworldly and this-worldy, Ascending and Descending, Emptiness and Form, Nirvana and Samsara, as equal gestures of One Taste."[73]

Buber's hits on some of the basic criticisms of mysticism, and he is quite right that these shadow elements would make us poor communicators, which is especially problematic if we think we are mystically minded and dialoguing but are actually monologuing by projecting ourselves onto others. We will explore these potential downsides when exploring New Age and Ascension thought and practices, as well as interspecies communication, which often claim to be responses to ecocrisis.

It is also not difficult to find criticisms of a transcendent dimension. Susan Griffin's influential *Woman and Nature* provides an ecofeminist perspective on transcendent traditions, in which she catalogues the historical roots of ecocrisis prior to modern dissociation. Buber and Griffin's seminal texts—*I and Thou* and *Woman and Nature*—were marker moments in my reading quest to find support for the overcoming of the anthropocentric, or in Griffin's case, the androcentric, bias of communication study. Despite the fact that I embrace mysticism and a transcendent dimension, although not uncritically, Buber and Griffin go a long way, or point the way, toward dialogue as a basis for life.

Griffin argues that a transcendent dimension—whether expressed via Platonism or monotheistic religions like Christianity—negates the fact that we are born of and immersed within earthly relations, leading to dualism and dominance, not nondual embrace. She further links transcendental thought to objective scientific inquiry, which also negates nature, treating it as an unrelated other. Thus, the traditions of philosophical and religious idealism and scientific materialism, which are usually understood as opposites, have something in common: they all dualistically deny the natural world via monologic practices. By traversing the androcentric shadow side of ancient Greek thought, monotheism, and science, Griffin articulates a thorough ecofeminist and dialogic critique; in other words, if we are to practice transversal rationality, participating in transcendent traditions by gathering fitting insights, we must, as always, also practice distanciation and discern the unfit.

Griffin utilizes commanding and calculative prose—mimicking the monologic language of a masculinist master narrative—to catalogue traditions of patriarchal knowledge-formation, and poetic prose to give the ecofeminist counterresponse. She writes in the prologue: "He says that woman speaks with nature. That she hears voices from under the earth. That wind blows in her ears and trees whisper to her. That the dead sing through her mouth and the cries of infants are clear to her. But for him this dialogue is over. He says he is not part of this world, that he was set on this world as a stranger. He sets himself apart from woman and nature. . . . And he says that he cannot hear us speak. But we hear."[74] Griffin then begins Book One, "Matter: How Man Regards and Makes Use of Woman and Nature," by critiquing the transcendentalist legacy bequeathed by Plato: "It is decided that matter is transitory and illusory like the shadows on a wall cast by firelight; that we dwell in a cave, in the cave of our flesh, which is also matter, also illusory; it is decided that what is real is outside the cave, in a light brighter than we can imagine, that matter traps us in darkness. That the idea of matter existed before matter and is more perfect, ideal."[75]

Griffin deconstructs the patriarchal privileging of abstract thought by problematizing three degrees of abstraction that supposedly lead to higher truths: "The scientist peels away uniqueness; the mathematician peels away sensual fact, revealing number; the meta-physician peels away even number and reveals the fruit of pure being."[76] She further criticizes the abstract notion that angels "live above the moon and aid God in the movement of celestial spheres." Such angels disregard temporal matters, and matter generally, as that is where the demons dwell. Thus, "it is decided" that "the demon resides in the earth . . . in Hell, under our feet." Of course, it is also decided that "women are closer to the earth," opening the door to the long hellish history of devaluing them both.[77]

Buber woke me up to the profound difference between monologue and dialogue, as did Griffin, but while he theoretically explores the dangers of living in a world dominated by an I-It attitude, she more specifically deconstructs the damage done, to both women and nature, when this attitude is ensconced in philosophy, religion, and science. Thus, Griffin further woke me up to the fact that we can read history, which is inevitably a history of relationships, through the lens of communication. Communicative metaphors, insights, and experiences, and associated narratives, are everywhere—or, like transversal rationality, they cut across all things—and they may be used to embrace or deny a fuller humanity and fuller response.

Griffin goes on in page after page of historically documented and monologic "he saids" to critically move from Platonic idealism to Christianity to numerous abstract, detached, and objective expressions of scientific materialism. For Griffin, dualism and dominance—expressed via abstract rationality historically associated with these manmade traditions—has limited our responses such that emotion and the body were dissociated long before the modern period, and she shows that it isn't difficult to traverse this legacy to gather examples. Griffin continues by referencing the power of language to disparage and silence: "The word "hysterical" is taken from the word hyster, meaning womb, because it is observed that the womb is the seat of the emotions (and women are more emotional than men). . . . That crying is womanish, it is observed, and that dramatic poetry, since it causes crying, ought to be avoided, that it "has a most formidable power of corrupting even men of high character."[78]

Griffin's reference to the etymological roots of "hysterical" and its derogatory association with women is obviously a tradition deserving of distanciation, yet was used to denounce Rachel Carson after the publication of *Silent Spring* in 1962. Carson was called a hysterical woman by chemical industry executives—male, of course—in an attempt to discredit her scientific, and poetic, narrative warning against the indiscriminate use of biocides, which killed life, not just pests. And this tradition marches on. More recently, Naomi Oreskes, a historian of science, has taken on ExxonMobil's spin on warming, showing that they long knew the facts but denied or distorted them; in response, she gets hate email and has been vilified, labeled a communist instead of hysterical. Other women scientists speaking out on climate have received death and rape threats, disclosing more examples of pre-rational Mythic-level thinking rejecting Rational and Pluralistic awareness.[79]

Griffin's roar catalogues the history of detached calculation and discovery, despite science itself revealing relationship: "Under the gaze of science, it is said, all the basic units of matter shed their substance," and it's found that electrons cannot be specifically located, now a particle, now a wave, and that the observer effects the observed.[80] Prediction continues, however, in the form of probabilities and tendencies, and the same predictive eye was turned to women: less objective, less superego, less ego, less libido, less sense of justice, more emotion, more clinging to home, tradition, and sameness. Men, it is said, are "responsible for civilization," while women identify with nature turned into a dualistically opposed It.[81]

Ecofeminists have deconstructed biological claims of closeness to nature, but historical and cultural affinity led Griffin to valorize women's capacity to listen to nature's voices. Jane Goodall, of course, famously listened by practicing I-Thou science, which may be tellingly contrasted to Harry Harlow's I-It experiments in which he took Rhesus Monkey babies away from their mothers, substituting artificial ones made of wire and cloth, and later spikes, to test the effects of maternal deprivation, and, he argued, love; that's right, that's how I-It science tries to learn about love. Harlow, who said he did not care for the monkeys or like animals, also did studies of isolation, placing monkeys in partial isolation for up to fifteen years and total isolation for up to two years; surprise, the monkeys emerged deeply shattered, many turning to self-mutilation.[82] Goodall, in contrast, studied wild Chimpanzees in Tanzania, learning via close calculative observation but also genuine sympathy and love, rather than torture, of their social and family relationships.

Still, while reading *Woman and Nature* one is inevitably struck by the list of amazing discoveries produced by abstract calculation and prediction, even though observer and observed are ultimately not-two, and even though the destructive downsides when applied to the fullness of life are made abundantly clear. After all, climate science is an abstract and predictive affair, which we desperately need and yet it's not enough. Also, Griffin's interpretation of Plato's "Allegory of the Cave," which she reads as dismissive of matter and the body in favor of a more perfect ideal is both instructive and restrictive; Plato's allegory, arguing that we must move from shadowy one-dimensional material thinking into the light of a wider awareness, remains relevant in an age of habitual overconsumption. And, it's worth noting that men, including Michael Mann and Bill McKibben, have also received death threats for climate research and activism, although, very much worth noting, not with gendered rhetoric.[83]

Griffin makes clear the links among transcendent idealism, androcentrism, and abstract science in relation to unfitting responses to ecocrisis. And again, she is not alone in her criticisms of transcendent traditions. Shepard argues that transcendental thinking is linked with the birth of sky god agriculture and the demise of hunter-gatherer life, ultimately resulting in ecocrisis, and deep ecologists and ecopsychologists argue that duality and dominance are linked to privileging heaven over our sacred, earthly home. Wilber is not ignorant of these criticisms, and responds with nondual embrace of transcendent and immanent, or descent and ascent, equating the descent tradition with agape and involution and the ascent tradition with eros and evolution. In other words, Spirit's involution, or descent, as manifest existence reflects self-giving agape love, and the evolution, or ascent, of Spirit toward higher unities reflects the self-organizing relational drive of eros. Wilber considers the dualism between these trajectories the "Great Dualism of all dualisms, manifesting as pure otherworldly and pure this-worldly spiritual perspectives."[84] Still, despite his emphasis on nondualism and integral mapping of the four quadrants, he may display bias toward the ascent tradition.

Andy Fisher, in *Radical Ecopsychology*, while mostly agreeing with Wilber's work, makes this argument, stating that modern Western culture has "little idea of what an embrace of the earth truly looks like" and what it "demands of us," which is why he turns to Shepard and indigenous wisdom.[85] He further argues that despite Wilber's four-quadrant approach, external societal structures are not fully explored in relation to ecocrisis, which leads to uncritical participation rather than critical distanciation. In particular, Fisher argues that Wilber, and those who apply his meta-theory in books on integral business and economics, are "soft on capitalism," neglecting the fact that "greenhouse gases accumulate in the atmosphere as capital accumulates on the earth" and "unless capital is accumulating it is not capitalism."[86] Said differently, he is concerned that Wilber and those influenced by him perceive the dignities of capitalism, including growth, technology, and progress transcending the cruelties of feudalism while delivering the material goods, but not enough of the cruelties of capital accumulation, whether the outrageous gap between rich and poor, the sixth great period of species extinction, or general destruction of the biosphere. To Fisher, these horrors are not fully addressed in his work, and one of the reasons is because bias toward the ascent tradition does not fully keep the earth in mind.

Fisher argues that Wilber and integral followers embrace moral or conscious capitalism, dismissing Marxism as a historical artifact instead of seeing it as the most thorough deconstruction of capitalism available, filled with critical insights needed in an age of ecocrisis.[87] In contrast, he advocates for Marxist critical theory and ecosocialism, which would include, of course, free-market processes, and be more democratic, not less, with governmental oversight ensuring material and spiritual freedoms for all species, like the right to dwell within a life-giving environment. But ascent bias displayed as faith in the spiritual marketplace leads to the failure to fully explore interior growth in relation to exterior socio-cultural structures, especially the blocks that Ramsey addresses via interpellation and the habit body or a systemically distorted communicative system.

Whether or not capitalism can be reformed is a key question of our time. For Fisher, the answer is, by definition, an obvious "no," and I suspect it would be "no" for the endless array of communities suffering from what amounts to "planned ecocide."[88] One need only pay attention to daily stories cataloguing more of the litany, with capital accumulation leading to misery accumulation. Where is the morality, the consciousness, the dignity for the more-than-270,000 debt-ridden farmers in India who have committed suicide, numbers increasing as climate disruption worsens? Or children in Malaysia suffering from respiratory disease from burning imported coal? Or residents of Flint, Michigan, facing financial crisis and poisoned water partnered with environmental racism? Or citizens of Puerto Rico, before and after Hurricane Maria, traumatized by disaster capitalism designed to increase inequality and reduce democratic processes? Or indigenous peoples of Brazil losing their homes due to clearcutting of rainforests for monoculture crops, or orangutans who have lost their homes in Indonesia due to clearcutting of forests for monoculture palm trees for palm oil, or thousands of other species who have left this home, having gone extinct as humans continue to expand and extract?

None of these besieged people are able to move up the developmental holarchy, and none are empowered to communicate their despair to the powers-that-be, although many indigenous peoples are fighting the good fight, demanding their voices are heard. Still, too many fellow human beings, and fellow beings, like orangutans but also so many other species, are considered voiceless and disposable. And a huge portion of the voices on the planet having no voice is built into the system.

Wilber is also not ignorant of economic injustices, and while not tackling the merits of ecosocialism versus moral capitalism, he has stated that the loss of jobs due to AI and automation may lead to guaranteed annual income.[89]

And, in a recent coauthored book with Alan Watkins, they have stated that the so-called free market is not free, nor should it be; that Rational-level economics is partial, fragmented, and broken; and that climate crisis may lead to class warfare over resources that even Marx could not imagine.[90] Furthermore, Wilber himself is critical of the focus on one quadrant over another, calling it "quadrant absolutism," and he has also thoroughly explored the shadow side of transcendent-focused traditions, including Buber's criticisms regarding a disturbing detachment from this-worldly concern and Griffin's historical recording of abstract rationality and damaging dissociation from nature.[91] All stages have potential dysfunctions, whether pre-rational, rational, or transrational, and Wilber thoroughly catalogues them so we may engage the process of "cleaning up."[92]

Still, the claim can be made that Wilber is biased toward nondualism, although for obvious reasons. His holarchic model of development is based on wider identifications, and thus it makes sense for the as-of-now ultimate to be third tier nondual embrace of transcendent and immanent Mystery or the Supreme Identity. After all, dualism and separation generally are underlying causes of ecocrisis, and Wilber provides a response: a developmental path of stage-steps toward increased wholeness. For Zimmerman and Esbjorn-Hargens, this response can be simplified into a kind of slogan: Things are getting worse, better, and are perfect.[93] In other words, we are simultaneously despoiling our own nest and waking up—no end of problems, lots of potential solutions—but either way, from a Super-Integral perspective all manifestation is the perfect expression of Spirit. The perfect part may sound off-putting given the undermining of ecosystem services due to our many imperfections, but perceiving perfection reminds us of what could be, seeing beyond gloom and doom by embracing the evolving play of Spirit. And we need to play, not just lament human fallibility, to productively serve, showing up by doing what needs to be done, including letting go in order to liberate fuller embrace.

Transcendent bias, especially in postmodern times, deserves criticism. Many, like Schrag, are rightly disturbed by the positing of a transcendent dimension signified by capital letters and claims for ultimate Truth, given that such claims have been a historical force for oppression, not to mention the deconstructive insight that all we have are interpretations. But Eastern philosophy and religion and philosophy and esoteric religion in the West, or the many traditions of "waking up" in search of some ultimate, add to a wider responsiveness, or, said differently, the spiritual enlightenment of the East and esoteric West must be integrated with the rational enlightenment of the West. For Wilber, the Great Liberation, with capital letters indicating breadth and depth, reflects a greater awakening that includes many paths, many perspectives and interpretations, as well as many transcending and including liberations at stages of development. Liberation itself develops—stages unfold, spiritual truths grow and evolve—but a greater, more integral liberation might be described as experience and expression of the groundless Ground of Being, while acknowledging that we always remain on shifting ground and literally, the ground, the alive and diverse earth. While I prefer dialogic interplay among differing perspectives, for Wilber integral simply means more inclusive and thus more responsive.

Schrag, of course, does not go as far as Wilber; the gift event, while a transcendent ideal immanently expressed, does not reflect nondualism. Self and other remain asymmetrical, seeking "convergence without coincidence," which protects against the danger of projecting oneness onto others and thus the assumption of mutual understanding that Peters warns against. And Ramsey argues for areligious religiosity marked by transcendence without the transcendent. Still, Schrag and Ramsey, in response to a purely deconstructive postmodernism, salvage religious feeling, whether as agape or eros, albeit interpreted differently than Wilber; both recognize that if logos is in force, after the deconstruction of religious claims for ultimate Truth, we must creatively construct by gathering insights in pursuit of fitting responses. For Schrag, this is manifested as the grace of gift-giving, especially when confronting injustice; and for Ramsey, as a fitting responsiveness to mundane and but especially world-historical events.[94] Again, their obvious difference from Wilber is his embrace of spiritual experiences, which, as we have begun to explore—it will be explored more—ultimately must be rationally interpreted given that unfit interpretations abound.

Nonetheless, nondualism, rather than reflecting fusion or projection of self onto other, is more akin to a gift event stimulated by will, or spiritual practice, and characterized by grace. We cannot wish away such awareness, in which distance between self and other is transcended while physical distance remains, due to poor interpretations. Nondual awareness as ideal is marked by all modes of communication being available, and this widened awareness leads us to perceive particular situations with more clarity, which then allows the gift to move, or come through, in the form of right action, and thus the pursuit of justice for all species. Yet, we remain thoroughly human and in the world, susceptible to misperception and misinterpretation. And, as Ramsey and Fisher have pointed out, we are not going to develop very far unless we are cognizant of societal blocks, and our capacity to create strategies to change and overcome them. Or, as Sartre wrote: "Freedom is what you do with what has been done to you."[95]

Criticism of a transcendent dimension and therefore spiritual communication will likely dog this project due to its seeming impossibility even if possible; however, the transcendent is inevitably part and parcel of our imaginative life. And such imaginative work, or play, simply entails contemplation of the nameless something, or nothing, depending on the tradition, which all religious and spiritual traditions have named, including God, Godhead, G-d, Brahman, Allah, Tao, Great Mystery, Spirit, Emptiness, the Infinite, the Absolute, the Divine, etc.—from which the manifest world has emerged. Differing traditions and devotees have differing life-denying and life-affirming interpretations, but the imaginative formulation of a transcendent dimension remains. Out of nothing came something. We are here; how did we get here?

This unavoidable, fundamental question gave rise to a transcendent dimension. And thus we should embrace this question, positing inconclusive answers informed by contemplation and spiritual communication, which beget more questions that provide direction in our transient lives, moving from "a culture of no-truth to a deliberately developmental culture."[96]

TRANSCENDENTALISM AS EXEMPLAR: SYMPATHY WITH INTELLIGENCE

Not surprisingly, a salient exemplar of a transcendental worldview embracing spiritual communication comes from the transcendentalists. As a philosophy, transcendentalism also provides a salient example of transversal rationality, as it gathers insights from numerous traditions: the panentheism of Hinduism; Plato's focus on ideas behind forms; the life of Jesus rather than institutionalized Christianity; Goethe's vision of a unified but continually changing nature; Kant's privileging of subjective knowledge and perception beyond the senses; Romantics like Coleridge, Wordsworth, and Carlyle, who furthered Kant's inquiry; and Emerson's call for self-reliance, exploring voice and what one is meant to do on earth.[97]

Thoreau embodied these ideals, living the transcendentalist vision. In early years, he spent his spare time in the Concord woods and exploring rivers and ponds, already gaining a reputation as "the one who did not fear mud or water."[98] The summer prior to attending Harvard, he spent time floating on Walden Pond on his first handmade boat, paddling to the middle then lying on the seat, gazing skyward, allowing the fates to take him where they will, finding that "idleness was the most attractive and productive industry."[99]

Productivity. Industry. Signs of the times that would shape him, and his defiance. He and other young men of the early to mid-1800s were asked to find their place amid burgeoning industrialization and westward "Manifest Destiny" expansion and exploitation; no longer would sons be restricted to following footsteps of farmer fathers or staying put. Daughters, while permitted to attend the Concord Academy, were expected to return to and run households. Yet the battle for liberation was brewing.[100]

It was a revolutionary time of industry and expansion, but also women's rights, Native rights, and most combustibly, abolitionism and the basic human right to not be someone's property. Interest in science was growing, with the question of how things came to be culminating in Darwin's 1859 *The Origin of Species*, which showed that all, including different races, were related and thus not so different. The seeds of ecocrisis were also planted, as were the ethical responses of the rights of nature and animal rights, thanks to the Thoreau he would become. And the gap between rich and poor was increasing, with potato famine Irish immigrants being the latest to suffer poverty and prejudice.[101]

Religion, too, was challenged. Unitarian minister William Ellery Channing preached that humans were defined by "likeness to God" and "inner light," not sin. Transcendentalists like Emerson would run with that idea, championing self-culture, or the cultivation of our inner spiritual nature, or soul within the world soul, and, by extension, our collective responsibilities when such light was dimmed by social institutions and those in power.[102] Transcendentalism became the first intellectual movement in the United States, its heyday a brief decade between the mid-1830s and 1840s, with Concord, home of the first Revolutionary War battle, and prior home to native Algonquians, at the center of the movement. Or more specifically, Emerson's home, where a handful of early transcendentalists would meet, leading scholar Lawrence Buell to argue that transcendentalism is best understood as "an outpouring of radiant energy" rather than "an organized enterprise."[103]

Quite a place, and time, for Thoreau to grow, with Emerson, his mentor, calling for a revolution of intellect and Spirit, an "original relation to the universe," not following England, not following tradition, not following the crowd.[104] In other words, for the transcendentalists, the question is not what traditions we should fully participate in or distance ourselves from, but what aspects of traditions are fitting and unfitting, gathering inspiration while creating new responses, and new men and women. One such woman was Margaret Fuller, the editor of *The Dial*, a transcendentalist journal. Fuller, a feminist pioneer, journalist, educator, political revolutionary, and revolutionary writer, influenced both Emerson and Thoreau; their discussions on Goethe were particularly important, persuading them to ground transcendental leanings more fully in the natural world, or the "law of the leaf," in which we perceive evolving patterns of intelligence.[105]

In "The Great Lawsuit," an essay published in *The Dial* and later extended as the groundbreaking *Woman in the Nineteenth Century*, she argued that women must have inward and outward freedom to grow, as there is "but one law for all souls."[106] Like all transcendentalists, she focused on the liberation of the divine spark within; for women that meant arbitrary barriers must be struck down such that all paths were open. Not seeing the destiny of women was a defect in man's moral development, as women were "taught to learn their rule from without, not to unfold it from within."[107] Education was focused on making them better companions and mothers of men, which actually made them like slaves, yet for Fuller, masculine and feminine were fluid, passing into each other.[108]

Fuller struggled to find her voice in a world that did not value women's abilities, or the fact that they too had souls. She wrote of her despondency in her journal; feeling pressured by church to conform and afraid of disappointing her father, who had educated her as if a son, she was weary from not having her gifts recognized by society. Emerson, mentoring Thoreau, saved him from a similar fate, stating he was "spiced through with rebellion" after their first meeting. A mentor may not have saved Fuller from her lostness, but a nondual spiritual experience did: she "saw there was no self, that selfishness was all folly, and the result of circumstance; that it was only because I thought self real that I suffered; that I only had to live in the idea of the All, and all was mine. . . . My earthly pain at not being recognized never went deep after this hour."[109]

Fuller went on to teach at Bronson Alcott's Temple School in Boston. Alcott, a prominent transcendentalist and the father of Louisa May Alcott, used Socratic dialogue to draw out insights from young students, until scandal from subject matter on religion and sexuality led to his school closing. Fuller loved reading, and teaching allowed her to share that love; Barry Andrews, in his book on transcendentalism, states: "Without doubt, she was the most well-read woman in America. "[110] She later taught at the Greene Street School in Rhode Island, and then elite women in Boston at Elizabeth Peabody's bookstore and salon. Peabody, another early transcendentalist guided by the fire of soul, assisted Alcott at the Temple School, started the *Dial* with Emerson, where she was the first to translate a Buddhist text for an American audience, advocated for early education for children, and called the mistake of confusing ego and soul "egotheism."[111]

Thoreau, Fuller, Alcott, and others who met at Emerson's home were philosophical idealists but did not discuss ideas detached from the world, gathering instead because they were dissatisfied with philosophy, religion, and literature in regard to how they influenced social and political life. And their transcendentalist vision, extolling self-reliance, intuition, and a fuller rationality, or logos, in pursuit of spiritual freedom, led to radical activities in the abolitionist movement, the women's movement, Native American rights, elementary education, journalism, labor, and at Brook Farm, a short-lived eco-socialist commune.[112]

And much of this activism was rooted in their embrace of the spiritual communication of higher laws or principles.

Thoreau was willing to call himself a mystic, and he stumbled upon mystical experience. During a time when agriculture was already growing in the direction of large farms and stultifying labor, Thoreau hoed his beans at Walden Pond as if conducting a concert: "When my hoe tinkled against the stones, that music echoed to the woods and the sky, and was an accompaniment to my labor which yielded an instant and immeasurable crop." Lost in his work, in reverie, he cultivated a loss of separate self: "It was no longer beans that I hoed, nor I that hoed the beans."[113]

Emerson, of course, had his own moments of spiritual communication. Not surprisingly, the most well-known occurred in the woods, "bathed by the blithe air, and uplifted into infinite space" such that "all mean egotism" vanishes and he became a "transparent eye ball; I am nothing; I see all; the currents of Universal Being circulate through me; I am part or particle of God."[114] Such experiences of unity-in-diversity led him to realize that we are much more than our individual selves, more than mere mortal coils, and thus nature always speaks of Spirit, while Thoreau extolled the language that all things speak.[115] Despite double consciousness, both practiced deep listening in sympathy with intelligence, and thus were called to respond to beauty, and the conditions that diminish beauty and cause suffering.

Thoreau shared plenty of practices to elicit such grace: "simplify, simplify, simplify," mentally and materially, living deliberately; read and write; saunter amid nature, attending to the senses and voices of nature; and aesthetic-poetic contemplation of Spirit, of the voice of nature perpetually shining through diverse nature. Even Peters acknowledged that communication is the ultimate border-crossing concept, and these practices elicited many profound crossings via spiritual communication, but also aesthetic and mythic-animistic communication.

In Thoreau's time, with enlightenment rationality becoming the leading edge of consciousness, and crises like slavery stirring souls, principles like freedom, rights, and equality emerged as a force for change, taking us to the modern world. In the 1960s, the fight for civil rights, women's rights, and clean air, water, and soil stirred souls, and enlightenment rationality evolved to include postmodern Pluralistic principles, with differing voices heard and differing movements formed as dominant narratives were deconstructed, including Mythic pre-rational religious views. The next stage, transcending and including a variety of views, has the potential to lead us to a new vantage point: The first time in history in which a significant percentage perceive and integrate partial truths, leading to more depth of consciousness and less deep-seated conflict. Or, picture this: a common humanity on a common home guided by Integral rather than Mythic-level spirituality, honoring all sentient beings, and finding rational solutions to eco-social problems. Sounds good, and sounds like the transcendentalists were already pushing the leading edge of consciousness back in the mid-1800s.

Reading the transcendentalists may grate on postmodern ears, since they speak so much of truth, of higher law, rather than interpretations of partial truths, but they were prescient, and faced early versions of the eco-social problems we face today. Perfectly imperfect, they worked, and played, in all four quadrants, exploring our liberatory potential by integrating existential and spiritual dimensions of life. They focused on growing up, waking up, cleaning up, and showing up, knowing that the more responsive we are, the more gifts we share.

NOTES

1. See Ken Wilber, *Sex, Ecology, Spirituality: The Spirit of Evolution* (Boston: Shambhala Publications, 1995). However, most of his books explore the ascent and decent of Spirit.
2. Integral World, http://www.integralworld.net/.
3. The quotation from Robert Kegan is on the back of Wilber's *The Religion of Tomorrow*. Kegan is also a founding member of Wilber's Integral Institute and conversations between them can be found on YouTube.
4. Ken Wilber, *The Religion of Tomorrow: A Vision of the Future of the Great Traditions* (Boston: Shambhala Publications, 2017), 197–250.
5. Ibid., 186–187.
6. Ibid., 166–179.
7. Ibid., 559–569.
8. Ibid., 36–45. See Jean Gebser's *The Ever-Present Origin*; many of Wilber's terms come from Gebser. Also, for the 100 theories supporting Wilber's meta-theory, see the appendix in his *Integral Psychology*.
9. Ken Wilber, *A Brief History of Everything* (Boston: Shambhala, 1996), 158–163.
10. Wilber, *The Religion of Tomorrow*, 199-200.
11. Ibid., 200. See also *A Brief History of Everything*, 174–185.
12. Wilber, *The Religion of Tomorrow*, 314–321.
13. Ibid., 202–203.
14. Wilber, *A Brief History of Everything*, 185–190.
15. Wilber, *The Religion of Tomorrow*, 203–205.
16. Ken Wilber, *Integral Meditation: Mindfulness as a Way to Wake Up, Grow Up, and Show Up in Your Life* (Boston: Shambhala, 2016), 61–71. See also Wilber, *A Brief History of Everything*, 185–190. Wilber often repeats the levels/stages of development in his books; both *Integral Meditation* and *Brief History* are good reviews.
17. Wilber, *The Religion of Tomorrow*, 44–45.
18. Wilber, *A Brief History of Everything*, 190–201.
19. Wilber, *The Religion of Tomorrow*, 205–209.

20. See Thomas Berry, *The Dream of the Earth* for insights into a universe comprised of a communion of subjects rather than a collection of objects.

21. Wilber, *The Religion of Tomorrow*, 321.

22. Ibid., 326.

23. Ken Wilber, *Trump and a Post-Truth World* (Boston: Shambhala Publications, 2017), 29–31.

24. Ibid., 3–11.

25. Ibid., 99–111.

26. Wilber, *The Religion of Tomorrow*, 321. See also Alan Watkins and Ken Wilber, *Wicked and Wise: How to Solve the World's Toughest Problems* (Great Britain: Urbane Publications, 2016), 134–138.

27. Wilber, *Trump and a Post-Truth World*, 137-145.

28. Alan Watkins and Ken Wilber, *Wicked and Wise: How to Solve the World's Toughest Problems.*

29. Wilber provides a good review of lines of development in *Integral Meditation*, 165–195. However, lines, which he sometimes calls waves or streams, are explored in many of his books.

30. Wilber reviews James Fowler's stages of faith in relation to spiritual intelligence in *The Religion of Tomorrow*, 199–209.

31. Ibid.

32. Vatican, 3, http://www.vatican.va/content/francesco/en/encyclicals/documents/papa-francesco_20150524_enciclica-laudato-si.html.

33. Ibid.

34. Yale Climate Communication, https://climatecommunication.yale.edu/publications/the-francis-effect/.

35. Sean Esbjorn-Hargens and Michael E. Zimmerman, *Integral Ecology: Uniting Multiple Perspectives on the Natural World* (Boston: Shambhala/Integral Books, 2009), 226–227.

36. Watkins and Wilber, *Wicked and Wise*, 94.

37. Wilber, *Trump and a Post-Truth World*, 54.

38. See Watkins and Wilber, *Wicked and Wise*, for insights on strategic communication, 260-265.

39. Chris Hedges, "Onward Christian Fascists," Truthdig, December 30, 2019, https://www.truthdig.com/articles/onward-christian-fascists/.

40. Wilber, *The Religion of Tomorrow*, 45, 517.

41. Ibid., 210.

42. Esbjorn-Hargens and Zimmerman, *Integral Ecology*, 226–258.

43. Wilber, *Integral Meditation*, 8.

44. For Underhill, see *Mysticism: The Preeminent Study on the Nature and Development of Spiritual Consciousness*; for James, see *The Varieties of Religious Experience.*

45. Wilber, *The Religion of Tomorrow*, 124–179.

46. Ibid, 130–131.

47. Quoted in Wilber, *Sex, Ecology, Spirituality*, 285.

48. Barry Andrews, *Transcendentalism and the Cultivation of the Soul* (Amherst: The University of Massachusetts Press, 2017). 30.

49. Ibid., 33, 39.

50. Wilber, *A Brief History of Everything*, 211.

51. Wilber, *The Religion of Tomorrow*, 135–145.

52. Wilber, *Sex, Ecology, Spirituality*, 306.

53. Wilber, *A Brief History of Everything*, 228–229.

54. Ibid., 240.

55. Wilber, *Sex, Ecology, Spirituality*, 302.

56. Wilber, *The Religion of Tomorrow*, 177.

57. Ken Wilber, *The Eye of Spirit: An Integral Vision in a World Gone Slightly Mad* (Boston: Shambhala, 2001), 300.

58. Ibid., 85, 87–88.

59. Wilber, *The Religion of Tomorrow*, 174–177.

60. Wilber, *The Eye of Spirit*, 12–13.

61. Wilber, *The Religion of Tomorrow*, 172–173.

62. See Parker J. Palmer, *Let Your Life Speak: Listening for the Voice of Vocation* (San Francisco, CA: 2000).

63. See Wilber's *Sex, Ecology, and Spirituality* for an early and thorough description of his four quadrant approach, but many later books also explore the four quadrants.

64. Wilber, *The Religion of Tomorrow*, 263–270.

65. Ibid., 600.

66. Esbjorn-Hargens and Zimmerman, *Integral Ecology*, 46–47.

67. Quoted in Duane Elgin, *Promise Ahead: A Vision of Hope and Action for Humanity's Future* (New York: HarperCollins, 1999), 173.

68. Ken Wilber, *A Theory of* Everything (Boston: Shambhala, 2000), 116.

69. Wilber, *Sex, Ecology, Spirituality,* 497.

70. Martin Buber, *Pointing the Way* (New York: Harper & Row, 1963), xv-xvi.

71. Ibid., 28.

72. Wilber, *Sex, Ecology, Spirituality*, 326.

73. Ken Wilber, *One Taste* (Boston: Shambhala, 1999), 134.

74. Susan Griffin, *Woman and Nature: The Roaring Inside Her* (New York: Harper & Row, 1978), 1.

75. Ibid., 5.

76. Ibid., 6.

77. Ibid., 7.

78. Ibid., 13.

79. Scott Waldman and Nijna Heikkenen, "As Climate Scientists Speak Out, Sexist Attacks Are on the Rise," *E&E News*, August 22, 2018, https://www.scientificamerican.com/article/as-climate-scientists-speak-out-sexist-attacks-are-on-the-rise/.

80. Griffin, *Woman and Nature*, 42–43.

81. Ibid., 38.

82. Lauren Slater, "Monkey Love," *The Boston Globe*, Boston.com news, March 21, 2004, http://archive.boston.com/news/globe/ideas/articles/2004/03/21/monkey_love/.

83. Bill McKibben, "Let's Agree Not to Kill Each Other," *The New York Times*, October 20, 2018, https://www.nytimes.com/2018/10/20/opinion/sunday/lets-agree-not-to-kill-one-another.html.

84. Wilber, *Sex, Ecology, Spirituality*, 338–339.

85. Andy Fisher, *Radical Ecopyschology: Psychology in the Service of Life*, 2nd edition (New York, SUNY, 2013), 234.

86. Ibid., 235.

87. Ibid., 333.

88. Ibid., 235.

89. Wilber, *Trump and a Post-Truth World*, 115.

90. Watkins and Wilber, *Wicked and Wise*, 266–275,

91. Wilber, *The Religion of Tomorrow*, 560–562, 352–412.

92. Ibid., 264–266.

93. Esbjorn-Hargens and Zimmerman, *Integral Ecology*, 307.

94. Calvin O. Schrag interviewed in *Experiences Between Philosophy and Communication,* edited by Ramsey Eric Ramsey & David James Miller (New York: SUNY Press, 2003), 45.

95. Carolyn Gregorie, "Living Well, According to Some of the Wisest People Who Ever Lived," HuffPost, December 6, 2017, https://www.huffpost.com/entry/the-truest-things-ever.

96. Wilber, *Trump and a Post-Truth World*, 142.

97. Richardson Jr., *Henry Thoreau*, 72–73.

98. Walter Harding, *The Days of Henry Thoreau: A Biography* (NJ: Princeton University Press, 1992 (1962), 31.

99. Ibid., 31.

100. Richard Lebeaux, Young Man Thoreau (MA: University of Massachusetts Press, 1975), 18-27.

101. Lawrence Buell, *The American Transcendentalists: Essential Writings* (New York: Modern Library, 2006), xiii.

102. Ibid., xix.

103. Ibid., xxvii.

104. Ibid., xiii.

105. Andrews, *Transcendentalism and the Cultivation of the Soul*, 45–46.

106. Margaret Fuller, "The Great Lawsuit," in Lawrence Buell, ed., *The American Transcendentalists: Essential Writings* (New York: Modern Library, 2006), 307.

107. Ibid., 309.

108. Ibid., 313, 317, 319.

109. Margaret Fuller, "Recollection of Mystical Experiences," in Lawrence Buell, ed., *The American Transcendentalists: Essential Writings* (New York: Modern Library, 2006), 160.

110. Andrews, *Transcendentalism and the Cultivation of the Soul*, 81.

111. Ibid., 49.
112. Buell, *The American Transcendentalists*, xxvi–xxviii.
113. Henry David Thoreau, *Walden* in *The Portable Thoreau*, edited by Carl Bode (New York: Penguin Books, 1975), 408.
114. Ralph Waldo Emerson, "Nature," in *Nature, Walking*, edited by John Elder (Boston: Beacon Press, 1991), 8.
115. Thoreau, *Walden* in *The Portable Thoreau,* 363.

Chapter 3

To Learn but Not Return

Paul Shepard and Mythic-Animistic Communication

When I was an environmental studies graduate student at the University of Montana, one of my professors seemed surprised to learn that I was a big fan of Paul Shepard's work. I was confused at first, but then realized he knew of my interest in mysticism and probably thought Shepard's grounded arguments conflicted with my more transcendent explorations. But he did not know about another interest: the communicative insights, metaphors, and experiences that are abundant in his writing. Plus, while Shepard and Wilber have differing views, they explore two much-needed sides of the same more-than-human world. Or rather, while Wilber warns against regression to our pre-rational past, Shepard warns against the repression of hunter-gatherer lifeways specific to oral cultures. In other words, we need to transcend and include, and to fully include we must realize that before the emergence of ancient Greek logos there was the experience of mythos, and it too has been ignored in modern times. We turn to Shepard, then, to traverse a mythic-animistic life-world filled with nonhuman voices, finding an mytho-historical, evolutionary, and ecological context for rethinking ecocrisis.

In *Nature and Madness*, Shepard begins with a question: why do humans persist in destroying their habitat? He considers numerous possibilities, including inadequate philosophies and value systems, but these seem too academic, reliant on the history of ideas rather than a deeper look at our psycho-history as hunter-gatherers and the pathologies that occur when a mythic-animistic view of creation replete with presences and voices becomes mute. Shepard, like Wilber, focuses on development, but for him this muting is an irrational stepping away from who we are rather than a rational step forward to who we must become. Thus, a "surround of living plants, rich in texture, smell, and motion," and "the unfiltered, unpolluted air, the flicker of wild birds, real sunshine and rain, mud to be tasted and tree bark to grasp, the sounds of wind and water, the calls of animals and insects as well as human voices" are not "vague or pleasant amenities for the infant," but essential to healthy growth in which "the clues to the meaning of life are embodied in natural things."[1]

For Shepard, two of the many wisdoms expressed by hunter-gatherers are the quality of their attention and the significance of place, and what is noticed, disregarded, and expected via overlapping senses reveals a world. Of course, a high level of attentiveness demands skilled listening, which, for Shepard, means recognizing that "all sound is a voice—dynamic, revealing, and communicating."[2] But with the burgeoning of agriculture, domestication, and urban life, farmers and citizens "ceased to listen to a million secret tongues in the wilderness."[3] Specific places lost their significance, ceasing to become "distinct, though unconscious, elements of the self, enhanced by mythology and ceremony, generating a network of deep emotional attachments that cements the personality."[4]

Shepard argues that we must learn from the past by salvaging the dignity of hunter-gatherer existence, especially since the modern world, and instrumental rational consciousness, reflects a mere cipher in time compared to our long history as hunter-gatherers.[5] Thus, respecting pre-Neolithic (before 10,000 years ago) modes of thought and communication is not a matter of return, but a realization that we've never left the experiences within which the human mind was formed. The human genome emerged from millions of years of evolution, and thus our biological and ecological heritage holds clues to our deepest psychological needs and traits, with many formed via the complex dynamics of predator-prey reciprocity.[6] Cooperation was needed among tribal members to be successful in the hunt, as was conceptualization, forethought, and planning. Hunters watched other animals, both predator and prey, and then mimicked their keen sensitivity to their environment. And children learned through perceiving the otherness of animals, trees, rivers, and plants, leading them to develop kinship bonds within a diverse natural world. Tribes owned little, wasted little, and their natural ecstasy was habitually aroused accompanied by an adrenaline and endorphin rush.[7]

Shepard does not deny aggressive and competitive traits or examples of brutality and ecological destruction among early hunter-gatherers.[8] But, generally speaking, he argues that the small social structure of their lifeway led to mutual dependence manifested as non-hierarchical gender roles, sharing, and communicative skill.[9] Wilber wisely warns against romanticizing tribalism, but Shepard's main point is difficult to deny: human traits and needs were initially formed during our long history as hunter-gatherers, and these needs and traits emerged via a chancy game of rewards, pitfalls, and numerous unexpected encounters. Hunting elicited humility and reverence, since the hunter, despite developed skills, was ultimately dependent on ritualized converse with animal spirits and the grace of the game. Hunting also demanded risk, which stimulated growth, ecstatic experience, and storytelling within the context of place and a gifting earth.[10]

Like Griffin, Shepard articulates the dangers of disembodied, abstract thinking. He is particularly harsh in his assessment of otherworldly orientations, since agriculture replaced the hunter-gatherer ideal, bringing with it transcendent sky-gods worshipped for good growing weather and abundant crops. And as agriculture and the belief in transcendent sky-gods grew, so did population, cities, the hoarding of food, domestication, disease, toil, violence, hierarchical societal institutions and social relations, and environmental degradation.[11] Humans had less leisure time, less time for gaming and game, and fewer experiences of timelessness in which feeling and thinking merged with the cycles of nature. The use of our sensuous imaginations, connecting predator and prey in communicative reciprocity, decreased in favor of the abstract imagination with its gods and particular mode of thought. Our identity shifted along with our imaginative life, and mind was removed from its embeddedness within the natural world.

In *Beyond Geography: The Western Spirit Against the Wilderness*, Frederick Turner is more upfront than Shepard concerning the dysfunctions of pre-Neolithic tribal groups, admitting that fear and terror before unseen powers played a significant role in governing behavior and that propitiation before these powers was ritualized as mutilation and sacrifice—in other words, Magic and Mythic-level behavior. He further states that brutish realities should not be dismissed via a "crude antirationalistic counteroffensive."[12] However, despite these rational admissions, for Turner it is also rational to explore the core of myth, which remains the celebration of life. The intersubjective "extension of mind into the world" and "attribution of soul life ("animism") to nonhuman aspects of the phenomenal world" that is typically condemned by objective rational thinkers fails to consider that "over the long course, Life stripped of its numinous dimensions may be phylogenetically insupportable."[13]

For Turner, these numinous dimensions are conveyed and cherished via stories that attempt to connect its hearers to a meaning-saturated reality disclosed via humility, respect for mystery, kinship bonds, and reciprocal converse with other-than-humans. For Shepard, stories reflecting this animate experiential life-world cause "forms to spring to life in the mind, re-presented in consciousness, training the capacity to imagine."[14]

Wilber, in contrast, argues that without recognizing the step forward from animism to abstract thought, we inevitably devalue rationality. The result is a damaging regression that celebrates instincts over contemplation. He levels this charge against the ecophilosophy of deep ecology and ecopsychology, both influenced by Shepard, arguing that their bias toward biospheric immersion over transcendent contemplation is expressed in a tendency to champion pre-Neolithic hunter-gatherer life as a lost golden age. Wilber admits that such immersion, along with permeable boundaries between self and other, was sometimes more attuned to ecological wisdom, but this stance too often ignores that magic-mythic tribes are mainly at an early level of development in which whole and part are pre-rationally confused rather than transrationally integrated.[15] According to Wilber, pre-Neolithic tribes should be honored, because they are literally our roots, but their lifeways do not reflect eco-utopias: in fact, tribes were undeveloped and often brutal, the horrors including short life expectancy, one in three infants being killed for population control, tribal conflicts, and eco-destruction due to the inability to practice long-term thinking and perceive the consequences of actions. Put simply, Wilber argues that it is a mistake to compare the worst aspects of modernity with the best aspects of early hunter-gatherer tribal existence.[16]

Wilber may be right that Shepard, and others, make this mistake, but his argument must be balanced with Shepard's claim that hunter-gatherer lifeways acted as a catalyst to growth, including the development of the human mind. Stimulation from predator-prey reciprocity dramatically increased brain size, as imitation of animal others gave way to forethought, cooperation, and communication among hunters. Shepard writes: "At this point in the growth of the mind it became fully human and brought into play various complex forms of cognition: the pantomime, the mimicked reference, sharing the idea of an animal by imitating its calls, the way it kicks, stamps, tosses its head, or ritually fights its own kind in stylized performances." For Shepard, a "huge repertoire" of communicative skills emerged from these activities.[17]

At first, I was confused by the opposing views of thinkers that I admired—like Wilber, Shepard explained a lot, as does deep ecology and ecopsychology—but my dissonance was a catalyst to more reading and a realization: differing perspectives temper each other, although it takes the proper transversal temperament to learn from this dialogic interplay. Such is how I view the debate, or dialogue, over hunter-gatherer existence, which I have been exploring for many years. Thus, it is worthwhile to continue briefly reviewing differing views, especially since they inform possibilities for fitting responses.

The anthropologist Claude Levi-Strauss writes that there are two common mistakes when discussing early humans: claiming that their thought is solely determined by survival, making it inferior to "civilized" thought, and claiming that it is fundamentally different kind of thinking. Oral cultures are without writing, but still capable of "disinterested thinking" and the desire to understand the world around them. However, Levi-Strauss does detect differences between mythic-animistic and scientific thinking, with the former desiring to understand everything and the latter proceeding in steps, giving explanations for limited phenomena before proposing larger theories. For Levi-Strauss, "the human mind is everywhere one and the same" with "the same capacities," but certain capacities are not developed when they are not needed. Oral peoples did not develop scientific thinking; they did develop sensory perceptions, precise knowledge of their environment, and myths, although these myths often gave the illusion of understanding.[18]

In *Saving the Appearances*, Owen Barfield also praises the mind of "primitive" humans, stating that they practiced an original participation in which the separation between self and nature was not as rigid as it is in modern society. To Barfield, we make a grave error if we assume that the "primitive" mind only practiced confused, superstitious, and magical thinking. Rather, he states that tribal members often exhibited an extrasensory link, or participatory consciousness, which is not wholly detached from the other in the act of perception. However, Barfield also recognizes the unfolding spiritual depth of the cosmos, asserting that our continued evolution demanded that humans more fully differentiate between mind and nature so that rationality could fully emerge.[19]

In *In the Spirit of the Earth: Rethinking History and Time*, Calvin Luther Martin argues that early humans are defined by their propensity for speech and tool making, as well as their ability to fashion images making them "part of it." For Martin, we remain "talkers" and "tinkers," but we have lost the images that connect us to the other-than-human, leaving few constraints on what our words and hands bring to our relationships. For Bateson, whose systems thinking was inspired by anthropological study of indigenous cultures, ecocrisis is a crisis of perception, but Martin adds that it is also a failure of imagination. We no longer participate with the obvious intelligence of nature and other beings, regenerating the system and vivifying relationships via our sensuous imagination and communication. Martin writes: "Humans learned their human powers and abilities from these other-than-human beings. They strove zealously to do so, in order that humans might fully converse with these other beings, and, in doing so, keep the world running properly. Hunting, fishing, shellfishing, plant gathering, tool making, house building, the making of clothes and adornments, feasting—all were part of that conversation."[20]

Like Shepard, Martin broods over why early hunter-gatherers gradually left their more sustainable and joyful lifeway, changing their image of themselves and their communicative practice. Instead of the evolution of a fuller rationality, he suggests that it was the Neolithic fear of want and death brought on by the slow turn to agriculture. Contrary to popular belief, pre-Neolithic hunter-gatherers were not destitute, always fretting over their next meal or the eventuality of death. Rather, inherent in the hunter-gatherer lifeway is the understanding that nature provides sufficient food and that death should be experienced naturally, without anxiety. Death anxiety is the product of Neolithic concern with an afterlife, adding concepts like heaven and hell, which only emerge with agriculture and transcendent sky-gods and genuflection over amenable weather for crops. When crops were fallow, there was fear, when crops were plenty, there was increased population, hoarding, and more fear. This cycle of fear and dependency began gradually but then picked up steam as agricultural norms displaced hunter-gatherer conversation, muting nature in favor of an increased desire for control.[21] However, Turner argued that pre-Neolithic hunter-gatherers were also influenced by fear, although expressed differently as propitiation before unseen powers via rituals of mutilation and sacrifice.

And, in David Abram's influential *The Spell of the Sensuous,* more insights are gathered from our mythic-animistic past, especially on bodily based converse within a more-than-human world and the limits of language. Abram, while not citing Bateson, also claims that we exist within a larger system of communication, and while his only reference to Shepard is an epigram lamenting the connection between written history and the rejection of habitat, he further articulates the mythic-animistic view of nonhumans as communicating others. Both moves lead him to celebrate the sensuous imagination and communicative habits of oral peoples, for whom "all could speak, articulating in gesture and whistle and sign a shifting web of meanings that we felt on our skin or inhaled through our nostrils or focused with our listening ears, and to which we replied—whether with sounds or through movements, or minute shades of mood."[22]

For Abram, perception, when open and spontaneous, allows us to creatively respond to specific, changing environments and nonhuman others via a reciprocal exchange of presences. This ongoing silent conversation, or always occurring dance of reciprocity, happens bodily and not always consciously as a "continuous dialogue" unfolding "far below" or even "independent" of verbal awareness.[23] Abram argues that this commingling of presences affirms an animistic attitude, in which we learn the value of nonverbal communication. Thus, we must learn from oral cultures that honor nonverbals and thus use language to "encourage and augment the participatory life of the senses."[24]

Humans developed in response to their surrounding environment, and this gave birth to language as a means of responding. One can easily imagine the awe and wonder elicited by animal others and the night sky, or senses heightened in the hunt or from feeling wind on skin, all stimulating the desire to speak. But Abram argues that literate Western civilization promotes language use that seems to "deny or deaden this life," leading to a "massive distrust" of the senses in favor of abstraction beyond appearances.[25] When language leads us to perceive nature as resources rather than the source from which we have evolved, we cut ourselves off from primal energies that sustain us, continuing the destructive monologue mistakenly labeled as progress.

Still more: in *Collapse: How Societies Choose to Fail or Succeed*, Jared Diamond adds an important element to this debate, arguing that we should not champion or denigrate past peoples, as they are not superior or inferior. Diamond investigated environmental collapse in earlier cultures, writing that managing resources has always been difficult, "ever since Homo sapiens developed modern inventiveness, efficiency, and hunting skills by around 50,000 years ago," and "any people can fall into the trap of overexploiting environmental resources." This trap has many roots: resources seeming inexhaustible, signs of depletion masked by natural variation, the difficulty in getting humans to exercise restraint, and the complexity of ecosystems generally which make damaging disturbances hard to predict, even for a "professional ecologist." Diamond argues that past peoples were neither ignorant and primitive nor environmental saints; instead, they were similar to us facing similar problems.[26]

That early humans were, well, human, falling into existential and ecological traps, while also learning from trial and error, is undoubtedly true. Shepard does not disagree, stating that hunter-gatherers are subject to "human shortcomings" and "incompetence" and are not always "happy and well-fed"[27] Still, Wilber, along with Zimmerman and Esbjorn-Hargens, point out many misinterpretations from romanticizing hunter-gatherers: along with the aforementioned comparing of the best of hunter-gatherer life with the worst of modernity, or the "dignity/disaster fallacy," we should have knowledge of and compare differing tribes rather than conflating them, and also should not conflate the individual state consciousness of a shaman with the average stage-consciousness within a tribe. Furthermore, we must consider an important distinction: pre-Neolithic tribalism should not be conflated with indigenous cultures that have evolved to our current world, like Native Americans; there are similarities, but many differences that make a difference, including Native American tribes positing a transcendent dimension expressed as the Creator or Great Mystery.[28]

Wilber, along with Zimmerman and Esbjorn-Hargens, also argue that earth-friendly cosmologies should not be conflated with earth-friendly behaviors, and indigenous peoples that were destructive of their environment would have done more damage if they had the technological means. While evidence supports that this is also undoubtedly true, studies of traditional ecological knowledge, or TEK, disclose that despite differences between pre-Neolithic hunter-gatherers and indigenous peoples evolving to today, they display characteristics of a consistent world view of belief, knowledge, and practice, including a nondominant, respectful ethic expressed as a sacred ecology. In other words, while earth-friendly cosmologies do not always lead to earth-friendly behaviors, it sometimes, or often, does, depending on where and when.[29]

Fikret Berkes, in *Sacred Ecology*, writes that this TEK ethic is commonly practiced as local knowledge of place via close observation over time—or Shepard's focus on the quality of attentiveness within specific places—which may be partnered with science to address eco-social problems, including climate crisis.[30] The Inuit, for example, notice things, like the alterations of species distributions and thawing permafrost in comparison to previous years, or decades. Not all scientists are on board, as belief in a "land that listens" and "glaciers who may be offended by human disrespect" do not fit well with geophysical science. And yet, indigenous myths about glaciers reveal "dramatic geophysical upheavals" during the eighteenth and nineteenth centuries.[31]

Berkes is also critical of romanticizing indigenous peoples, especially as the noble savage and exotic other, which leads to adulation followed by denigration after Natives fall from a perch they never claimed for themselves. However, while indigenous peoples are people, and have not developed to the level of Eco-Sage—who has?—TEK has something of value to offer, and this offering is often rooted in mythic-animistic communication in which there is a "sacred, personal relationship between humans and other living beings."[32] Wilber, while sympathetic, always reminds that we must not confuse pre-rational, rational, and transrational experience, and we must learn from past mistakes; after all, that is what makes rationality, at least potentially, an evolutionary step forward. Those who embrace logos as transversal rationality can integrate insights from differing time periods and world views, expressing a world-centric awareness, which early hunter-gatherers could not do.

Still, we have the question of Wilber delegating animism to pre-rational levels, while many environmentalists, especially deep ecologists and ecopsychologists, embrace animism as a step forward within a Gaian living system that has been systematically reduced to dead matter. Fisher, in *Radical Ecopsychology*, suggests a kind of middle path that may satisfy both sides: make animism a line of development with differing expressions at differing levels.[33] This move, along with making a distinction between earlier hunter-gatherer consciousness and indigenous awareness that has evolved to today, and thus animistic expressions through time, further opens the door to interesting integrations between indigenous knowledge and science, as well as the celebration of deep listening as a practice crucial to all ages. Consider, for example, the lessons gifted to Thoreau from Joe Polis, and his efforts to integrate them with the burgeoning natural history and science of his day.

In Donal Carbaugh's studies of Blackfeet "just listening," tribal members widen their current awareness via long-held animistic practices, respecting ancestors and mythic traditions, conversing with the spirits within places and the spirit of place.[34] Environmental communication scholarship of this type is critical of the rational focus on the social construction of nature via discourse rather than the continual construction of an embodied and fluid and listening self attuned to the natural world that then discourses. Or rather, criticizing the power of dominant discourses—like bigger, better, and more—to shape monologic perceptions is obviously needed, but more steps are necessary; or, as Carbaugh puts it, we must focus on "representations of nature" but also on its "presentations," becoming attuned to other "expressive systems."[35] Put simply, we should connect with nature and then speak rather than constantly speaking without such connection.

Berkes gained insight into Cree expressive systems after ten years of academic study gathering abstract data on sustainable hunting and fishing, and when invited to attend meetings among elders, tasked with writing down practices to be passed on to the young. Berkes just listened, absorbing discussion on spirituality, rituals, the cycling of animal populations, and dealings with missionaries. A world view pulsating with life emerged, with one elder telling a story of a missionary repeatedly telling the Cree that "there were no spirits in the bush," while knowing that "the land was sacred and full of spirits." Such fullness led to ethics of respect and reciprocity; when violated, one likely came back empty from the hunt, as animals determined success. But did hunters experience telepathy with animal minds or simply get into the right frame of mind to be successful? Such mythic-animistic possibilities may certainly be critically questioned, or violated, and were by some younger Cree, yet for Berkes ecological wisdom may be studied via science *and* "danced or told as myth" After all, he began his studies by asking: Why, with no government regulation, did the Cree not overfish? And why wasn't their resource base degraded by the tragedy of the commons?[36]

In an interview with Derrick Jensen in *Listening to the Land*, Shepard lists TEK's many wisdoms, as well as modernity's frequent lack of wisdom, perhaps falling prey to the dignity-disaster fallacy, yet his insights are instructive. Shepard begins by stating our "historical experiences with the natural environment condition our responses to it," which is why hunter-gatherer responses are better than modern responses of division, domination, destruction, and despair.[37] These four "ds" feed upon themselves, abusing the planet. However, the deeper root is a fifth "d": domestication. Put it all together and today's children develop within tame, human-controlled environs rather than hunter-gatherer wildness and wild animal others, and the result is the expectation of control, which is ultimately an illusion. Base a civilization on an illusion and crisis is what you will get, if not collapse.[38]

The domestication of plants and animals produced a "different cosmology, value system, and cultural set of assumptions" and reduced wild others that kept humans "sane and small."[39] Yet, our Pleistocene bodies, psyches, and physiologies remain, and while we can't go back—Shepard states he's heard that refrain for twenty-five years—we can recover aspects of who we are.[40] To begin, we must learn our heritage, and then add exposure to wildness, especially for our kids, realizing we are one species among many. Richard Louv has more recently taken on this task, writing of "nature deficit disorder" and the cure of getting outside and learning from nature. To Shepard, such learning should include close observation of species, face-to-face small group decision-making that protects against authoritarianism, and art and music inspired by nature. Not surprisingly, he also advocates for hunting and eating wild meat, while also stating that the turn to vegetarianism is understandable given unhealthy domestic meats. And he challenges the claim that nomadic tribal living inevitably led to tribalism and war, stating that without rootedness in land, there is less to defend. Own land, you get war. Hunter-gatherer conflict is typically between individuals; tribes often welcomed strangers, assuming others to also be welcoming. What they often found instead is colonization and enslavement.[41]

Shepard speaks of hope, circa 1995, calling it "cheap and available" and a "last resort," while stating that ecological disaster is not a future reality but an already happening "creeping thing" that includes poverty, mental illness, suicide, and ultimately the "disintegration of natural systems."[42] He died in 1996, and one can only wonder what he would say as climate crisis has crept forward into climate emergency. We expect Hollywood-like Armageddon and have gotten our share via volatile storms and flooding, pillaging of rainforests, and out-of-control bushfires, among many other eco-disasters, but the roots are the daily undermining of ecosystems and mental health.[43] Interestingly, for Shepard, the solution of caring for nature and growing within small communities not only leads to overcoming dualism or separation but respecting and learning from the mystery and "unbridgeable gap" of otherness—he is even critical of Buber's I-Thou ontology, fearing that it presumes that the other will be like us.[44] Peters would be pleased, Wilber somewhat chagrined.

Shepard celebrates commonalities and common home among species, but the presence of otherness provides much-needed humility and limits: we are neither "masters" nor "voyeurs," but part of food chains.[45] He further argues that we are not human without others, as we emerged "enacting, dreaming, and thinking animals." Wonder before accepted differences stirred the growth of mind and self, as did making sense of existence via myth, especially the contradiction of loving and killing animals. Of course, we all live within myths, whether acknowledged or not, which provide models of behavior and, ultimately, response. Myths, for Shepard, do not have to explain everything, like attempts at metaphysics, but relate us to life and death rather than happy illusions of heaven: our sanity resides in landscapes. And plants, too, play powerful roles as others from which we learn; their presence may be limited by immobility, at least compared to animals, but they elicit "instincts for cover and comfort and protected observation," not to mention gifts as food, tools, and medicine.[46]

Shepard's views may be challenged—for one, there is debate over whether human physiology is more designed for eating meat or plants—but what stands out is his emphasis on otherness, especially since, for Wilber, mythic-animistic awareness is marked by pre-rational confusion between self and other or projection of self onto other. How pre-Neolithic hunter-gatherers responded to otherness and their environment, and how they recognized it via thinking, is not easily settled—I wasn't there and don't know anyone who was—yet it is likely that both are right or that they provide partial truths that temper each other. Shepard further argues that spontaneous response to nature's details stimulated categorical thought. Ironically, this resonates with the evolution to scientific categorization, which Shepard celebrates; he is not against all things modern, especially ecological science.[47]

Again, for Shepard, two core wisdoms expressed by hunter-gatherers are the quality of their attention and the significance of place, especially in regard to healthy development. These qualities are much less debatable and worth carrying forward. Yet, he also documents other favorable characteristics of early tribal experience, including an "eco-predicated logos" in which we "read the world as the hunter-gatherer reads tracks"; respect for nonhuman life, and death, via predator-prey reciprocity; a small-scale joyful subsistence existence in which tools are vocational instruments supplying a "gestural response to life, subordinate to thought, art, and religious forms"; respect for numinous experience as a sign of sacred presence; the celebration of wildness instead of domestication; and attentiveness to the "Voice of life," continually recounted and awakened via rituals and stories."[48]

And, for Shepard and other environmental thinkers, all this was lost with the shift to agriculture. But while agriculture obviously changed human life, it is equally obvious that we are currently dependent on agriculture. So while we cannot fully go back—nor would we want to, according to Wilber—we can learn from the above list, transforming modes of living even as debates over hunter-gatherer life continues. Many practice sustainable agriculture, selling local and organic produce, by attending to the wisdom of specific places and listening to and telling stories of the land, or, as articulated by Wendell Berry, listening to the genius of place in the spirit of dialogue.[49] Shepard would be on-board with such listening, and they both deconstruct the insanity of agribusiness, but he questions Berry's focus on Christian social morality and stewardship as response, which, for him, is focused more on use rather than respect.

And so, the debates on what is most fitting keep coming. Again, practicing transversal rationality discloses that what is unfit may repeat—like the destructiveness of agribusiness—and thus, while we must be open to new insights, what is unfit may be more final and we are better able to make this determination by learning from multiple modes of knowledge and communication. Sustainable agriculture, for example, does not preclude abstract rationality—successful farming is filled with mathematical figuring—or other forms of rationality, like contemplation, which may act as stepping stone to an increased aesthetic awareness and perhaps nondual experience or spiritual communication. After all, Thoreau had a mystical experience while hoeing his bean field, transcending the distance between self and other.

For Schrag, decentered subjects dwelling within the space of communicative praxis must display an ethos that embraces both "ethical and aesthetic forms of life" if we are to approach fitting responses.[50] Such an ethos demands responsiveness to past, present, and future times, but also specific places. The practice of mythic-animistic communication engages with specific places, but also the wider, natural world within which we dwell, as we are thinking animals dependent upon a biosphere. We dwell communicatively, socio-historically, and mythically via narrative, but we also dwell on a vibrant earth, and thus mythic-animistic communication is a first step toward an ecological ethos that discloses the good and the beautiful in support of the flourishing of life. However, the extending of subjectivity, and thus listening, to the other-than-human exposes us to new dangers, especially the faulty use of imagination; but this extension is necessary, while we remain wary of the unfit, if we are to more fully respond to a myriad of eco-social challenges.

By practicing transversal rationality, then, we can say that Shepard, Wilber, Turner, Levi-Strauss, Barfield, Martin, Abram, Diamond, and Berkes articulate interpretations based on anthropological and historical research that seek to disclose the experiential reality of our hunter-gatherer past. In doing so, we must question what is being concealed as they reveal new insights. The point, again, is that dialogic transversal rationality, in contrast to monologic instrumental rationality, traverses various arguments, acknowledging possibilities for learning from hunter-gatherer traditions while remaining critical via distanciation. For me, such traversing reveals that humans are fearful and foolish in all ages, a conclusion supported by Diamond and Shepard, but also Peters' when emphasizing miscommunication and existential walls rather than dialogic bridges. However, there is also wisdom to each age, and we would not want to be more foolish then is existentially our lot by refraining from exploring this possible wisdom.

Despite worth-exploring concerns about Shepard and romanticism—I can imagine Peters cautioning against the dream of communication and rising expectations while applauding his emphasis on otherness—there is much to learn from trying to understand how pre-Neolithic hunter-gatherers, as well as hunter-gatherers through history, communicate within a dialogic life-world filled with messages. But those lessons, Wilber is quick to point out, are only gathered through rationality, and yet such messages, meant to keep the world running properly, are only heard if we practice deep listening. Attentiveness is paramount, marked by recognizing the aliveness, reciprocity, and gifts of nonhuman others and specific places. Telling stories and participating in rituals are also key qualities of the dialogic openness that constitutes mythic-animistic communication.

At the very least, we should respect stories of reciprocity, developing deeper sensitivity to the messages that make up *our* world, including signs that reveal the madness of a monologic mindset, like melting ice flooding coastal communities; wildfires burning forests, homes, and millions of species, driving some to the brink of extinction; and storms made far worse by human expansion and the loss of wetlands. Deep ecology and ecopsychology, for example, maintain that the force of life persists, or repeats, and rooting ourselves in sensual experience sensitizes us to life-denying messages, but also life-affirming ones. Post-metaphysical theorizing must embrace this evolving physical experience, gathering insights from our past participatory consciousness and conversation for our frequently dissociated present world. However, the differences between Wilber and Shepard disclose both shadow-sides and light-filled guidance, and thus they deserve further exploration.

SHADOWS AND LIGHT, UNFITTING AND FITTING: DEEP ECOLOGY

While Wilber and Shepard's opposing views on animism created dissonance for me, it was more pronounced when I read Wilber's cutting criticisms of deep ecology, especially since I saw little incongruence between Wilber's work and this ecophilosophy. But Wilber perceives several shadow elements in Shepard-inspired deep ecological identity, thought, and activism. This debate goes back twenty-plus years, yet we still have not figured out a more fitting, and therefore more comprehensive, response to ecocrisis. Thus, it is also worth exploring to articulate why we need an ecology of communication.

Deep ecology is inspired by many traditions, including a bit of mysticism, but, generally speaking, it does not recognize a transcendent dimension, mixing, at best, mythic-animistic, rational, and aesthetic communication with an emphasis on mythic-animism. Thus, for deep ecologists, the most fitting response to ecocrisis is identification with the earth and fellow earth-sharers via an ecological self that fights to preserve wilderness. The ecological self also asks critical questions revealing how thoroughly societal institutions are grounded in the nature-culture dualism and anthropocentric attitudes in which only humans have intrinsic value; the result, of course, being ecological destruction.

Arne Naess, a Norwegian philosopher and the intellectual founder of deep ecology, explicates the ecological self by relating it to joy rather than foundational ethics and moralizing, acknowledging, like Schrag, that fitting responses are not known in advance of praxis. Naess, countering the doom and gloom eco-activist stereotype, focuses on "sources of joy which are available to people through an increased sensitivity toward the richness and diversity of life, and the landscapes of free nature." Such sensitivity becomes a powerful motivator, calling forth an eco-self that realizes joy resides not in the subject or object but in their relation. We begin to participate in "something bigger than our ego; something which has endured for millions of years, and is worth continued life for many more millions of years." For Naess, care "flows naturally if the "self" is widened and deepened," making moralizing and asking for sacrifice unnecessary. We protect nature because we are protecting ourselves.[51] This naturally flowing joy, care, and protection makes us more ethically responsible, but instead of responsibility being a chore demanded from an external source, it "comes through" via eco-selves responding to local and global eco-existential conditions.

For deep ecologists, such coming through may be practiced by doing as Aldo Leopold suggested: delimit thought and imagination by learning to "think like a mountain." Leopold learned this lesson when a young trigger-itch hunter shooting into a pack of wolves, killing a mother wolf and then watching a "fierce green fire die in her eyes." Pained and humbled, he realized that while fewer wolves meant more deer for hunting, more deer also meant destruction from overgrazing, as if "God took pruning shears" to the mountain landscape.[52] Thinking like a mountain, then, restores widened awareness of our larger ecosystem as a communion of subjects.

Deep ecologists turn to Shepard for early examples of such communion, leading to the claim that the modern period is characterized by a disembodied philosophical outlook in denial of the undeniable: we dwell within a larger earth community. Thus, embracing our immersion within the biosphere, which Wilber interprets as pre-rational mythic-animism that must be included but also transcended, is key to being fully human. When we repress our animal nature, too much of our experience ends up being artificially mediated by various forms of technology, leading to out-of-touch anthropocentric arrogance that represses relationships experienced via mythic-animistic conviviality and proven by ecological systems science. To combat this distortion, deep ecologists argue for an ecocentric, or biocentric egalitarian, attitude in which the unique qualities of all species are valued. Dialogic relations with nature are celebrated, with humble humans providing for their vital needs by living a life that is simple in means, rich in ends.[53]

Deep ecology, then, provides a compelling counter-narrative to the myths of linear progress, unlimited growth, and technology as savior. Environmentalism that participates with these modern myths without critical distanciation is shallow, responding to ecocrisis with an emergency room mentality without going to the deeper roots that cause ecological emergencies in the first place.

The ecological self also leads to various forms of nonviolent activism. A dramatic example comes from the deep ecologist John Seed, founder and director of the Rainforest Information Center, in his work to save the forests of Australia. In responding to the question of how he sustains the struggle against giant lumber interests, he declared: "I try to remember that it's not me, John Seed, trying to protect the rainforest. Rather, I am that part of the rainforest protecting itself. I am that part of the rainforest recently emerged into human thinking." Joanna Macy, also influenced by deep ecology, writes that practicing our ecological self to this degree evokes a kind of grace. We experience ourselves being "acted through and sustained by something larger than ourselves."[54] Such grace may emerge as rational response to erotically charged world-historical events, including eco-social crises, which is how Ramsey articulated "coming through"; however, as we have already begun to explore, it may also emerge from biospheric immersion expressed as mythic-animistic communication.

Sounds good, right? Deep ecology focuses on the common good of all species, not just humans. What could Wilber possibly find disagreeable?

For deep ecologists, the ecological self is already within us waiting to be realized and expressed, and yet Wilber is highly critical of the eco-self, as well as activism inspired by it, even though those inspired by deep ecology are among the most passionate lovers and defenders of the earth. Again, at first glance, there does not seem to be much incongruence between deep ecology and Wilber's model of development. The ecological self, while embracing mythic-animism, is also influenced by systems science and nature mysticism. The only criticism would seem to be that deep ecology stops its process of identification too soon by not recognizing Super Integral levels and subtle, causal, and nondual states. However, for Wilber, the absence of these levels and states is a crucial error, because it leaves out a transcendent dimension, resulting in a flattened spirituality in which the earth is seen as ultimate reality. In other words, deep ecologists fail to fully gather insights from the great mystics, causing them to collapse the holarchy of matter, life, mind, and Spirit. This theoretical collapse has practical results: poor interpretations of spiritual progress, with deep ecologists attempting to regress to pre-rational levels where mind and nature are not yet fully differentiated rather than progressing to transrational realms where they are integrated—or the pre-trans fallacy—and this regression may lead to unfit solutions and activism in response to ecocrisis.[55]

Wilber argues that even if the ecological self did reside at the Integral level of nature mysticism, deep ecology still points toward a goal—which he calls the eco-noetic self—without a real path to get there, since such a path requires an actualizing holarchy. Regrettably, many deep ecologists reject all forms of hierarchy and thus are left without a genuine developmental model, leading them to conflate biocentric egalitarianism, in which each species has equal intrinsic value, with unity-in-diversity nature mysticism. Wilber responds that all species have equal Ground value in Spirit, but this does not mean that all species have *equal* intrinsic value. In fact, some species have instrumental value—we need to eat—even though all species has emerged from Spirit unfolding, and thus all have value. In other words, everything has equal Ground value when considered from the perspective of the absolute plane, but differing intrinsic and instrumental value on the relative plane. Equal from one perspective, different from other perspectives, and we need to consider all perspectives when making ethical decisions.[56]

Thus, by not recognizing a transcendent dimension and the involutionary-evolutionary ontology of physiosphere/matter, biosphere/life, noosphere/mind, theosphere/Spirit, deep ecologists fail to recognize differences that make a difference. Biocentric egalitarianism sounds progressive, but extends Pluralistic level dysfunctions to nonhumans, leading to a relativizing of value in which there is no difference between a mosquito and human child. Thus, eco-romantics, which are what Wilber calls deep ecologists, are actually advocating for magic-mythic level indissociation in response to modernist dissociation rather than higher level differentiation and integration.[57]

To Wilber, deep ecology offers immersion within an alive biosphere and web-of-life systems science (Gaia theory, ecological science, systems theory) as a new spiritual paradigm replacing mechanistic science. But, for him, animism at higher levels is not really animism anymore, but Spirit in the world, and systems science takes us no higher than the Rational level of development: again, systems science, as wonderful as it is, is still science and does not integrate higher spiritual realms of awareness. Thus, deep ecologists need to honor human potential by acknowledging Spirit unfolding as both the highest level in the developmental holarchy and the holarchy itself. By doing so, they would discover a transrational path that leads us to experience a relationship of oneness with Spirit beyond and within all things.[58]

Ironically, Wilber argues that such a path is significantly deeper than deep ecology's return to instinctual fusion or joy at being parts in the web of nature: genuine transformation is catalyzed when we realize that we simultaneously express Spirit and work toward our full potential in Spirit. In fact, Wilber argues that humans represent a higher stage of evolution since we express more of the holarchy of matter, life, mind and Spirit, a claim that needs critical unpacking given that our dissociation, justified by thinking ourselves the crown of evolution, has led to endless destruction. Still, he argues that human self-consciousness, marked by rationality, language, and technology, are more complex than the unique qualities of other species because we transcend and include physical matter and biological life, participating in the intersubjective creation of culture and contemplation and experience of Spirit, which are all needed for a fuller responsiveness.[59]

Wilber also disputes the charge of anthropocentrism, arguing that rather than being an ecological evil, it is a stage of development reflecting the emergence of human rationality, which is absolutely necessary if we are to solve irrational eco-social crises. Again, developing our full rational capabilities to vision-logic allows us to reflect upon the long-term effects of our actions, a quality that mythic-animistic hunter-gatherers lacked and most today lack due to our reptilian brains; Wilber agrees with Shepard that our long history as hunter-gatherers remains with us, but in mostly negative rather than positive ways. Also, he argues that as humans we cannot help but perceive the world as humans do. Other species evolve within their unique perspective and so do humans. The problem, therefore, is egocenteredness, which is mostly overcome at rational levels, especially healthy Pluralistic care for others, rather than anthropocentrism.[60] Yet, for deep ecologists, anthropocentrism reflects a distorted human perspective in which the value and voices of nonhumans are disregarded.

Again, the above criticisms reflect the pre-trans fallacy in which the prerational (go back to nature) is mistaken for the transrational. The main forms of this regression include a tendency toward anti-rationality, the relativizing of value via biocentric egalitarianism, a romantic celebration of pre-Neolithic hunter-gatherer cultures as spiritually advanced, and the lack of a genuine model of development. To Wilber, there are many negative results of these regressions, with the most extreme being a doom and gloom environmentalism that tends toward ecofascism. In other words, since deep ecologists, inspired by Shepard's analysis of hunter-gatherer life, feel that the eco-golden age was in the past, the only solution to ecocrisis is to de-industrialize the world back to beginnings of human emergence. As a result, he claims that eco-romantics, who have fallen prey to the dignity-disaster fallacy, attempt to force a confused eco-ideology down the throats of the modern miscreants who are destroying the planet.[61]

All of this kind of exploded my brain when I first read it. What followed was picking up the pieces, first my brain and then the differing strands of the arguments. Transversal rationality cuts across times, places, texts, but also experiences; insights from my Appalachian Trail hike supported the interpretations of both deep ecology and Wilber, and so I had plenty of unpacking to do.

There is obviously much disagreement between Wilber and deep ecology, with Shepard and deep ecologists rightly concerned with the repression of body and biosphere. Wilber honors the good intentions of deep ecology, while stating that he is equally concerned with repression; that is why his model reflects an ever-increasing process of differentiation and integration. However, even Zimmerman has stated that he may be biased in favor of the ascent tradition.[62] This bias could be a reaction to the regressive bias of deep ecology, but whether Wilber has a problematic bias does not change the fact that his developmental model provides context for his criticisms of deep ecology. Still, he does not seem to recognize that this ecophilosophy has opened many to a deeper sense of self and experience; I am consistently enthralled by Wilber's intellectual discernment, but some assertions are overstated. For example, he writes: "the worshipers of nature are the destroyers of Nature [Spirit]."[63] On a theoretical level, this statement is true. Deep ecology leaves a transcendent dimension out of its philosophical theorizing. However, on a practical level, deep ecology-inspired activists have consistently worked to protect Spirit expressed as nature.

The exploration of the ecological self has helped many to integrate body and mind, so often dissociated, opening them to eco-religious experiences of Spirit's immanence or the world soul, even if not articulated as such. These experiences are a prime motivator for all kinds of needed activism. And, at its best, the ecological self expresses not only mythic-animistic communication but also rational and aesthetic communication, while also opening us up to the possibility of spiritual communication. For example, my hike—which, once again, while life-changing for me, was not the longest or wildest compared to the trips of many others—led to a different self-experience that included the widened identifications of the eco-self, which then primed me for more experiences. Or, mythic-animistic communication is a stepping stone to and often interwoven with the other forms of communication.

Yet, Wilber's criticisms disclose the potential for unfit responses. Zimmerman utilizes Wilber's work to interpret Nazism's racist, yet nature-centered, blood and soil "complex and contradictory movement" mixing tribalism, technology, and nature worship, among other pro and anti-modern biases. What concerns Zimmerman is the rhetorical similarity between Nazism and deep ecology, including the desire for identity with nature, or nature's laws and the "mysteries of the blood," as well as a rejection of Judeo-Christian transcendentalism. Thus, deep ecology must be clear that "a strong identification with and concern for non-human life" does not lead to "authoritarian and misanthropic attitudes." Zimmerman states that the link between Nazism and nature reverence was so strong that a whole generation had to pass before German Greens could emerge as a political force during the 1970s.[64]

Zimmerman acknowledges that some similar rhetoric does not make deep ecology fascistic. However, this example reveals why good theory, or good meta-theory, is so important: pre-rational interpretations may lead to tribalistic dismissal of human others. After all, tribalism, like nationalism, is a prime cause of warfare, providing mesmerizing myths that disparage other tribes and their myths and meanings. Shepard, of course, argues that small-scale tribal living has many positives and does not necessarily lead to tribalism. Still, Wilber warns against the links between magic-mythic animism and tribalism, especially since identification with the earth and nonhumans does not guarantee identification with the human family at Rational and Integral levels. For example, some *individual* activists inspired by deep ecology have made incredibly thoughtless statements, like AIDS is good because it reduces population, which reflects the regressive, biospheric focus that Wilber criticizes.[65] Many years later, similar sentiments are bubbling under the surface of the coronavirus pandemic; but, overpopulation, while a deep concern, is not the main issue when it is specific wealthy and overconsuming populations, and the structures that support wealth accumulation and consumption, that are doing the most damage. After all, a study has revealed that 100 corporations are responsible for 71 percent of greenhouse emissions.[66] And that, of course, is why development is needed in all four quadrants.

Wilber's argument that identifications with the biosphere may lead to antirationality also has merit. When I taught at Purdue University, I attended weekly meetings of an environmental discussion group, aptly named "A Gathering of the Tribe." A biology professor who taught classes in deep ecology and wilderness protection facilitated the meetings, and discussions were often lively with much critical thought. However, a student once derided rationality as a mode of knowing, blaming it for ecocrisis while celebrating instinct and feelings. He made no distinction between instrumental rationality, which has situational worth despite its damaging effects, and higher forms of rationality. He also failed to recognize what mode of knowing he was using, and using poorly, to make his claims.

This example may seem minor, but eco-activism is done based on similar interpretations of rationality and rational communication. Radical environmental groups like Earth First! (EF!), the Earth Liberation Front (ELF), and the Animal Liberation Front (ALF) tend to privilege mythic-animistic identifications and communication, seeing little worth in modernist institutions. That, of course, is not to say that these groups are not rational, but they may fall prey to the pre-trans fallacy. More recently, Deep Green Resistance's (DGR) platform details the insanity of industrial ecocide and asks whether we should rid ourselves of civilization. Their response is a plan, Decisive Ecological Warfare (DEW), which includes aboveground organizing and nonviolent direct action and underground tactics such as networking and mobilization, sabotage, systems disruption, and dismantling infrastructure. DEW is obviously not for the meek, but for the fed up who believe that warlike resistance is not only necessary but overdue. Following feminist ethics, non-hierarchical respect and care for others is highlighted, as is rejection of the myth of human, and male, supremacy.[67]

A thorough exploration of these eco-activist groups is beyond the scope of this project, but there is a tendency toward return to pre-industrial, and in some cases hunter-gatherer, lifeways rather than learning from the past by perceiving dignities and dysfunctions. Radical environmentalists, responding to the irrationality of destroying our own home, are mostly Eco-Guardians and Eco-Warriors, marked by pre-rational mythic-animistic identification with nature and aggressive defensive tactics, as well as Pluralistic Eco-Radicals who deconstruct the dysfunctions and destruction of modernity, without perceiving dignities, while fighting for eco-social justice. Rational-level Eco-Managers and Eco-Strategists who work on passing laws and conserving resources would be too beholden to status quo systems, Pluralistic-Integral Eco-Holists who use vision-logic to map out ecosystem complexity and wicked problems would be too slow to act, Eco-Integralists who see value in all perspectives would be accused of over-including conflicting views that also slow down response, and Eco-Sages would be too focused on inner over outer work, and too focused on other worlds rather than "the only world we've got," to quote Shepard.[68]

Derrick Jensen, a core advocate of DGR and DEW, argues that fighting for the health of the landbase is all that matters. Future generations will not care about how spiritually enlightened we are, or whether we recycled, or our many excuses for ignorance and not acting: all they will care about is whether there is air to breath and water to drink. Pollution causes degradation of habitat, burning oil and gas causes climate disruption; the fitting response is to stop doing what causes damage to health and home. Simple, understandable to a child. To Jensen, then, active resistance to industrial civilization is the only sane, and rational, response to an insane system of 97 percent of native forests destroyed, 98 percent of native grasslands destroyed, 200 species driven extinct daily, more plastic than plankton in parts of the ocean, more ocean acidification, more melting ice, more climate refugees, more overconsumption rather than a life simple in means, rich in ends. Jensen argues industrial civilization destroys landbases and won't stop "because we ask nicely," and thus fighting back, and going back, in a sense, is the only way forward.[69]

To me, all of this makes sense. However, thinking your view is the only view usually does not end well, which is why Schrag champions transversal rationality and Wilber's meta-theory seeks to integrate multiple views, creating a more encompassing and therefore better, but not final, view. And radical environmental groups, since they are comprised of humans, will most likely express a combustible mix of unfitting and fitting responses, which is why discerning the unfit is so important.

Despite the need to be wary of the unfit, Zimmerman sees promise in immanently focused radical environmentalism, including deep ecology and ecofeminism, arguing that these ecophilosophies, along with New Ageism, are rooted in 1960s counterculture ideals, including envisioning a "new age" characterized by "dramatic improvements in personal, political, ecological, and spiritual conditions." Counterculturalists rejected the idea that a new world could arise in the "context of modernity's dualistic, control-oriented, hyperrational outlook," leaving them to gather insights and practices "drawn from premodern tribal peoples, from the world's wisdom traditions, from contemporary visionaries attuned to the complex relations between humankind and nature." Sounds like transversal rationality at work, especially since Zimmerman further argues that counterculturalists also gathered insights from the dignities of modernity's democratic traditions, including equality and rights. Thus, thoughtful radical environmentalists "criticize some aspects of modernity, while appropriating and transforming other elements of its emancipatory vision."[70]

Those inspired by Shepard and deep ecology, then, get us closer to fitting responses by practicing transversal rationality, distancing from modernist dissociation while participating in enlightenment rationality expressed as ecological science, rights, and liberation movements, as well as progress thoroughly redefined from a dialogic and sustainable perspective, with degrowth and technology as tools becoming part of new narratives.

However, according to the ecofeminist philosopher Val Plumwood, deep ecology also needs to distance itself from incorporation of the other, an attribute of modernist master-slave dualisms. She writes of Seed's call for identification: "What John Seed seems to have in mind here is that once one has realized that one is indistinguishable from the rainforest, its needs will become one's own. But there is nothing to guarantee this—one could equally well take one's own needs for its."[71] Plumwood warns of psychic inflation, and speaking for others without listening, a concern for New Ageism and interspecies communication. To counter incorporation, she calls for a relational self within human and nonhuman worlds; there must be differentiation between self and other in order to express relational care. She also calls for the deconstruction of dominating hierarchies, which, like Griffin, she claims is the result of androcentrism more than anthropocentrism.

Naess would respond that identification does not mean incorporation, but the discovery of our preexisting relationship with diverse others, while Wilber would agree with Plumwood's critique of deep ecology's theoretical flaws, arguing that the process of identification requires differentiation and integration leading to transrational experience. Of course, Wilber would also be critical of Plumwood's more immanently focused treatment of relationship and otherness. Such debates seem endless but do not merely reflect intellectual nitpicking; rather, they facilitate the discernment of unfitting responses in particular eco-social situations. This is precisely why an ecology of communication constitutes a fitting responsiveness to ecocrisis: instead of kneeling before moral absolutes or drifting like tumbleweed among relativistic landscapes, we are called forth by dwelling within and embodying a widened communicative awareness.

Despite Wilber wisely pointing out shadow-sides, deep ecology is another piece of the puzzle informing the practice of an ecology of communication, especially mythic-animistic communication, while providing insight into monologic narratives that constrain the relational coming through of joy. The ecological self remains inspiring when stripped of potentially unfit expressions, especially in regard to wilderness experience. However, not everyone has the time and access to explore wilderness and wildness. While deep ecology explores some of the factors that restrict such privilege, like industrialization, its analysis is incomplete.

SHADOWS AND LIGHT, UNFITTING AND FITTING: ECOPSYCHOLOGY

Like deep ecology, ecopsychology also deserves attention in relation to mythic-animistic communication, although, not surprisingly, Wilber places them in the same leaky transcendent-ignoring, flattened spirituality, pre-trans fallacy boat. However, since Wilber argues that the underlying evil is the divorce between transcendent freedom and immanent embrace—the dualism of all dualisms—ascent and descent must be integrated if we are to express sympathy with intelligence; and that means that the descent tradition is half of the cure, and, as we saw, Andy Fisher argues that we do not understand "what embrace of the earth truly looks like." Due to Wilber's seeming ascent bias, he could be accused of misunderstanding, while deep ecology and ecopsychology, which have Shepard-inspiration in common, do understand.

Fisher opens *Radical Ecopsychology* with a Shepard epigram stating that humans need nature not only for food, water, air, and energy but also for sanity-supporting connections that we have scarcely explored. He cites Shepard's core concern: "If environmental crisis signifies a crippled state of consciousness as much as it does damaged habitat, then that is perhaps where we should begin." And so ecopsychology begins, exploring "eco," taken from the Greek "oikos," meaning home, and the logos, or order and meaning, of the psyche. Combine the two and it only makes sense to study psychological well-being in relation to our earthly abode.[72]

Fisher points to psychoanalyst Harold F. Searles' *The Nonhuman Environment,* published around the time of Carson's 1962 *Silent Spring* but receiving far less attention, as an early core text exploring how the larger human-nonhuman matrix has a great influence on the psyche, which we ignore at our peril. Other early influences include Leopold's late 1940s arguments for thinking like a mountain and being plain members of the land community and "protoecopsychologist" Gregory Bateson's 1970s writings arguing that our individual minds are part and parcel of the pattern that connects. Beauty, wonder, and awe, stimulated by diverse and flourishing nature, are essential to a healthy psyche and more encompassing perception integrating the senses, rationality, and imagination. Ecopsychology rejects the Cartesian dualism of mind and extension, following Bateson by exploring the extended mind and the extension of communication to the other-than-human world. Thus, like deep ecology, an ecopsychological response to ecocrisis counters modernist dissociation by liberating and listening to the voices of nature.

Robert Greenway, who took students on hiking trips for thirty years in association with a course taught at Sonoma State University, was the first to propose a name, psychoecology, which later morphed into ecopsychology. Hikers exposed to extended periods of wilderness—usually three weeks, long enough to leave cultural selves behind—questioned their habitual lives and were often cured of addictions like alcohol and cigarettes, discovering the ecological self as well as experience approaching the nondual. He writes in a journal entry: "After a time we gravitated toward a large flat space on top of one of the rocks next to a pool and formed a circle, our habit over the past weeks. And then, without knowing how it happened, distance disappeared and there was an openness into ourselves that was an openness to each other, that embraced the pool, the river, and farther out into the wilderness, the "other world," the whole Earth, the universe."[73]

Greenway calls such transformation the "wilderness effect," in which we cross the boundary from physically inhabiting wilderness to the psychological experience of wildness, typically taking three to four days; such crossing leads us to discover a primordial sense of community with nature and fellow hikers that seems to already be within us. Greenway draws on Bateson's systems ecology, and Joanna Macy's and Wilber's interpretations of Buddhism, among other sources, to describe the wilderness's profound effect on the psyche.[74] Of course, the permeable boundaries discovered during extensive stays in the wilderness are not modern discoveries, as the mythic-animistic past and present of numerous hunter-gatherer cultures were the first to provide ecopsychological insights. However, turning to Wilber, Greenway also sought to legitimize the ineffable transrational spiritual experiences stimulated in wilderness, while acknowledging diversity of experience and interpretation.

Greenway laments our psychological disconnection, arguing that separation leads to domination and exploitation of habitat, undermining species survival.[75] Such domination occurs from being dualistically divorced from nature, but dominating attitudes may also arise during wilderness excursions. Those who hike in wilderness inevitably push physical boundaries but may not cross psychological ones; while being in the wilderness is bound to have some effect, humility is required to lessen ego-isolation. In fact, wilderness trips have been used, particularly via corporate empowerment programs, to solidify the competitive ego and develop skills useful in the work place, not to call forth the sensuous imagination and reciprocal dialogue with the other-than-human world. The nature-culture dualism is not overcome, but enforced, making wilderness an instrumental tool for corporate ends.

Greenway is also concerned with the understandable attraction of entering wilderness for psychological healing. Wilderness areas are often overused and abused, as there are just too many of us, and thus desire for healing may result in more destruction. Greenway suggests using meditation and yoga instead of wilderness trips to experience the extended mind, stating that they "open to the same awareness that occurred in the wilderness." However, he has also added meditation and yoga as pre- and post-trip practices, hoping to heighten and maintain the effects of wilderness on the psyche. He concludes: "Let that which serves the culture—at this point in our history at least—be done in culture. And if we do use wilderness, let us use it in ways that further its rehabilitation as well as our own. Let us use it for those healing processes that cannot take place anywhere else."[76]

In 1992, Theodore Roszak took another step in articulating ecopsychology with the publication of *The Voice of the Earth*, turning to—you guessed it—archaic hunter-gatherer cultures, arguing that animistic reciprocity represents a moral world view. Roszak defends the image of the "noble savage," although not from the usual charge of romanticism. Instead, he is critical of those who grant nobility to "savages" without acknowledging competence, claiming "savages" were deemed noble despite "ignorance, backwardness, poverty." However, "as the empire of cities crowds the wilderness toward obsolescence and threatens all traditional societies with extinction," nobility is reconceived: "*savagery as ecological wisdom*, the secret of survival and permanent well-being."[77] This mirrors Shepard, who writes that "the idea of the inherent 'nobility' of the individual savage was laughed out of school a century ago, and properly so," at the same time that he argues for the dignity of hunter-gatherer lifeways.[78]

Like Shepard, Roszak admits that hunter-gatherers were not always ecologically wise, referencing tribes and nomadic peoples from the Mediterranean basin who "overcut and overgrazed the land so severely that the scars of the resulting erosion can still be seen." Still, he argues that this cannot be fairly compared to current ecocrisis: "One finds nothing in the historical record to compare with the wanton assault upon the global environment we witness in our own time—and this on the part of industrial societies that arrogantly claim to know better." Roszak states that we need the "greatness of soul" that will allow us to "seek instruction from nonliterate, nonurban societies."[79] On the other hand, Wilber, Zimmerman, and Esbjorn-Hargens would accuse Roszak of the dignity-disaster fallacy and not factoring in the lack of technological means to amass destructive ends.

Roszak's turn to the past, however, goes deeper than this debate. He argues that Freud's id, rather than merely reflecting the unconscious within a libidinal context, must be understood within an ecological one: the id is "the very protohuman psychic core that our evolution has spent millions of years molding to fit the planetary environment," and this wisdom exists within our unconscious as the voice of the earth. We need to listen to this wisdom to have any hope of preserving the world.[80]

For Roszak, the id's larger context is the self-organizing universe from which life and mind emerged, characterized by "environmental adaptations" and the "basic properties of living things and of the planet as a whole." He pronounces: "*Seen from this perspective, what the id conserves from its long maturing process is our treasury of ecological intelligence*."[81] Bateson wrote that Freud turned within to expand communication to include the unconscious, while he expanded communication outward to include larger mind. Jung also turned within to expand the communication system via our collective cultural unconscious, but Roszak expands it still more by adding the ecological unconscious. These expansions of mind reveal that modernity's dualistic opposition between inner and outer worlds is a monologic mistake. Inner and outer are dialogically intertwined, which, according to Roszak, is how they were experienced by archaic peoples: the most significant image of the "noble savage" includes the id as "Earth's ally in the preservation of the biosphere."[82] When the human mind is placed within the context of the id's "intuitive environmental knowledge," what dwells within may be accessed to harmoniously dwell without.[83]

Roszak argues that this intelligence is easily seen in children, or at least in children who have not had their sense of wonder systematically destroyed by degraded landscapes. The earth is everyone's "primary care giver," and parents owe their young "a warm and trusting connection to the Earth that accounts for our evolutionary history." Referencing Searles' study of the relations between childhood development and nonhuman environments, Roszak states that healthy children greet the natural world with an instinctively animistic response: "It is alive and personal for them. It has a voice."[84] The unconscious, then, is rooted in animism; and so, like Shepard, he argues our mythic-animistic past, or "the distant preverbal origins of our species," remains with us despite being taught that the world is "an accumulation of dead and purposeless stuff."[85]

Mythic-animism is also found in the world's folklore and fairy tales, or the stories that enthrall children because they reflect their experience of a sentient world in which "animals seem to be going about what they do with intelligence and intention." Such stories should not be confused with clothed and talking Disney characters, which reflect a dissociated adult awareness; instead, mythic-animistic storytelling discloses an intuitive admiration of animals and nature, an admiration that may well extend into the perception of personality, for it is what the "unbiased eye sees in their conduct." Roszak defends what many deem anthropomorphism, arguing that perceiving personable animals is not merely mistaken projection, at least when done with intuitive care. He further states that we "educate" children out of this view, making them "realistic." But we must learn from children and childhood, becoming "childlike," in the good sense of sensing aliveness and wonder, in response to the world.[86] Scientists like Jane Goodall have put this responsive attitude to good use, discovering knowledge where less sensitive and attuned researchers have found none.

Roszak argues that taking sentience and intelligence out of nature denies the id its inborn ecological wisdom. He clearly follows Shepard's lead by valorizing the intellectual capabilities and mental processing of animistic hunter-gatherers, despite the fact that they were not always ecologically wise. Wilber, of course, argues that we must resist the temptation to respond to damaging modernist dissociation by romanticizing hunter-gatherer indissociation rather than evolving to leading-edge vision-logic differentiation and integration. It is noble and necessary to include pre-rational levels of development, integrating body-knowing and unconscious id with the rational mind, but it is not noble to become childish rather than childlike. However, the crucial elements that Shepard, Roszak, and ecopsychology generally bring to the discussion are the depth of our psychological connection to earth and other species and the need for a deep, rather than shallow, environmentalism. And both elements emerge when practicing mythic-animistic communication.

Roszak furthers the project of ecopsychology by taking us inward, prodding us to listen to the voice of the earth via the ecological unconscious. This communicative turn within makes us more responsive to an animistic world without, or at least an alive world, where the extended mind and dialogic relations with nature are experienced. A deaf inner world and deadened outer world do not lead to responsibility and fitting responses. Roszak's ecopsychological depiction of ecocrisis—and our lack of sanity—discloses that we are currently transcending too much and including too little of what makes us fully human.

All of this may be hard to hear if you grew up in or are living in an urban environment, or specific urban environs with little to no green space. Jonah Lehrer, in "How the City Hurts your Brain," writes that harsh, brain-taxing environs impair mental processing and memory, as well as reduce self-control. So much stimuli leads to constant cognitive choice of what to attend to and what to ignore. Any kind of nature break provides focus, including for studying students, and may aid healing, like hospital patients who have a window overlooking courtyards.[87] Given Shepard's arguments, this should be no surprise, as our brains developed while living in small groups, not overcrowded metropolises.

During my hike, I passed a gaggle of urban-dwelling teenage Outward Bounders out in the wilderness for the first time. They were only a couple of days removed from their cultural lives, and it showed: they peppered me with questions about who my favorite baseball team was and other cultural interests that had faded from my focus. By nightfall, however, they walked slowly and silently into the shelter area, putting up their tents without a peep. They were clearly exhausted, but perhaps they had also exhausted some of their internal chatter, finding a bit of a quiet mind. Today, my urban-dwelling students often arrive at our nature-immersed campus with discomfort and some fear. Taking a walk on well-worn trails adjacent to campus, which, to me, is one of the perks of being there, does not inspire, at least not initially. So many trees, they say, and what of fierce and threatening wild animals. Many are surprised to see stars in the night sky, which Thomas Berry has called a soul loss.

Cities, of course, supply plenty of inspiration for great art and great conversations, but there is a reason that Shepard's first book was called *Nature and Madness*. But the madness is structural, with winners and losers, sometimes big losers; we all know whose neighborhoods will have the polluting factories and toxic waste and garbage dumps. If the physical doesn't get you, the psychological will, at least to some degree, especially if you are poor or a person of color. Deep ecology does not really have a fitting response, and neither does ecopsychology, although Fisher has attempted to radicalize it, disclosing that mythic-animistic communication needs the critical skills of rational communication.

To Fisher, ecocrisis is most certainly madness, or more specifically: "Modern society is in an extreme, pathological state of rupture from the reality of the natural world."[88] And, not surprisingly, he turns to Shepard to support that claim, since "perhaps no one has done more to reconceive other-than human beings as psychological counterplayers." Shepard, before Roszak, focused on childhood, and then adolescence, arguing that immersion within nature, along with rituals that support communicative reciprocity, is imperative for healthy development. But he does not insist that we become childlike in response; instead his focus on repression led him to assert that we have become "childish adults" who degrade the earth that sustains us, striking back at a "natural world we dimly perceive as having failed us."[89]

We may respond with various eco-therapies, including within city parks, and we should, especially if therapeutic practices are designed to support the real work of creating a sustainable society. And spiritual practices like meditation, yoga, and group rituals not only allow hikers to bring the wisdom of wilderness trips back to their cultures but also support activists, making them less susceptible to burnout from confronting a world of wounds. But we all face degrees of wounding on a daily basis, and thus we all need daily therapies and practices. But therapies and practices fall short if they attempt to help us fit into the insanity. After all, one can imagine clients complaining to therapists about not getting the Hummer or SUV they desired. This is not to disparage psychoanalytic dialogue, but our conversations must extend beyond the personal. Or as Fisher puts it: "As witnesses to the psychic injuries wrought by our society, psychotherapists are uniquely positioned to be social critics."[90]

And that brings us to the necessity of a radical response, which Fisher calls the critical task of ecopsychology, in which we shift the therapeutic focus from individual inner states to collective outer techno-industrial capitalist structures that do violence to inner states. Once again, Fisher, like Ramsey, turns our attention to Wilber's interobjective societal structure quadrant, which too often inhibits our subjective and intersubjective responses. For him, unlimited growth and technological dominance defines our notions of progress but also sanity, reasoned discourse, and respectable behavior. Thus, those who fight the power will be deemed crazy, or hysterical, or communist, by the powers-that-be. The great work, then, will be played out in rhetorical debates and rational discourse, and in whether we move up the developmental holarchy, leaving Magic and Mythic thinking and take steps to Rational and Pluralistic dialogue, and hopefully Integral communicative competence, but that work must be grounded within ecopsychological insights that disclose what is sane and insane. And the practice of mythic-animistic communication does just that.

And so, mythic-animistic communication is essential but not sufficient. And like all things, and all forms of communication, it has its shadow-side, especially when not balanced with other forms of communication. But if Shepard is right that we cannot really leave our hunter-gatherer past because it remains so much a part of us, and if Greenway is right that the wilderness effect can reawaken dormant mental-functioning, and if Roszak is right about the existence of an ecological unconscious, then we are much more than we think, especially if we live in cities. In my *The Voice of Nature* course, I sometimes ask students to write a narrative describing experiences in nature or the absence of nature when growing up. A student from Boston wrote of protecting his tiny patch of front yard, defending it against those who would trash it and cleaning it up after it was. To me, that is the beginning of sanity, wherever we live, and enables us to respond more fittingly to our eco-social predicament, or simply what we face moving forward.

But again, to move forward, including the creation of sustainable cities designed with lots of green space, we need to reclaim our mythic past by transcending *and* including, growing into a widened awareness. Shepard does not advocate for a hunter-gatherer twenty-first century, but he does argue that we must incorporate the mythic-animistic practices of hunter-gatherer lifeways—the practices interwoven with the human genome—and then "the culture will reshape itself."[91]

Perhaps no current thinker has valorized such practices, and mythic-animistic communication, more than David Abram.

In his 1996 *The Spell of the Sensuous*, an influential ecopsychological text, Abram carried forward many of Shepard's views at the year of his death, exploring all-can-speak bodily based converse within a more-than-human world. Writing more poetically than Bateson, he also makes the claim that ecocrisis reflects a crisis of perception: modernity has narrowed mind such that our dominant activity is the collection and classification of objective facts, which then gets used as a tool of domination. But Abram returns to our senses, exploring perception as continual conversation with sensuous surroundings: "The color of sky, the rush of waves—every aspect of the earthly sensuous could draw us into a relationship fed with curiosity and spiced with danger. Every sound was a voice, every scrape or blunder was a meeting—with Thunder, with Oak, with Dragonfly."[92]

While animals, plants, and insects express gene-embedded instructions, they are also expressive, responding to ever-changing environs. This includes human interactions, with communication as perception creative, open, and receptive, and all animate species reoriented within specific locales.[93] This reciprocal exchange of presences invites us to respectfully attend to others and then respond in turn, each moment a different moment demanding a different response. For Abram, this dance occurs bodily, and not always consciously, via "a sort of silent communication."[94] Of course, this commingling of bodily presences and silent communication affirms an animistic attitude: "Prior to all our verbal reflections, at the level of our spontaneous, sensorial engagement with the world around us, we are all animists."[95]

Like Shepard, Abram praises participatory consciousness in which "inanimate objects like stones or mountains are often thought to be alive, for whom certain names, spoken aloud, may be felt to influence at a distance the things or beings that they name, for whom particular plants, particular animals, particular places and persons and powers may all be felt to *participate* in one another's existence, influencing each other and being influenced in turn."[96] However, as with the discourse of New Ageism and interspecies communication, we must interrogate the rhetoric here, asking whether aliveness extends to stones and if telepathically influencing things and beings at a distance reflects an advanced awareness or pre-rational and undeveloped Magic level create-your-own-reality wishful thinking.

In his 2010 *Becoming Animal*, Abram directly addresses these concerns. Oral cultures live within many-voiced, no distinction between animate and inanimate local environments. A pack of wolves, spider webs, thunderstorms, and yes, rocks, all speak, as oral languages, in contrast to abstract alphabets, kept one close to earth. Rocks express themselves nonverbally; for example, depending on our attentiveness, we perceive that boulders, while expressing solidity, change over long time scales. Everything interacts at different speeds and with different styles, yet all are part of an animate community, and respecting that community was key to success in the hunt, as words, connected to earth, had impact. If all are alive then all are sensitive, and we must be sensitive in return, quieting chatter, sensing every movement, every nonverbal cue, while making decisions on which way to move amid a moving matrix. If others can hear at a distance, we must be attuned and not offend. Such animistic awareness, while alien to alienated verbal minds who mostly converse with themselves, with their own symbols, marks the vast majority of our history as communicators, and without it we would not have survived.[97]

Abram also responds to the charge of romanticism, arguing that those critical of looking backward to hunter-gatherer life have no trouble returning to ancient Greece or Rome to gather insights. The difference, of course, is demarcations between so-called civilized and uncivilized conduct, often centered on language; yet such a stance fails to perceive that oral cultures have a different experience of language that includes silence, reticence, and slowing down while being attuned to the body and senses. Language, then, is not only for humans; all species participate via signs and gestures, expressing their unique song. Literacy, as wonderful as it is, removed words from land and put them into scripts, then books, now internet, leading humans out of their bodies and into their heads. We learned to communicate as if the earth does not matter or as if matter was subservient to mind.[98] Abram, again like Shepard, argues that such detachment birthed transcendent sky-gods and transcendence generally, positing, and depositing, creative intelligence beyond the wild, earthly intelligence from which language emerged.[99]

Abram has been criticized for the irony of using rationality and language to disparage rationality and language, but that's a gross misreading; not only does his poetic prose make it obvious that he loves the music of the written word, and he obviously reads a lot, but he also argues that reading alphabetic symbols is similar to reading the marks and indents and lines on rocks and nature generally.[100] We remain animists no matter how we read, as voices emerge in our heads when in relationship, whether to ink on the page or rocks and twigs and bark, or animal others. Of course, the printed page only connects us to human voices, but voices are everywhere.[101]

In *Becoming Animal*, Abram also somewhat surprisingly acknowledges the merits of modernity and monotheism. Like Griffin, he links this unlikely duo in their earthly detachment, yet admits to noble advances in abstract thought, science, technology, philosophy, religion, and the arts—quite an impressive list—of course, he also laments the loss of indigenous intuitions and practices.[102]

And so, while Abram acknowledges the problematic nature of listing universalized TEK wisdoms, he does so anyway, stating they are inevitably expressed differently among differing tribal groups. They include the intensely local focus of oral awareness resulting in place-based knowledge; perception as reciprocity, in which everything is sensitive and responds; everything as animate, and thus everything has the power of speech without words; everything as part of stories situated within land, evoking dreams and the felt-sense of existing within the Earth's imagination; a nonlinear sense of time marked by the sun's daily disappearance into the earth; and sensing bodies providing partial perspectives amid an uncertain world, yet, inhaling and exhaling, influencing the wild mind. Thus, intelligence, rather than transcendent to the earth, displaced solely onto God, or soullessly reduced to random natural selection, is creative, alive, and ultimately, participatory. To Abram, we desperately need to reclaim this regenerating multivoiced reality as survival response, as the wonders of monotheism and modernity also emerged with shadowy blinders producing endless blundering and plundering.[103]

Abram tells a story told by a variety of oral peoples of the disappearing sun that appears to enter the earth. Or rather, it does enter, resting within the center before firing up the sky once again. The story has little to do with facts, other than the fact that the arcing sun and arc of the story match visual perception, and for Abram that is its value: such stories return us to our senses and bodily based awareness. There are obvious shadowy blinders here as well; in fact, Wilber is critical of Joseph Campbell's hero's journey interpretations of archaic myths, claiming that magic-mythic level tribes interpreted them literally while rational-level moderns find profound metaphoric meanings. But Abram points out that "literal" derives from "letter," which early oral peoples did not have, nor did metaphoric meaning make sense: literal and metaphoric are markers of literacy, and thus these terms do not apply to stories that reflect a different way of thinking that does not withdraw from living landscapes. Stories emerged to provide practical information to listeners regardless of truth and falsity, which, like literal and metaphoric, were not yet fully imagined into existence.[104]

Plato's allegory, of course, disclosed the shadow-side of such visually based thinking, which continues to this day via flat-earthers chained within caves of one-dimensionality. Yet so many who think they have left the cave, transcending limited perception, have also transcended, in their heads, the vibrant earth, labeling it as fallen and sinful. But the religious are not the only ones to have fallen prey to this sin. Abram also tells a much more modern story of perusing the shelves of a city bookstore, amazed by amassed wisdom of art, science, philosophy, literature, and spiritual traditions, not to mention books on TEK and indigenous lifeways (like his own) increasingly available to all via digital downloads, yet forests are decimated, wetlands paved-over, oceans plastic-ridden, acidified, and dead-zoned. How, he wondered, could there be so much wisdom yet so much destruction? And then, a realization: "*We have written all these wisdoms down on the page*, effectively divorcing these many teachings from the living land that once held and embodied these teaching."[105]

Abram does not renunciate the printed word or digital discovery but argues for the renewal of oral culture and listening as a sensory skill. We need not turn back the clock but it would be wise to turn it off, rediscovering timeless experiences by attending to spirits in specific places.[106] For an animist such spirits are not transcendent but embedded, akin to the etymological root of spirit as breath or wind within a sensuous earth from which we draw our awareness.[107] However, in his brief praise of monotheism when not monologue but dialogue, he writes of transcendent Spirit expressed as immanent voices, the "one" animating many earthly inhabitants. And he is quite fond of shamans, or medicine persons, who "remember the primordial sacred language, and who are thus able to slip, at will, out of the purely human discourse in order to converse directly with the other powers."[108]

Such appeals to the transcendent and transcendence would seem to be contrary to Abram's embodied arguments concerning communication. Shamans, for example, temporarily transcend the body in order to enter the spirit world, which may reflect subtle state spiritual experience. Interestingly, Shepard disparaged shamanistic questing, claiming that shamans were more prominent with the birth of agriculture.[109] Regardless, despite these nods to a wider and deeper sympathy with intelligence and spiritual journeying, Abram's focus is thoroughly grounded: earth, senses, body, all alive and ritualized in stories of reciprocity and locality. The Gaian earthbody, and our receptive bodies, operate quite well without conscious interference. And while shamans induce non-ordinary trance states catalyzed by the primal beat of rhythmic drumming, Abram highlights the embodied experience of the shaman, expressed as an immense openness and willingness to recognize the other as a communicating subject, coupled with an intimacy that views animals, plants, rocks, and all fellow earth-sharers as family.[110]

Wilber cautions that shamanic experiences, which go back 40,000 years, are interpreted by archaic peoples within magic-mythic cultural space, disclosing insights but also superstitions that were in need of rational analysis not yet available. But Abram does not explicitly enter this ongoing debate; instead, like Shepard and Roszak, his focus is not on returning but learning from our long history as hunter-gatherers by exploring practices concealed by abstract rationality. But one of the lessons of shamanism is that body-based experiences may open us to spiritual insight. In fact, we deny spiritual affinities by denying bodily empathy, lost in constructs and silencing nonhuman voices in order to privilege our own human ones. For Abram, we deny the sacred via abstraction but also vision when disconnected from other senses. We ignore bodily empathy and spiritual affinities when we "see" what we have been socialized to see—everything as separate. To move beyond this separative perceptual bias we must invite synesthesia, the visual becoming visceral as senses intercommunicate.[111]

For postmodern selves, then, practicing mythic-animistic communication might look like this. We open ourselves to the mystery of the other while directing attention within. We are silent and behold, we feel and ponder. We drop grasping egos, give ourselves to grace, and become receptive to messages emerging within us. Such is how we use our sensuous imaginations. We open ourselves to the possibility of relationship and this acts as a conscious meditation in which our internal voice speaks, our individual unconscious beginning to connect with its evolutionary source as we resonate with the songs of other beings. So when we say that rocks and trees and animals speak to us, we refer to the silent communication of our perceptual bodies, as well as feelings and thoughts that are imparted to us from within when in sympathy with the spirited wind or breath that animates every earthly expression. We discover sense-based affinities with diverse forms of life, synesthesia awakening shared biological heritage.

If we go further into aesthetic communication, we attune to panerotic energies and discover psychological affinities for beauty and unity, awakening our shared heritage within larger mind; and if we go still further into spiritual communication, adding insights from esoteric religious traditions and affinity with Spirit unfolding, one in many, many in one, we reclaim our shared spiritual heritage by perceiving divinity in all things.

Put it all together, and we transcend habitual defenses, embodying a responsiveness characterized by biophilia and eros and agape, and thus an array of wisdoms that we too often ignore due to definitions of "normal" consciousness.

Ecopsychology does not go this far, and for Shepard, and no doubt Abram, this goes too far, too far away from the sensuous earth. Regardless, this hopeful description of communicative possibilities inspired by ecopsychology, aesthetic responsiveness, and spiritual traditions must be tempered by another possibility: the self-deception that may occur when rational communication is not highlighted. Are we being spoken to by nonhuman others or are we projecting onto others our own thoughts and desires? In other words, the failure to also practice transversal rationality may lead to the belief that unfit responses are fitting.

Roszak and Abram do not emphasize the need for rational communication as a response to ecocrisis, nor does Greenway, although he acknowledges the marvel of distinction making as "evolutionarily crucial." However, he further argues that we have taken the mental function of distinction making and pushed it until "distinctions become disjunctions" as Western psychological processing is marked by differences being placed in completely different categories. This narrow perception cannot solve problems of ecocrisis, but it can and does create them. Thus, Greenway follows Griffin by arguing that our mental processes are marked by "radical dualism" constituted by a "complete divergence of realities."[112] Ecopsychology, then, in the hands of Roszak, Abram, and Greenway, tends to emphasize the downsides of rationality more than calling for higher-order rationality and world-centric dialogue. This has the potential to be problematic, but they also have much to offer to the discernment of fitting responses.

Roszak and Abram highlight the necessity of mythic-animistic communication for grounding us within diverse relationships that make us human, with Roszak focusing on listening to the voice of the earth via the ecological unconscious, and Abram focusing on preverbal bodily empathy and perception as conversation. Greenway discloses the transformative potential of wilderness trips, and responds to ecocrisis by highlighting nondualistic modes of ego processing experienced on these trips, and practices like meditation and yoga, all as a means for creating relational habits of mind, and thus his focus includes spiritual communication.

However, Roszak, but especially Abram, can be, and has been, criticized for emphasizing mythic-animistic communication at the expense of other forms of communication. As stated, this is mostly unfair, as recovering mythic-animistic converse is his focus. On the other hand, Wilber may be, and has been, criticized for not fully grasping the devastating repression of oral cultures and their communicative skill within a more-than-human world. Still, from his perspective one could argue that Roszak and Abram, and most certainly Shepard, over-respond, in the sense that they wish to reclaim animistic aliveness while failing to be critical of the magic-mythic cultural space occupied by early animistic cultures.

And Fisher argues that we must widen our psychological concern by mining our hunter-gatherer past to undermine the sociocultural structures that currently harm all species. Both must be addressed if we are to counter modernity's monologic diminishment of the psyche and earth. Thus, focusing on our interior psychological life in relation to eco-social processes is a radical act. And Abram is a fan, praising Fisher in the foreword to *Radical Ecopsychology*, while also extolling Shepard's "audacious writings," which, despite being little known, has influenced so many environmental thinkers.[113]

THE COUNCIL OF ALL BEINGS AS EXEMPLAR

In *The Others*, Shepard makes our communicative debt to nonhumans explicit: "Our species and our best observers emerged in watching the Others, participating in their world by eating and being eaten by them, suffering them as parasites, wearing their furs and skins, making tools of their bones and antlers, and communicating their significance by dancing, sculpting, performing, imaging, narrating, and thinking them." Carson's *Silent Spring*, he continues, was not only prescient because it warned of the indiscriminant use of pesticides, but a deafened world emptied of wild voices. But despite praising Carson's book in his own, he disparages nature writing, and art, when it substitutes for adventuring among the Others, a mistake primal peoples did not make.[114]

A modern-day example of such mythic-animistic adventuring is The Council of All Beings ritual, influenced by Naess and deep ecology and largely designed by John Seed and Joanna Macy to create a dialogic rapport among humans, other species, and the earth that actualizes the ecological self and inspires activism. In arguing for the ecological self, Naess called for "therapies which heal our relations with the widest community, that of all living beings," and Seed and Macy responded with the Council ritual, an instinctual, emotional, intuitive, and imaginative response to the eco-state of the world, with some rationality thrown in there too.[115]

Seed borrows a phrase from the Buddhist monk Thich Nhat Hanh, stating the specific purpose of the ritual is to allow humans "to hear within themselves the sound of the earth crying," and then to let the earth and other species speak through them. Thus, Leopold's "world of wounds" is fully entered in an effort to heal the rift between mind and nature, supplying energy for nonviolent resistance. For Seed and Macy, the destruction of our life-support systems is the "deepest and most pervasive source of anxiety in our time." They further state that people sense and feel this destruction on an "inchoate level, in their bodies" but that the "enormity of the threat makes it harder to talk about it or confront it squarely."[116] The Council of All Beings is designed to provide a ritualized space within which we may confront what we habitually deny, turning repressed despair into a conscious and creative response to ecocrisis.

The Council of All Beings is informed by rationality in its utilization of ecological science and evolutionary theory, but the ritual is clearly focused on heightening senses and emotions. Seed argues that our problem is not an absence of information, but numbness and apathy: "This refusal of feeling takes a heavy toll on us, impoverishing both our emotional and our sensory lives. It also impedes our capacity to process and respond to information as we screen out and filter anxiety-provoking data. But such feedback is precisely what we need to adapt and survive."[117] This passage thoughtfully recalls Griffin's and Abram's celebration of body-knowing and Ramsey's concern with widening our awareness and increasing our capacity to act, and, as we will see, it's consistent with Bateson's explication of communication dysfunction as it relates to ecocrisis and survival. However, as always, we must question whether the Council ritual problematically privileges mythic-animistic communication and imagination over rational communication and rigorous thought.

The Council of All Beings has been performed all over the world in indoor and outdoor locales, and either in one-day or three-day sessions. However, a three-day natural setting is considered ideal. A Council workshop includes practices of mourning ecological destruction and remembering our evolutionary heritage, preparing participants for the ritual in which they speak from the perspective of nonhumans. Other practices are also included, like sharing personal stories, poetic readings, breath work, and group exercises that lessen boundaries between self and other, but mourning, remembering, and speaking as an other-than-human species are the core practices utilized to elicit the ecological self, making deep ecology's ecophilosophical insights experientially real.

Mourning can be expressed in any way one needs to mourn what is happening in the world, but workshop leaders specifically direct participants to grieve the loss of species. Macy has written "Bestiary," a poetic lament listing numerous extirpated species that may be presented to the group for reflection. Macy and Pat Flemming, who worked with Seed on the Council from its beginning in Australia in 1985, write that mourning "erodes the culturally conditioned ego, the fictions that "I" am or should be in control, that I can hold aloof from what befalls others."[118] Mourning, then, while typically expressed as anger and grief, also elicits connectedness and caring; otherwise, there would be no reason to mourn. Thus, mourning is an empowering step, re-establishing feelings and connections with nonhumans.

Remembering takes the form of a narrative depicting the evolution of life. In *The Universe Story*, the eco-theologian Thomas Berry and the cosmologist Brian Swimme poeticize the scientific narrative of evolution, calling the big bang the "primordial flaring forth" or the "great radiance" in an effort to provide a sense of belonging. Everything has descended from the same primordial source, making every life form our cousin.[119] And E. O. Wilson has argued that science provides an evolutionary epic from which we may derive meaning.[120] The evolutionary remembering of Seed and Flemming does similar work, connecting participants to the "mystery of the universe coming into being," as well as human emergence: "The first cell was born. You were there. I was there. For every cell in our bodies is descended in an unbroken chain from that event. Through this cell, our common ancestor, we are related to every plant and animal on the earth." Thus the ecological self, identifying with all life, is scientifically supported. From an ecopsychological perspective, the eco-self and epic of evolution are also discovered within our bodies and ecological unconscious.[121]

Speaking from the perspective of other life-forms, then, has a rational basis; however, Macy and Flemming acknowledge the imaginative element of the Council of All Beings, which they admit is novel and "risks appearing contrived or silly."[122] They also provide a sample introductory narrative of the Council process: "This morning we engaged in a number of group exercises to help make us more conscious of our embeddedness in the web of life. They helped us remember our bio-ecological history, as our species and its forebears evolved through four-and-a-half billion years of this planet's life. They helped us relax into our bodies, into our intuitive knowings, and our trust in each other. Now . . . we assemble to prepare ourselves for the promised ritual of the Council of All Beings."[123]

Instead of immediately beginning the Council, participants are instructed to find a place that calls to them, and then be still and wait for a further call from a life-form, a life-form "for whom you will speak." A sample call from a nonhuman is recalled, with a female participant being chosen by mountain: "I relax and breathe in, breathe in Mountain . . . I feel my rock-roots go deep down to where the Earth herself is very hot. My base is wide, very wide and solid. Storms come and go over my surface, leaving but a ruffle in my tree-skin. Even the occasional quake of Earth only causes me to shiver and sense more keenly my vitality. I am ancient." This imaginative identification leads her to feel great strength within herself, as well as deep peace, "immovable, beyond time." Strength and peace are the gifts she, as mountain, will offer "to all who rest on me." But humans abuse these gifts: "They gouge my bones, strip flesh, bring danger to my inhabitants. I must speak of this at the Council this afternoon."[124]

The drum sounds and she is called back to the group, with participants making masks representing the life-forms through which they will speak. The masks are made and the drum sounds again: The Council of All Beings is convened. Mirroring native traditions, an invocation calls forth the powers of the four directions and the ritual begins: "We will, as the life-forms we have assumed, speak spontaneously, letting be said what needs to be said." There are three stages of speaking within the Council. First, participants will spontaneously speak as other life-forms, stating why they have come to the Council while expressing their confusion, anger, and grief. Second, small groups will remove their masks, moving to the middle of the circle to listen as humans to the laments and concerns of other species. And lastly, masked participants will offer humans insights needed to stop ecological destruction.

The workshop leader acknowledges ecocrisis and calls forth the wisdom of other-than-human ancestors. Mountain speaks: "I am Mountain. I am ancient and strong and solid, built to endure. But now I am being dynamited and mined, my forest skin being torn off me, my topsoil washed away, my streams and rivers choked. I have a great deal to address to humans today." Numerous other life-forms also speak—Woompoo Pigeon, Cow, Wild Goose, Slug, Red Kangaroo. Then the workshop leader, speaking as Weeds, calls for masked humans to speak to their brethren: "This is our world, too. And we've been here a lot longer than you. For millions of years we've been raising our young, rich in our ways and wisdom. Yet now our days are numbered because of what you are doing. Be still for once, and listen to us." Anthropocentrism is clearly negated here, traded in for the ecological self if one finds wisdom in the ritual, and anthropomorphism if one is more critical. One wonders what Weeds thinks of invasive species, although invasive species, which conservation biologists especially lament, is anthropogenic, a human-generated problem. Perhaps if we listened to the wisdom of Weeds and the other life-forms, such problems would never have emerged.

Many life-forms speak, and then Mountain erupts: "Humans! I, Mountain, am speaking. You cannot ignore me! I have been with you since your very beginnings and long before. For millennia your ancestors venerated my holy places, found wisdom in my heights. I gave you shelter and far vision. Now, in return, you ravage me. You dig and gouge for the jewel in the stone, for the ore in my veins. Stripping my forests, you take away my capacity to hold water and to release it slowly. See the silted rivers? See the floods? Can't you see? In destroying me you are destroying yourselves. For Gaia's sake, wake up!"[125]

After the workshop leader temporarily takes off their mask and gives human thanks to all fellow beings for their honesty and insights, life-forms continue to offer their wisdom to humans. Slug offers the wisdom of slowness and knowing the "ground you travel on." Water offers the wisdom of flow, persistence, and flexibility. Condor offers a keen farsightedness. Lichen offers patience, Rainforest offers balance and harmony, and Dead Leaf offers freedom from the fear of death. Wildflower speaks: "I offer my fragrance and sweet face to call you back to life's beauty. Take time to notice me and I'll let you fall in love again with life." And then, Mountain wants to speak through again, in a kind of *The Giving Tree* offering: "Humans, I offer you my deep peace. Come to see me at any time to rest, to dream. Without dreams you may lose your vision and your hope. Come, too, for my strength and steadfastness, whenever you need them."[126]

The Council of All Beings draws upon and blends different Native traditions, including aspects of vision quests. The Council participates with these traditions, gathering insights from archaic and indigenous peoples in search of fitting responses to present-day problems. Such gathering may provide wisdom, but some would view it as appropriation of traditional rituals, and thus distanciation must also be elicited. In other words, instead of a gathering of insights, it is a taking, and not only a taking from Native peoples, but a taking out of context, with ritualized meanings connected to particular peoples and specific places co-opted and perverted. New Ageism is often guilty of such theft, especially in the material marketing of "spiritual" products like dreamcatchers. However, to call all such gathering appropriation is to negate the possibility of learning from earlier peoples and differing cultures. How one gathers insights and for what purpose—the how and what of communication—must be considered. The Council of All Beings also draws upon science, and has unique features, and thus perhaps its creators may be forgiven if the gathering from oral traditions is done with respect and if meaningful insights are accrued by participants; in an age of ecocrisis, we need all the help we can get.

The Council also draws upon the mythic-animistic imagination via the practice of identification. The ecological self is called forth and participants "think like a mountain," although one wonders if Leopold meant for this evocative phrase to be interpreted so literally, with "Mountain" speaking out forcefully against our world of wounds. Thus, literalness must also be critically considered—once again, the how of communication—especially if it reflects solipsistic imaginings with communicative action happening solely within our heads. However, the Council's literalness reflects a conscious extension of imagination to connect with emotions and other life-forms, fostering healing and a critical appraisal of ecocrisis that leads to activism. And anthropomorphism, while an understandable charge, may be understood as a situational response or projection for a specific purpose, with consideration of the quality of our projections as the main concern.

Still, it would not surprise me if some participants feel silly, and if the experience is not life-changing. But this does not mean that the Council of All Beings reflects a misuse of mythic-animistic communication or disdain for rational communication. Rather, it could be argued that the hunter practice of mimicking nonhuman others to learn their qualities is put to good use, the ritual designed to reclaim the mythic past from which we have emerged. I have not participated in the Council, but I have participated in deep ecology-inspired workshops and social protest, and any silliness I may have felt from unfamiliar rituals or being a small voice among other small voices facing monologic powers-that-be is always countered by the insanity of ecocrisis. Speaking as Mountain is not nearly as silly as mountaintop removal for coal.

The Council of All Beings leads us to reflect on the reasons for ritual and the insights of mythic-animistic communication. Shepard argues that nature has been aligned with myth and culture with history. Meaningful nature rituals connect us to the wisdom of mythos, which we must integrate with the logos of culture and history; but Shepard's point is that history, although a narrative of rational inquiry, reflects a mode of knowing that is inevitably limited. We are born "anti-historians" because "our secret desire" is the wisdom of mythos, which calls to us in the form of "a yearning, a nostalgia in the bone, an intuition of the self as other selves, perhaps other animals, a shadow of something significant that haunts us, a need for exemplary events as they occur in myth rather than History."[127] In other words, we are drawn to meaningful narratives that we may experience and consult via ritual and the practice of mythic-animistic communication, narratives that connect us to the earth and other-than-humans, narratives that elicit the call to responsibility and activism.

For Shepard, the key question confronted by moderns is how can we become native to the land. Fitting responses to this question require rational *and* imaginative traversing of the mythic past. The Council workshop's inclusion of evolutionary remembering is designed to tap into this past, this history of unfolding mind, but the telling, when ritualized and made real in our bodies, draws us "like fire": "We cannot explain it, but it is there, made fragile in our psyche and hearts, drowned perhaps in logic, but unquenchable."[128] For Shepard, and deep ecology and ecopsychology, remembering our mythic-animistic past and then projecting ourselves forward via the evolutionary epic makes us belong, and then defend this belonging against the forces that make us forget deeper identity expressed as the ecological self, like the hubris of linear progress with humans at its pinnacle, transcending without including.

The Council of All Beings' practice of imaginatively speaking through the perspective of other life-forms should not be equated with New Age channeling, as Council participants invite panerotic energies and the ecological self to "come through" rather than channeling actual animals, plants, or earth. Thus, the erotic energy of evolution, or eros, may speak through and as us, as may "biophilia," coined by E. O. Wilson, with bio, or life, reflecting four billion years of co-evolution, and philia, from the ancient Greek, referring to loving friendship. Of course, our reactions to other forms of life are not always loving; there are situations in which fear may be the best response. However, both eros and biophilia remind that we are not fully human unless we are cultivating relations with what is not human, both rationally and imaginatively, especially when we consider the loss of biodiversity and ecosystem health upon which human survival depends. Again, the potential silliness of the Council of All Beings pales in comparison to species extirpation and our lack of a loving response. We must rationally consider what is true foolishness and a true misuse of imagination.

For Seed, deep ecology recognizes that only a "total revolution in consciousness will be of lasting use in preserving the life-support systems of the planet," and the Council ritual is meant to be a step toward this revolution, which may take form as revolutionary activism.[129] The collective authors of *Thinking Like a Mountain* provide the activist example of Graham Innes, or rather, the testimony of Graham Innes. Innes, along with others, seemed to connect with the wisdom of our mythic-animistic past by burying himself in the present earth to block bulldozers from "developing" the Daintree Rainforest: "This nonviolent action had been forced upon us as a last resort to save the Daintree Rainforest from the blades of men unhinged by greed, prestige, and authority."[130] Being buried in the earth, with only his head above ground, led Innes to "Earth bonding": "Her pulse became mine, and the vessel, my body, became the vehicle for her expression." However, a back-hoe operator brought in to remove protestors, who were linked via chains fastened to underground logs, did not share the same awareness.[131]

During a break while police tried to figure out their next move, Innes confronted the back-hoe operator: "This is the opportunity I had been waiting for. Unknowingly preordained. The language was mine and I was speaking from the heart—and yet it was not of me. It was as though nature had overtaken my consciousness to speak on her behalf." Like the Council of All Beings, Innes proceeded to speak imaginatively as other, this time earth itself, with, it would seem, some degree of channeling. He, or the earth, spoke in a commanding voice: "Sir. You are stripping the earth of her mantle and she will die. She will die as surely as the naked baby left unclothed on the beach in the mid-day sun. Dying slowly but certainly of exposure. Stripped of her mantle and laid bare to the harsh unfiltered rays of the sun, the earth slowly sterilized. No longer allowing nature's vitality and fullness to work the miracle of creation."[132]

Innes continued giving the earth's soliloquy, not blaming the back-hoe operator but trying to reach him: "Sir. I ask you to desist from this act of madness, from taking part in this vandalism. Please withdraw. Go home now with honor. A hero whose praises of courage will be sung by all. There will be no shame, no stigma attached to such a decision. I know that in your heart you know it to be so. Act now and withdraw a hero. Listen to your heart. I see you hesitate. This is not the time for hesitation when your heart speaks to you of right action."[133] Innes writes that a crowd of nearly one-hundred fell silent during his oration, becoming immobilized by his words. The tension mounted to such a degree that police and the back-hoe operator withdrew for a two-hour lunch. They returned with a renewed determination to not be dissuaded by Innes' words, and soon dislodged him.[134]

This was clearly a powerful experience, one that most would have trouble imagining. Personally, I am amazed by the commitment of activists who put their bodies on the line. And the Council of All Beings clearly prepares activists for such activism. But we must return to the question of interpretation. Was the earth speaking? Was Innes connecting to the mythic past, the animistic fire within and without, the ecological unconscious, or the eros of the evolutionary epic? Was he expressing his ecological self, or biophilia, or more, having aesthetic or spiritual experience? What, if anything, was Innes tapping into, other than his own heart and mind? Most importantly, is such activism a fitting response to ecocrisis?

Practicing transversal rationality to explore mythic-animistic, aesthetic, and spiritual communication, while attempting to discern fitting and unfitting responses to ecocrisis, makes for a combustible mix rather than easy answers to questions like those provoked by Innes' experience. However, as I will continue to show, an ecology of communication provides an interpretive framework for thinking through complex questions that confront us, bringing our life-world into relief and relieving us from more, but most certainly not all, of the unfit. To me, Innes spoke from more than mythic-animistic identity or pre-rational biospheric immersion even though his was literally immersed. He likely had a unity experience of some type, while also calling forth the mythic past by expressing our unconscious and unquenchable secret desire to speak from dimensions of the psyche that are connected to earth and other species. An animistic line of development expressed at higher levels is another possible interpretation. Still, we do not have enough detail to critically consider whether his activism constitutes a situationally fitting response. We can say that being dislodged was inevitable, but the "coming through" resonated in the moment and beyond, for him and those that witnessed it, and those who have heard about it, and perhaps in invisible ripples eventually made visible as future gifts.

Wilber, since he is critical of deep ecology, the ecological self, and the superstitions of the mythic past, may have a different interpretation; he would likely support activism that learns from the past but does not wish to return to it. For me, however, a nagging question remains: along with returning to our mythic past and imagining ourselves as nonhuman species, how about returning to the history of Manifest Destiny conquest of Native Americans, which has played itself out in the form of today's class and race divisions? Our experiences of nature are too often mediated by technology, but they are also mediated by economic exploitation and racism. Also, what of Griffin's long history of androcentrism?

Dorceta Taylor has researched how class, race, and gender have influenced the U.S. conservation movement, and the data discloses environmentalism-so-white and so-male, especially among the largest and most influential environmental organizations.[135] That is slowly changing, but would change faster if we focused on the people's history of environmentalism, not just wilderness protection, in which workers, women, and people of color fought to protect the health of their communities. But arguing that we need to take care of people first, rather than earth first, is a false dichotomy: we are people on a diverse, shared planet, and practicing mythic-animistic communication discloses the insanity of not being in sympathy with that existential reality. Also, imagining ourselves as other takes us to at least the Pluralistic stage, in which compassion and care more fully emerge, and that compassion and care, from the perspective of the evolutionary epic, extends across the board to all species.

Still, I can imagine many of my urban-dwelling students of color thinking the Council is crazy, yet they easily identify with Majora Carter when I show her TED talks describing her work to protect the Bronx, kicking out toxic polluters and creating more green space, but also the struggles of growing up in the city. But the desire to create green space in cities may come from practicing mythic-animistic communication, and once such an inviting space is there, that would invite diverse species, and that would invite more mythic-animistic communication, and that would call forth physical and psychological health.

From the perspective of mythic-animistic communication, which, again, may be problematic when separated from other forms of communication, we can also say this: Earth-based rituals of archaic hunter-gatherer and indigenous cultures evoke myths that heighten dialogical relations; ultimately, whether or not this heightening is experienced should measure the worth of The Council of All Beings.

NOTES

1. Paul Shepard, *Nature and Madness* (Athens: University of Georgia Press, 1998 (1982)), 6–7.
2. Ibid., 21.
3. Ibid., 23.
4. Ibid., 24.
5. Paul Shepard, *Coming Home to the Pleistocene* (Washington, DC: Island Press, 1997), 19.
6. Ibid., 154–155.
7. Ibid., 51–65.
8. Ibid., 69. See also Paul Shepard, "A Post-Historic Primitivism" in *Limited Wants, Unlimited Mean,* edited by John Gowdy (Washington, DC: Island Press, 1998), 289–290.
9. Shepard, *Coming Home*, 54–55.

10. Ibid., 52–54.
11. Ibid., 85–88.
12. Frederick Turner, *Beyond Geography: The Western Spirit Against the Wilderness* (New Brunswick: Rutgers University Press, 1992 (1980)), 15.
13. Ibid., 15–17.
14. Shepard, *Nature and Madness*, 7.
15. Wilber, *Sex, Ecology, Spirituality*, 166.
16. Ibid., 166–170, 452.
17. Shepard, *Coming Home to the Pleistocene*, 54.
18. Claude Levi-Strauss, *Myth and Meaning* (New York: Schocken Books, 1979), 15–19.
19. Owen Barfield, *Saving the Appearances: A Study in Idolatry* (London: Faber & Faber, 1957), 1mcmlvii.
20. Calvin Luther Martin, *In the Spirit of the Earth* (Baltimore: John Hopkins University Press, 1992), 18.
21. Shepard, *Coming Home to the Pleistocene*, 596–597.
22. David Abram, *The Spell of the Sensuous: Perception and Language in a More-Than Human World* (New York: Pantheon, 1996), ix.
23. Ibid., 53.
24. Ibid., 72.
25. Ibid., 72.
26. Jared Diamond, *Collapse: How Societies Choose to Fail or Succeed* (New York: Penguin, 2005), 9–10.
27. Shepard, *Coming Home to the Pleistocene*, 78.
28. Esbjorn-Hargens and Zimmerman, *Integral Ecology*, 296–297.
29. Fikret Berkes, *Sacred Ecology* (New York: Routledge, 2012), 266.
30. Ibid, 266.
31. Ibid., 173–174.
32. Ibid., 123.
33. Fisher, *Radical Ecopsychology*, 232.
34. Carbaugh, "Just Listen: 'Listening' and Landscape Among the Blackfeet."
35. Carbaugh, "Quoting 'The Environment': Touchstones on Earth," 68.
36. Berkes, *Sacred Ecology*, xv–xvii.
37. Paul Shepard interviewed in Derrick Jensen, ed., *Listening to the Land: Conversations about Nature, Culture, and Eros* (San Francisco, CA: Sierra Club Books, 1995), 248.
38. Ibid., 249–251, 257.
39. Ibid., 251.
40. Ibid., 253–254.
41. Ibid., 254–255.
42. Ibid., 256.
43. Ibid., 256.
44. Ibid., 258.
45. Ibid., 248–249.
46. Paul Shepard, *The Others: How Animals Made Us Human* (Washington DC: Island Press, 1996), 3–8.

47. Ibid., 10.

48. Shepard, "A Post-Historic Primitivism," 319–320.

49. Wendell Berry, *What are People For?* (San Francisco: North Point Press, 1990).

50. Calvin O. Schrag, *Communicative Praxis and the Space of Subjectivity* (Bloomington: Indiana University Press, 1986), 209.

51. Arne Naess, "Self-Realization in *Deep Ecology for the 21st Century*, edited by George Sessions (Boston: Shambhala, 1996), 236.

52. Aldo Leopold, *Sand County Almanac* (New York: Ballantine Books, 1970), 137–141.

53. Bill Devall and George Sessions, *Deep Ecology* (Utah: Gibbs-Smith, 1985).

54. John Seed, Joanna Macy, Pat Flemming, & Arne Naess, *Thinking Like a Mountain: Towards a Council of All Beings* (Philadelphia: New Society Publishers, 1988), 37–46.

55. Wilber, *Sex, Ecology, Spirituality*, 447–454.

56. Wilber, *Sex, Ecology, Spirituality*, 517–520. See also Wilber, *A Brief History of Everything*, 204–207.

57. Wilber, *Sex, Ecology, Spirituality*, 460–465.

58. Ibid., 447–472. See also *A Brief History of Everything*, 278–295.

59. Ibid.

60. Ibid.

61. Ibid.

62. Michael Zimmerman, *Contesting Earth's Future: Radical Ecology and Postmodernity* (Berkeley: The University of California Press, 1994), 196. See also Zimmerman, "A Transpersonal Diagnosis of the Ecological Crisis," in *Ken Wilber in Dialogue*, edited by Donald Rothenberg and Sean Kelly (Wheaton: Quest Books, 1998).

63. Wilber, *Sex, Ecology, Spirituality*, 288.

64. Zimmerman, *Contesting Earth's Future, 173–174.*

65. Ed Vulliamy, "Seattle Fears Green Rage," *The Guardian*, November 27, 1999, https://www.theguardian.com/world/1999/nov/28/wto.theobserver.

66. Tess Riley, "Just 100 Companies Responsible for 71% of Global Emissions, Study Says," *The Guardian*, July 10, 2017, https://www.theguardian.com/sustainable-business/2017/jul/10/100-fossil-fuel-companies-investors-responsible-71-global-emissions-cdp-study-climate-change.

67. See the Deep Green Resistance website, https://deepgreenresistance.org/en/.

68. See *The Only World We've Got: A Paul Shepard Reader*, edited by Paul Shepard (San Francisco: Sierra Club Books), 1996.

69. Aric McBay, Lierre Keith, and Derrick Jensen, *Deep Green Resistance* (New York: Seven Stories Press, 2011), 11–13.

70. Zimmerman, *Contesting Earth's Future, 3–4.*

71. Val Plumwood, *Feminism and the Mastery of Nature* (New York: Routledge, 1993), 178.

72. Fisher, *Radical Ecopsychology*, 3–4.

73. Robert Greenway, "The Wilderness Effect and Ecopsychology," in *Ecopsychology: Restoring the Earth, Healing the Mind,* edited by Theodore

Roszak, Mary E. Gomes, and Allen D. Kanner (San Francisco: Sierra Club Books, 1995), 126.

74. Ibid, 123.
75. Ibid, 130–131.
76. Ibid, 135.
77. Roszak, *The Voice of the Earth*, 226.
78. Shepard, *Coming Home to the Pleistocene*, 69.
79. Roszak, *The Voice of the Earth*, 226.
80. Ibid., 289.
81. Ibid., 291.
82. Ibid.
83. Ibid.
84. Ibid., 297.
85. Ibid.
86. Ibid., 299.
87. Jonah Lehrer, "How the City Hurts Your Brain," *Boston Globe*, Boston.com, January 2, 2009, http://archive.boston.com/bostonglobe/ideas/articles/2009/01/04/how_the_city_hurts_your_brain/.
88. Fisher, *Radical Ecopsychology*, 7.
89. Ibid., 9.
90. Ibid., 20.
91. Shepard, *Coming Home to the Pleistocene*, 173.
92. Abram, *The Spell of the Sensuous*, ix.
93. Ibid, 50.
94. Ibid, 53.
95. Ibid., 57.
96. Ibid.
97. David Abram, *Becoming Animal: An Earthly Cosmology* (New York: Pantheon Books, 2010), 269.
98. Ibid., 267.
99. Ibid., 178.
100. Fisher, *Radical Ecopsychology*, 235.
101. Abram, *Becoming Animal,* 269.
102. Ibid., 178.
103. Ibid., 268–272.
104. Ibid., 293–296.
105. Ibid., 281.
106. Ibid., 288–290.
107. Ibid., 149.
108. Abram, *The Spell of the Sensuous*, 88.
109. Shepard, *Coming Home to the Pleistocene*, 91–92.
110. Abram, *The Spell of the Sensuous*, 9.
111. Ibid., 61–62.
112. Robert Greenway, "Healing by the Wilderness Experience," in *Wild Ideas*, edited by David Rothenberg (Minneapolis: Univ. of Minnesota Press, 1995), 182–193.

113. Fisher, *Radical Ecopsychology*, xi.
114. Shepard, *The Others*, 11.
115. Seed, Macy, Flemming, & Naess, *Thinking Like a Mountain*, 11.
116. Ibid., 7.
117. Ibid., 8.
118. Ibid., 102.
119. Thomas Berry and Brian Swimme, *The Universe Story: From the Primordial Flaring Forth to the Ecozoic Era--A Celebration of the Unfolding of the Cosmos* (San Francisco: HarperCollins, 1992).
120. Connie Barlow, *Green Space, Green Time: The Way of Science* (New York: Springer Verlag, 1997).
121. Seed, Macy, Flemming, & Naess, *Thinking Like a Mountain, 45, 47.*
122. Ibid., 100.
123. Ibid., 80.
124. Ibid., 80–81.
125. Ibid., 88–89.
126. Ibid.
127. Shepard, "A Post-Historic Primitivism," 286.
128. Ibid.
129. Seed, Macy, Flemming, and Naess, *Thinking Like a Mountain*, 9.
130. Ibid., 91.
131. Ibid.
132. Ibid., 92–93.
133. Ibid., 93.
134. Ibid.
135. Dorceta Taylor, "Interview: How Green Groups Became So White," Yale Environment 360, https://e360.yale.edu/features/how-green-groups-became-so-white-and-what-to-do-about-it.

Chapter 4

The Pattern that Connects

Gregory Bateson and Aesthetic Communication

Mentoring is a gift event that keeps on giving, as lessons learned influence future actions. And sometimes you get more mentoring. Ten years after studying and meditating with Rafael, I felt an intuitive need to visit with him. His living room was filled with a library of books, both along the walls and jutting out in rows, and within an hour of arriving, I found myself perusing shelves and stumbling upon Gregory Bateson's *Mind and Nature*. I had been a Communication Studies undergraduate, and was teaching Communication Studies courses at Purdue University, but had never heard of him despite his argument that we live within larger communicative systems. Rafael walked around the corner, saw the book I was holding, and before I could ask many questions he said we should read it together.

At that time, I was active in the Ometeca Institute, going to conferences and helping Rafael out with editing the Ometeca journal. Ometeca has its roots in C. P. Snow's influential 1959 essay declaring that the sciences and humanities had split off into two separate cultures, which posed a great danger in Snow's time of growing nuclear armaments, but always poses a great danger when the myths of linear progress, technology as savior, and unlimited growth, rooted in scientific innovation, are not humbled and rethought via the traditional focus of the humanities on the true, good, and beautiful. It turned out that Bateson's writings provide a robust response to the dangerous divorce between the sciences and humanities, with the true of science and the ethical common good guided by pursuit of the beautiful.

Bateson began as an anthropologist, marrying Margaret Mead and studying totemism and art in New Guinea and Bali, and then moved on to other fields in the sciences and humanities—biology, psychology, cybernetics—taking steps in thought, and steps to an ecology of mind. And despite never encountering him in my formal education, he is also a seminal figure in the study of communication, with his writings on relationship pathologies and communicative double-binds remaining influential. However, his argument that we must live in sympathy with the intelligence of larger systems is largely ignored, even though such sympathy is central to communication study, and all study, and our collective sanity in the present and future age of ecocrisis.[1]

As we've seen, rational, spiritual, and mythic-animistic communication go a long way toward challenging anthropocentric biases; however, to fully counter dominant narratives and habit-bodies we need to further articulate an encompassing vision that replaces our monologic mindset with a dialogic one, recognizing that if our experience of communication changes, so too will our diverse relations to the earth and nonhuman species. Gregory Bateson adds an aesthetic dimension to this vision, integrating insights from mythic-animism and modern science on a path to immanent sacred unity.

According to Bateson, we are sadly unaware that we are immersed within an evolving communicative system of pattern and beauty. The ecological crisis, then, reflects a crisis of perception in which we are living a mistaken epistemology: our knowledge systems focus on parts, miss out on the whole. But Bateson's thought provides a corrective, especially for healing the damaging division between nature and culture. Although different logical types, or contexts, culture and nature both exist within the larger context of mind, or mental process characterized by difference as a constituent aspect of relationship and the continual exchange of information, expressed as breathing with the biosphere, sharing of energy, and nonverbal presence. Thus, our individual minds are subsystems within larger systems that extend mind, and thus intelligence, outward. For Bateson, this reduces "the scope of the conscious self," resulting in humility, as well as dignity and joy, as we are integral parts of "something much bigger."[2]

Bateson further argues that the evolution of "mind," or "the pattern that connects," is characterized by order and randomness, with randomness adding novelty to the evolutionary system. As a result, rigid knowing is antithetical to continual evolution because it closes down the dialogic interplay between these principles, which are both required for creative emergence. However, Bateson also asserts that basic scientific knowledge, or the collection and classification of facts, is necessary for cultural stability. And yet, such knowledge is often the product of purposive thought—his way of saying instrumental rationality—disconnected from intrinsic value and unresponsive to the ever-evolving pattern within which we are immersed. Purposive thought is the result of a one-way objective monologue with nature rather than an intersubjective dialogue within mind, and our failure to harmoniously participate within this communicative system inevitably results in double binds.[3]

Bateson first explored double binds in his studies of family communication and schizophrenia. In a classic example, he tells the story of a hospitalized son being visited by his mother. The son put his arm around his mother's shoulders, leading her to stiffen. Reading her body comportment, he removed his arm, and she responded, "Don't you love me anymore?" The son then blushed, causing his mother to further chastise him, "Dear, you must not be so easily embarrassed and afraid of your feelings." Bateson writes that he was able to stay with her for a few more minutes, and then assaulted an orderly after she left. The son is clearly stuck in a "damned if you do, damned if you don't" double bind situation, which is psychologically damaging when habitually repeated, especially if the contradictory communication is ignored and there is no escape from the relationship. In other words, the mother forbids discussion of her mixed messages, and the son must "deceive himself about his own internal state in order to support mother in her deception."[4]

Double binds occur when we lack larger shared meanings to make thoughtful, and thus fitting, decisions in particular situations. Said differently, since the process of perception is an act of attending to one context instead of another, double binds occur when communicative subjects are attending to different contextual referents. Imagine a conversation in which two people are discussing car directions while referring to different roadmaps. The lack of a shared referent would cause a downward spiral of confusion and miscommunication. Of course, as Peters would be quick to remind, miscommunication happens all the time when we refer to different cognitive maps. To Bateson, the only way to have some degree of mutual understanding in such situations is through metacommunication or communication about communication in which we discover a larger context of shared meanings. Without a larger context, the inevitable result is dysfunctional and unhealthy relationships, or for the son who lives without access to metacommunication, clinical care, and, hopefully, fewer visits from his mom.

The same analysis can be applied to ecocrisis: we have a dysfunctional communicative relationship with fellow nature because our cognitive maps deceive us by solely defining humankind within the context of culture, reducing, and in some cases obliterating, our communicative and perceptual intimacy with the plants, animals, sounds, textures, and colors that make up our earth community. And, just like dysfunctional family relations are difficult to escape, we are immersed within dysfunctional techno-industrial-capitalist systems. Bateson focused on Lake Erie, the "poster-child" of pollution at the time of his writing, to explain the effects of this crisis of perception: "When you narrow down your epistemology and act on the premise 'What interests me is me, or my organization, or my species,' you chop off consideration of other loops of the loop structure. You decide that you want to get rid of the by-products of human life and that Lake Erie will be a good place to put them." But, for Bateson, such decisions are unfitting because "the eco-mental system called Lake Erie is a part of *your* wider eco-mental system—and if Lake Erie is driven insane, its insanity is incorporated in the larger system of *your* thought and experience."[5]

There is a big difference between misperceptions from having different roadmaps and those involved in the pollution of Lake Erie. The root is the same—the absence of a larger context of meaning—but it probably won't take very long for metacommunication to make us aware of differing roadmaps. In contrast, dysfunctional relationship with Lake Erie was only addressed after many years of pollution and degradation. This dysfunction is the result of epistemological error—our dominant modes of knowing are monologic rather than dialogic, leading to damning decisions within reductionistic contexts—and the basis for making this judgment is a pragmatic one: the risk to human survival and the flourishing of all life.

Bateson's point is that we exist within recursive cycles of nature and culture expressed as mental process, or the constant exchange of information, and yet we act as if these cycles don't apply to us, constantly *attempting* to separate ourselves from the communicative system of mind. As Abram argued, we are restricted by isolated, verbal minds—and abstract language itself—and thus in everyday conversations we habitually speak as if nature is somewhere other than where we are right now or where we occasionally vacation rather than the source and sustenance of our being. Bateson puts the matter bluntly:

> As you arrogate all mind to yourself, you will see the world around you as mindless and therefore not entitled to moral or ethical consideration. The environment will seem to be yours to exploit. Your survival unit will be you and your folks or conspecifics against the environment or other social units, other races and the brutes and vegetables.[6]

He further exhorts that this restricted vision, when combined with advanced technology, makes our likelihood of survival equivalent to a "snowball in hell": "You will die either of the toxic by-products of your own hate, or, simply, of overpopulation and overgrazing. The raw materials of the world are finite."[7]

The above passages once again assert that our collective sanity is dependent upon restructuring our thinking such that we perceive that mental characteristics are immanent within larger mind rather than only our individual minds. Indeed, we do not create our thoughts in isolation, and thus we do not fully own our thoughts; instead, it is more accurate to say that thought emerges via the recognition of relationship and difference. To illustrate, Bateson gives the example of a person chopping wood. After each strike of the axe, a new difference is perceived and our relationship to this difference guides the next strike. Thus, relationship and difference are the sources of mental activity and information. And this includes moments in which we are simply sitting in a chair contemplating, since such thinking is the result of past relationships produced by the recognition of difference. Humans make sense of information produced by relationships of difference in a unique way—via language and rationality—but this does not separate us from the communicative system of mind.

Bateson's vision was influenced by his exposure to indigenous cultures, which disclosed a different way of thinking and living not wholly governed by purposive thought. Like many environmental thinkers, including deep ecologists, ecopsychologists, and ecofeminists, he demonized Descartes, or the Cartesian separation of mind and matter, which tears "the concept of the universe in which we live into rags."[8] Conscious purpose, when isolated from other modes of mental processing, reflects aberrant mind, which, he pointed out, is perpetuated by schooling divided by disciplines.[9] This argument puts him in full consort with the Ometeca Institute, which seeks to overcome an institutionalized factory assembly line approach to knowledge-production, with overspecialized scientists spending little to no time exploring ethical-aesthetic questions and humanists having little understanding of how the ecosystem works. In some instances, this has been overcome, with climate scientists like Katherine Hayhoe conveying data and ethical imperatives to broad audiences. But discipline division remains, and is necessary for focused study, yet interdisciplinary and transdisciplinary education is needed to fully apprehend and respond to our many eco-social crises. Bateson went as far as to ask, "Why do schools teach almost nothing of the pattern which connects?" And then wonder if the answer is that teachers, who have learned the lessons of aberrant mind all too well in their own overspecialized education, carry the "kiss of death" and therefore teach nothing of "real-life importance."[10]

For Bateson, language-mediated purposive thought, which cuts us off from living beings, is secondary, while nonverbal messages of "all our relations" are primary. This claim, of course, is a precursor to Abram's similar arguments. Mental process, then, does not require consciousness; rather, consciousness is a shallow layer, with depth and breadth coming from the unconscious and living embedded within mind or the pattern that connects, in which everything responds to stimuli, to exchange of information, albeit it in different ways. We are part of the mental system, not outside it, and thus attempts at control will lead to double binds. Without using the term ecological unconscious, he also clearly agreed with Roszak that evolution has its wisdom and it mostly resides within the unconscious. He also recognized that the "ancients who endowed forests and lakes with personality were not without wisdom."

For Bateson, living these wisdoms is an aesthetic adventure marked by the integration of unconscious, conscious, and larger mind. He describes aesthetic experience as "responsiveness to the pattern which connects" and meeting the other with "recognition and empathy."[11] Thus, asking how we are related, or what patterns connect self and other, is an aesthetic question challenging the thinking, perceiving, and communicating of the modernist self. Said differently, Bateson extols the experience of unity-in-diversity, affirming it as beautiful, and later in his thinking, sacred, yet our monologic mindset has marginalized this experience: "Most of us have lost that sense of unity of biosphere and humanity which would bind and reassure us all with an affirmation of beauty. Most of us do not today believe that whatever the ups and downs of detail within our limited experience, the larger whole is primarily beautiful . . . I hold to the presupposition that our loss of the sense of aesthetic unity was, quite simply, an epistemological mistake."[12]

Bateson was critical of fellow anthropologists, no doubt university trained, for too much left-brain interpretation via language of right-brain nonverbal phenomenon, losing the aesthetic sense.[13] A sacred unity is not disclosed by left-brain thinking, but through art, meditation, poetry, dreams, ritual, and looking at the living world.[14] When one does integrate unconscious and conscious, right brain and left brain, while recognizing a larger external system of mind, it is much easier to perceive the pathology of wrong thinking, or the unfit, like ecocrisis, or polluting the eco-mental system called Lake Erie, or putting a potted plant on a radiator, which, Bateson states, is bad biology, but also bad aesthetics and bad spirituality: "What we are trying to do is defend the sacred from being put on the radiator."[15]

For Bateson, sacredness reflects the real—unity-in diversity—and myths may be poetic ways of communicating the sacred, reminding us to remain humble, guided by beauty within patterns of specific places.[16] However, he was critical of the use of myths to displace what we don't know into the supernatural. Mythic-animistic thinking, then, is not without error, just like Cartesian habits of the modern mind are not without error, especially when supported by the myths of bigger, better, and more, which lead us to pay too little attention to how symptoms relate to systems. Mistaken epistemology, whether from mythic supernatural displacement or Cartesian dualism, results in solutions that don't solve. To Bateson, in the modern world, such error leads to somebody getting paid "to make the pathological trend more comfortable," like "more roads for more cars" and "faster cars for restless people." There is little preventive care, too much care after the fact, which mirrors deep ecology's criticism that environmental policy, rather than going to the roots of eco-destruction, is too often like an emergency room.[17]

Since Bateson was fluent in biology, he knew that if something is good, more of it is not necessarily better: "good things come in optima, not maxima." Beyond the optima, or what is most fitting, often lies toxicity, whether oxygen, calcium, food, entertainment, psychotherapy, rage, and "perhaps even love."[18] And yet, momentary enlightenments, or what he came to call grace, in which beauty, and no doubt love, lead us to perceive the larger gestalt, the larger whole, are possible.[19]

To Bateson, then, scientists should not only be counting things that are related, but perceiving relationships that disclose immanent beauty. Lists are okay, but lists divorced from a larger whole are not, especially when we have lost the understanding of language as "living organized pattern" within larger patterns.[20] Bateson understood that there are always more possibilities for disorder than order—an admission that Peters would applaud—but aesthetic engagement prepares us for the possibility of grace, or the coming through of fitting responses, in support of ecological well-being, and, in a time of ecocrisis, survival. By connecting beauty, grace, and sacredness with health, survival, and yes, love, as optima, expressed as agape or eros or philia, we may take steps to an ecology of mind.

Bateson came to the realization that his vision of unity-in-diversity needed religious metaphors to be comprehensible. However, while mysticism explores experiences of aesthetic unity, and Bateson is sympathetic to this tradition, his version of unity remains mostly scientific, although of a holistic variety. Mind is immanent within nature, expressed via relations among phenomena and information exchange, not as the pan*en*theism of the transcendentalists, nor as only mythic-animism. However, he did briefly mention that larger mind may be equated with God, in a pantheistic sense.[21] And Bateson's sympathy with intelligence reflects the Greek unity of the rational soul with a rational and ordered cosmos, or panpsychism, which can be found in other traditions, including the philosophies of Spinoza and Leibnitz. Thus, Bateson's conception of mind, while always emphasizing that the pattern that connects is characterized by both order and randomness in creative, dialogic reciprocity, harkens back to the ancient Greek conception of logos, writ large, supported with systems science and yet informed by indigenous knowing and religious feeling.

And all of this combines into aesthetic responsiveness, or what I am calling aesthetic communication, which most likely arises at the brink of Integral and finds its home at Integral, with beauty a guide to perceiving unity-in-diversity and then possibly having graced experiences of unity-in-diversity, although stopping short of nondual spiritual communication. We can have experiences of beauty at any level—aesthetics is also a line of development—but when we transcend and include mythic-animism and scientific rationality, we are better able to embrace larger wholes, and better able to receive messages that disclose the beautiful.

For Bateson, the turn to the religious language of sacredness and grace reminds that humility, rather than scientistic arrogance, reveals more of this reality. In fact, he argued that humility is needed to express aesthetic responsiveness, not as morality, but as a working scientific attitude that recognizes that mind can never fully know itself. This humble attitude elicits a "cybernetic self," or an unfolding self-sense embedded within and dependent upon larger complex systems. Aesthetic responsiveness, then, is a feature of mental process, and evolution, which acts as a corrective to deep-rooted atomistic pathology. Beauty guides us to what is most fitting because it emerges from relationship and unity, while ugliness emerges from duality and separation. Perceive beauty, perceive relationship, and vice-versa. Such self-understanding leads to the profound realization that "the unit of survival is a flexible organism in its environment" and "the creature that wins against its environment destroys itself."[22]

Bateson's focus on aesthetic unity and self-transformation does not mean that he is unaware of the difficulties of epistemological change or simply the messiness of life. After all, he thoroughly studied relationship pathologies and miscommunication. And he also admitted to his own difficulties in thinking via a cybernetic self: "The most important task today is, perhaps, to think in the new way. Let me say that *I* don't know how to think that way. Intellectually, I can stand here and give you a reasoned exposition of this matter; but if I am cutting down a tree, I still think "Gregory Bateson" is cutting down the tree. *I* am cutting down the tree. "Myself" is to me still an excessively concrete object, different from the rest of what I have been calling "mind."[23]

Buber would likely advise Bateson to be fully present; only then does relationship and presence come into being. Chopping wood, for example, is personal to me, and I approach the task as a I-Thou spiritual exercise. This does not mean that I completely forget myself or that I am chopping wood for instrumental purposes—to keep the house warm—but, when aesthetically attuned to the log as both other and self, there is a momentary forgetting as the maul strikes, and a quiet mind usually leads to a loud "whack" as the log splits despite modest effort, or because of effortless effort. The results are far less resounding when my mind is disconnected and cluttered. But either way, information is received with each strike, placing me in relationship.

This example takes Buber's I-Thou ontology out of the abstract and into the everyday. And the fact that it is an instrumental example makes it all the more salient. We are continually confronted with instrumental needs, but we fulfill these needs more fittingly when responses are informed by aesthetic experience, and even more fittingly, of course, when informed by an ecology of communication as a whole. It is most certainly difficult to both be and do—as Buber admits when he states that "every You [Thou] is doomed . . . to enter into thinghood again and again"—but that is the point of I-Thou practices: to make our life more harmonious, more aware and respectful of intimate relations, giving us the sense of being addressed.[24]

Peters would applaud admissions of difficulty, calling for us to acknowledge the "splendid otherness of all creatures that share our world without bemoaning our impotence to tap their interiority."[25] But Bateson does not deny the splendid otherness of others, as difference is a constituent aspect of relationship, nor does he seek to tap interiorities; rather, he argues that responsiveness demands aesthetic openness to exchange of information. Thus, admitting that he struggles to live a cybernetic self does not represent a damaging contradiction—saying one thing, doing another—but a dialogic, and very human, tension between thought and action that acknowledges our individual difficulties and collective challenge. We are interpellated by social structures and narratives that are exceedingly I-It, and thus it is a loving struggle to overcome the habit of instrumental decision-making; but systems thinking and aesthetic experience can continually awaken us to a joyful life as individuals-in-relationship, responsive and responsible to I-Thou modes of existence.

Despite Bateson's struggles, he provides a compelling rethinking and reimagining of our place within the pattern that connects, as well as the relations among communication, aesthetics, and self. His description of a larger, evolutionary metapattern is rich, open, complex, and relational, giving depth and meaning to particular acts of communication. In addition, his work on double binds helps us to understand that evolution will only proceed in a creative fashion when we realize that while we are cultural beings, we are also nature and must communicate and listen in a manner that acknowledges this wonder. When purposive thought reduces wondrous and wild nature to "environment," to something humans study to control the planet, the argument shifts to problems and solutions "out there." We ignore the dysfunctional perceptual bias that labels nature as separate from ourselves. We cannot, of course, solve eco-social crises with the same instrumental thought and language that has created the crises in the first place. There must be a deeper change in epistemology, and thus a deeper change in our thinking, perceiving, and communicating.

Like Schrag, Bateson counters our Cartesian legacy by questioning the production, uses, and effects of knowledge. There is no final knowledge, only learning within contexts of creative evolution, self-transformation, and survival in relation to aesthetic responsiveness. Monologic inquiry is at the root of our despair because it fails to perceive the dialogic intimacy between knowledge production and ecological integrity. Bateson, like many early ecological thinkers, made dire predictions. In 1967, he estimated that within ten to thirty years, ecological instability would reach major proportions, citing the risks of nuclear warfare, increasing levels of carbon dioxide in the atmosphere, and the population "bomb."[26] That no end-of-humanity catastrophe took place during that time frame does not lessen the worth of his vision, especially when you consider the Exxon Valdez in Alaska, the Bhopal chemical disaster, Chernobyl in Russia, deforestation in Indonesia and Brazil, and the sixth "great" period of species extirpation, among too many other examples of ecological instability, or, from the perspective of mental process, insanity. His championing of ecological systems thinking also anticipated the many eco-social crises we presently face, including, of course, the specter of climate. And, most importantly, his vision discloses a root cause of ecocrisis: our habitual persistence in understanding ourselves as communicating subjects separate and apart from the larger communicative system of mind.

Bateson does not devalue the knowledge gained from studies in human communication—he did such studies himself—or knowledge gained via studies in other disciplines. After all, our knowledge of climate crisis comes from consensus among numerous scientists. But he does question what our studies are contributing to, as well as what we study and how we interpret and ought to act in light of results. Bateson's writings should inspire wise discernment guided by aesthetic awareness of the pattern that connects. Such awareness cuts across all disciplines, overcoming the downsides of overspecialization, but Communication scholars should highlight, rather than run from, Bateson's use of communicative experiences, metaphors, and insights to foster a deeper integration of knowledge.

As an undergraduate, my education was thoroughly anthropocentric, with only humans having voices, displaying mental characteristics, and exchanging information. Classes on the Rutgers campus often took place in historic halls within ivy-covered brick buildings: inside we filled up our brains with disparate knowledge, outside we were covered in green. The ivy was a presence too often ignored but having its say, soothing student minds and bodies as we walked, or sat and talked on the grass, or read a book and pondered while leaning against a tree. Inside and outside need to consciously merge, because they are already merged within patterns not of our making. But I did not start to become conscious until conversations with Rafael on how the "two cultures" and dividing disciplines undermines dialogue and invites destruction. I became more conscious when I found *Mind and Nature* on Rafael's shelves, tucked away as if a text from the past, and yet increasingly relevant as mistaken epistemology continues to have its say. And I became still more conscious after experiences on the trail.

What is largely unspoken and yet resoundingly clear in Bateson's work is that ecocrisis is a crisis of perception because we do not listen beyond the human, or very well among humans. But we would listen better, and respond better, if we realized we exist within larger communicative systems; that is what entered my imagination and thinking so many years ago in Rafael's living room and that is what provides guidance today. If all things sing with information, if all things have voice, in their own way, it is wise to pay attention. We become more conscious by listening inside and outside the classroom: to professors, ivy-covered buildings, books, friends, grass and trees, and trail. And we become more conscious by perceiving the double binds that restrict our full communicative capacities, and thus our ability to fittingly respond within the evolving and ongoing dialogic interplay of life.

Bateson thought outside the academic box, especially in "metalogues," or imaginary conversations with his young daughter Mary Catherine, in which he played, and played out his later thinking on aesthetics, grace, and the sacred. Mary Catherine Bateson, who became an accomplished author and supporter of her father's work, including in their coauthored *Angels Fear: Toward an Epistemology of the Sacred*, wrote that humans are capable of "being wrong in rather creative ways—so wrong that this world we cannot understand may become one in which we cannot live."[27] Scientists, Bateson advised, must explore aesthetics, and despite trepidation that he also shared, they must enter religious territory of grace and sacredness to be less wrong by being more whole; otherwise, instead of taking steps to an ecology of mind, we will take further steps toward a runaway system with no governor, no means to regulate the speed of negative feedback, an increasing danger in a tipping point, climate crisis world. However, the larger system of mind, like Wilber's developmental holarchy, has a self-correcting capacity that includes living systems from cells to civilizations.[28]

In an introduction to a 2000 reissue of Bateson's *Steps to an Ecology of Mind*, Mary Catherine wrote that we had not yet changed our epistemology, and thus our attempts at self-correction do not correct or make matters worse. She admits that her father had a tendency to overdramatize his message—"snowball in hell," "kiss of death," "driven insane," etc.—but the illogic of mistaken epistemology has logical, and therefore predictable, results. His timetable may have been wrong but his warnings were based on what is obvious once one changes their thinking: "The habits of mind that he described can be seen in every newspaper or newscast: the search for short term solutions that worsen the problem over time (often by mirroring it, such as violence used to oppose violence); the focus on individual persons or organisms or even species, seen in isolation; the tendency to let technological possibility or economic indicators replace reflection; the effort to maximize single variables (like profit) rather than optimizing the relationship among a complex set of variables."[29]

Twenty-odd-years later, all of these symptoms, treated as if separate from larger systems, march on. The pattern that connects includes competition, of course, but when not balanced with cooperation the system breaks down, whether in forest and ocean ecosystems or human societies with absurd power differentials between rich and poor. Awareness of symptoms of disorder like climate emergency, wealth disparity, and systemic racism and sexism has increased—we listen more, becoming more conscious of patterns of health and ill-health—but change will only occur if the powers-that-be are challenged. Or, as Bateson presciently warned, "Never vote for a man who is neither a poet or an artist or a birdwatcher."[30]

Bateson championed the need for rigorous thought, but never at the expense of imagination, arguing that "advances in scientific thinking come from a combination of strict and loose thinking," and that to insist too hard on "operationalism" is to "lose something of the ability to think new thoughts."[31] As a result, he considered the practice of rigor by itself to be "paralytic death," while also stating that "imagination alone is insanity."[32] The interplay between rigor and imagination in the service of beauty and unity is the essence of aesthetic communication.

Thus, along with mind and nature, order and randomness, and stability and flexibility, we have another duality that must in dialogue, with rigor linked to logos and imagination to mythos and eros. Ultimately, however, what is most needed is not analysis of Bateson's vision for its own sake, but following the trajectory of his work and taking next steps, both in thought and in action, and thus in fitting responses in an age of ecocrisis.

SHADOWS AND LIGHT, UNFITTING AND FITTING: LEARNING, LANGUAGE, AND LIVING IT

I had already started reading Wilber when I discovered Bateson's writings, and once again did not find any incongruence. And when Wilber's *Sex, Ecology, Spirituality: The Spirit of Evolution* came out in 1995, three years after the World Scientists' Warning to Humanity and the failed Rio climate conference, I immediately noticed an obvious nod to Bateson: chapter two was titled "The Pattern that Connects." However, I also discovered that Wilber does not mention him in the chapter or the book, except in a footnote where he equates his "cybervision" with "flatland systems theory"—in other words, no transcendent dimension and thus a lack of Kosmic depth. It is in that same book that he waylays deep ecology, ecopsychology, and ecofeminism, while also giving them some props. And I guess he gives props to Bateson by using his core metaphor to explore involutionary givens like holons, or whole-parts, which make up and express evolution: holons within holons within holons.

Holons are equivalent to patterns, and thus life is comprised of a nested holarchy of patterns within larger patterns, all the way down and all the way up: subatomic particles, atoms, molecules, polymers, cells, and on and on.

However, since Wilber includes a transcendent dimension in his vision, he states that involution begins with some holons, and this takes his integral post-metaphysics into the realm of metaphysical pre-givens: but, again, he argues that some structure, some logos, is needed to get a world going, like form itself. Also, given that patterned holons are not wholes or parts but both and always unfolding, his meta-theory respects the postmodern turn because there is no claim for solid metaphysical ground.[33]

For Wilber, then, patterns that connect describes his grand developmental vision, but he considers Bateson's transcendent-less systems science flat, without sufficient depth, and without genuine sacredness expressed as graced experiences of the world soul. Yet, while Bateson's use of religious rhetoric is differently directed, his reconception of mind is marked by systems within systems and thus holarchy. His vision does not go far, or deep, enough, at least for Wilber, but it is not flat, and like Wilber, he does explore development marked by levels of learning.

For Bateson, the ability to embrace and learn from the dialogic interplay of unity and diversity, order and randomness, and rigor and imagination is characteristic of the cybernetic self, but, of course, such embrace occurs in stage-steps. What he calls Learning I or proto-learning requires a self engaged in solving a problem. In contrast, Learning II or deutero-learning is marked by the skill of solving problems in general; we recognize contexts that characterize multiple problems, and we learn how to learn. Learning II is a rational step forward, yet also characteristic of purposive thought, scientism, and a fixed, modernist self separate and apart from objective nature, and we then mold a reality to fit our self-expectations, participating in habitual patterns of behavior that monologic culture deems to be sane.[34]

Fortunately, Bateson also formulates the possibilities of Learning III, or simply learning about Learning II. In Learning III, we become aware of Learning II as a level, and this provides the critical distance needed to take another stage-step, transcending, including, and changing our monologic epistemology to a more dialogic one. The result is a profound re-organizing of our self-concept, as well as our vision of scientific knowing. Fact and value are no longer completely separate, but intermingle, because there is no pure objective place to stand, no perception free from values.[35] This cybernetic self has clearly stepped up to the Pluralistic level, expressing vision-logic and approaching Integral, as our sense of self merges with "the processes of relationship in some vast ecology or aesthetics."[36]

For Bateson, the core distinction between Learning II and Learning III is the difference between digital/left brain and analogic/right brain knowledge-production and communication. Digital knowledge is verbal and directed toward communicating content; analogic knowledge is nonverbal and directed toward communicating context.[37] Despite strong scientific leanings, Bateson knew that perceiving larger contexts of meaning is the beginning of wisdom and that analogic knowledge—whether through art, meditation, poetry, dreams, ritual, myth, religious experience, or simply being immersed within living and breathing ecosystems—cannot be fully expressed via digital communication. Again, digital/left brain thinking does not disclose beauty, and thus it does not disclose sacred unity. Attempts to do so end up being acts of communicative closure, and, ultimately, acts of destruction. Digital communication must serve the analogic, not vice-versa. A cybernetic self understands this relation, integrating digital and analogic to achieve Learning III, perceiving "the contexts of contexts" such that one's self no longer functions as a "nodal argument in the punctuation of experience."[38]

Wilber's criticism, then, should not be that Bateson's vision is flat, but that it does not fully take next steps into third tier Super Integral, despite his turn to religious rhetoric. Both Bateson and Wilber embrace holarchy, although Bateson's ecology of mind does not take further steps to an ecology of Spirit that fully includes insights from spiritual experiences. He was tempted to do so, having taken LSD, like many other researchers in the 1960s, experiencing no separation between himself and the music he listened to, the "perceiver and the thing perceived" becoming "strangely united into a single entity."[39] He also spent time giving talks and living at the Esalen Institute amid the beauty of Big Sur, CA. Esalen is famous for exploring non-ordinary states, and many friends urged him to "meet more practitioners of the improbable" and expose himself to more experiences, but he was steadfastly skeptical of any communication without a material pathway.[40] Bateson preferred the company of Esalen counterculture explorers over scientistic experimenters—at least they were open to changed epistemology—but he wanted his work to catalyze discovery in-between supernaturalism and mechanism, not support what, for him, relies on superstition more than evidence.[41]

Despite the often clouded spiritual versus material debate, Bateson's focus on mental process provides some clarity to another one: whether humans are the crown of evolution. On the one hand, Bateson's reconception of mind may seem similar to biocentric egalitarianism: all species exchange information and have equal value. On the other, he recognizes holarchy and differences that make a difference, including differing contexts, which suggests different levels of value. Yet, for Bateson, the biosphere is saturated with mind, and thus, compared to Wilber, he would seem to perceive less differentiation between biosphere and noosphere. Wilber, of course, argues that the biosphere and noosphere are manifestations of Spirit unfolding and are differentiated and integrated, but Bateson, while not arguing for equality, more fully discloses the damaging downsides of mistaken notions of human superiority. Humans do not merely dwell within the noosphere and nonhumans within the biosphere, as both are deeply interwoven within the pattern that connects.

As we've briefly explored via Wilber, one way to unpack both debates is to consider three kinds of value: ground value, intrinsic value, and instrumental value. Ground value refers to the absolute or transcendent Ground of all Being from which all things emanate. If everything comes from the same source—Spirit, Godhead, Divine Ground, etc.—then everything is sacred and has equal Ground value.[42] Of course, not all will agree with this religious perspective; yet, from a more philosophical one, contemplation on the fact that life exists at all suggests all things have value. For example, Thomas Berry extends value to the immanent or relative plane via rights: all species have a right to exist (since they are here, as part of an amazing evolutionary journey), a right to habitat (you need habitat to exist), and a right to flourish within the predator-prey dynamics of their ecosystem niche (cooperation and competition within habitats).[43] On the immanent/relative plane of existence, then, we must consider intrinsic value, which suggests that each species has a right to be, to habitat, and to thrive—we can't just do whatever we want without considering that all species have worth intrinsic to themselves—but also instrumental value, which reflects the obvious fact that we exist within a food web and need to eat to survive. Thus, while using resources instrumentally is often a bad thing, it is not bad when awareness of Ground value and intrinsic value adds an ethic of least harm. Instrumental value, rather than leading to exploitive resource use, may be expressed via thankfulness and respect for nature's services.[44]

The key to using these terms, and acknowledging these realities, is that all three types of value occur at the same time, with some species having more intrinsic value and some species having more instrumental value, even though all species have equal Ground value and are sacred. At first blush, this take on value seems to produce double binds: how can species be equal and unequal at the same time? However, Wilber's developmental holarchy, as well as Bateson's levels of learning, provides relief from this paradox, or what seems like a schizophrenic dilemma: we just need some metacommunication that recognizes differing levels or contexts of meaning.

For Wilber, then, it is more ethical to eat a carrot than a cow, as a carrot is lower on the food chain and thus has less intrinsic value and more instrumental value, or more value as a food source for more species. It also has less sentience, feeling less pain and suffering less, although this has become another debate, as there has been much recent research into plant sentience and how plants also feel and display mental characteristics, in their own way. On the other hand, a cow has more intrinsic value—cows eat plants, plants don't eat cows—and cows are more sentient, feeling more pain and suffering more. Of course, cows and other animals may have instrumental value for humans, or meat-eating species, but they also have intrinsic value and must be respected. Plants also have intrinsic value and must be respected, but, generally speaking, animal species have more sentience. However, going further and acknowledging Ground value assures that we are sensitive to the sacredness of all life, plant or animal, or insect for that matter.[45]

Wilber's tripartite take on value, while acknowledging equality on the absolute plane, ultimately places humans on an evolutionary pedestal. After all, only humans produce thought capable of comprehending three kinds of value. But Bateson's reconception of mind, while recognizing the distinctive value of rigorous thought, problematizes the too easy and too quick crowning of humans as superior, as mental process cuts across human and nonhuman agency. His focus was on the intelligence of evolution as a whole more than rationality as a marker of human supremacy, and when we are not in sympathy we are hardly supreme. Humans governed by mistaken epistemology, and thus with no governor, no wider awareness of mental process to regulate our actions, are often more like the clowns of evolution.

Speciesism most certainly creates problems, and yet humans may use their specialness, including transversal rationality and embrace of the dignities of mythic-animism integrated with ecological science and aesthetic responsiveness, to solve them. To begin, we must recognize that while there are differences that make a difference among species, we are not so different on the level of mind or the noosphere. Rationality, rather than the crown of evolution, may simply be our human crown with other species having their crowning specialness within larger systems of mental process. And to be fully rational, we must perceive and embrace this wonder.

In *Biomimicry*, for example, Jeanine Benyus argues that we should mimic nature's intelligence; or rather, nature as model, measure, and mentor. Nature as model leads to science and technology that takes inspiration from natural designs, such as a solar cell inspired by a leaf. Nature as measure causes us to ask whether new innovations are well adapted for the flourishing of life. And nature as mentor changes our primary relationship with the earth from a storehouse of resources to a source of learning. There are many benefits from *this* tripartite—changing how we make materials, harness energy, heal ourselves, store information, conduct business—but one change, interwoven with the rest, is key: how we grow food.[46] Benyus writes that industrial agriculture has removed "the land's native dress" and imposed "a pattern of our own making," including "exotic plants instead of indigenous ones, annuals instead of perennials, monocultures instead of polycultures."[47]

Wendell Berry, along with Benyus, argue that this imposition must be challenged by deep listening, writing that farmers perceiving nature as model, measure, and mentor should be conversationalists, asking questions like: What would nature allow them to do there? What would do least harm to the land but also human and nonhuman neighbors? What would nature help them to do? And then we listen for a response catalyzed by outer observation but emerging from within, transforming use of the land but also the user. Such reciprocity binds farmers to specific places and fellow inhabitants, "changing and growing to no end, no final accomplishment, that can be conceived or foreseen."[48] In other words, while attending to the genius of the place engenders reciprocity, the gifts received are mysterious and always unfolding. Like all dialogic relations, there is a flow of meaning with each participant making a contribution, yet this flow is tentative and led more by grace than will. This flow, this coming through, provides present moment insights but does not negate the possibility of different responses in the future. Above all, such conversation is a collective act of creativity sparked by spontaneous interplay, spiraling and building such that participants leave the encounter changed, if just a little, and better able to formulate fitting responses.[49]

As we saw earlier, to begin such a dialogue with nature—or from a Batesonian perspective, within larger mind—we first open ourselves to the mystery of the other, including asking questions without expectations of predetermined answers. The sharing of animistic presence then makes us receptive to inner insights that we ignore in "normal" definitions of mind, and we may discover an affinity with all life by experiencing our shared biological and psychological heritage, leading to awareness of our shared spiritual heritage. However, when dialogue is muddied and muddled by those who are unable or unwilling to participate in its creative unfolding, the result is monologue after monologue.

Not surprisingly, Native knowledge, or TEK, also provides insights into a reconception of mental processing in relation to the natural world and listening to the wisdom of all our relations. In *Earth's Mind*, Roger Dunsmore extends our understanding of mind by arguing that Native Americans, and Native peoples everywhere, often perceive the earth community as a communion of subjects, with nonhumans having intelligence and personhood. Chief Joseph, speaking at the last council between the Nez Perce Indians and the United States government in 1877, expressed this sentiment in a single phrase: "The earth and myself are of one mind."[50] According to Dunsmore, this phrase does not represent eco-romantic primitivism, but a different experience, a different way of being and thinking. He further quotes John Swanton, an Alabama Creek Indian and anthropologist, on the Native world view: "The world and all it contained were products of mind and bore everywhere the marks of mind." For Swanton, mind did not evolve out of matter, but was there in the first place—an involutionary pregiven—yet for moderns it is "quiescent" but may be "roused," if we realize that "mind was visibly manifested in the so-called "living things" as plants, and . . . animals."[51]

This passage clearly states that mind and nature are interwoven. In fact, Dunsmore traveled to a Nez Perce reservation in Idaho and discovered that there was no word in the Nez Perce language that translates mind as a separate entity.[52] As a result, there are few explicit articulations in Native literature of the definition of mind. Instead, mind interwoven with nature is implicit in Native knowing and communication, expressed, as we saw via Carbaugh's study of the Blackfeet of Montana, as kinship and gratitude that allows one to learn from the powers of nonhuman others, ancestors, and nature generally. Dunsmore, reflecting on thirty years of teaching Native American literature, and Native American students, writes that despite having their world "hammered" for centuries, there is a "continuance of a mind/heart that was both aware of its necessary linkage to mind in all its manifestations, and that contained disciplines, ritual structures, and ceremonies for generating and sustaining that mind capacity."[53]

For Dunsmore, a seminal literary expression of larger mind in relation to ritual is Leslie Marmon Silko's *Ceremony*. *Ceremony* tells the story of Tayo, a half-white/half-Laguna Indian, who upon finishing his term of service in the Philippines during World War II—and a spell in the mental ward of the army hospital—returns to the poverty and despair of the reservation. Tayo is utterly lost, and the novel takes us on his journey to make sense of the white world, the reservation, and Native traditions of his youth.

But before telling Tayo's story, Silko begins with a tale told by Thought-Woman, the spider, who spins the web of all stories:

Ts'its'tsi'nako, Thought-Woman
is sitting in her room
and whatever she thinks about
appears.

She thought of her sisters,
Nau'ts'ity'I and I'tcts'ity'I,
and together they created the Universe
this world
and the four worlds below.

Thought-Woman, the spider,
named things and
as she named them
they appeared.

She is sitting in her room
thinking of a story now

I'm telling you the story
she is thinking.[54]

In this creation myth, the world comes into being via the stories of *Thought*-Woman, which mirrors John Swanton's statement that mind is primordially interwoven with nature, and thus with the patterns that sustain life on earth. Silko weaves the stories of Thought-Woman throughout the novel, and the presence of her "thoughts" become the context within which we interpret Tayo's struggle to realize Native knowledge of pattern and reciprocity that is always present, but forgotten. In a powerful passage, Thought-Woman explains the ramifications of this forgetfulness, which she labels the "witchery":

Then they grow away from the earth
then they grow away from the sun
then they grow away from the plants and animals.
They see no life
When they look
they only see objects.
The world is a dead thing for them
the trees and rivers are not alive
the mountains and stones are not alive.
The deer and the bear are objects
They see no life.

They fear
They fear the world.
They destroy what they fear.
They fear themselves.[55]

Thought-Woman's web of stories weaves together the pattern that connects, which are more than myth and metaphor; they represent potential reality in which we express intelligence that is not restricted to the inside of our heads or the static objectivity of scientific experts. The witchery of war has caused Tayo to lose touch with this reality, and it is highly symbolic that this loss has landed him in a mental ward, where medical experts are unable to diagnose the deeper roots of his troubles. The dominant culture's stagnant institutions and army of professionals fail to understand a simple truth: when we realize that thought and intelligence emerge out of relationships, we naturally nourish these relationships; when we do not, we destroy the very relationships that produce our thought and intelligence.

It would seem that we need to lay the ladder of hierarchy down, especially when you consider the amazing abilities of nonhuman species. For Benyus, science's microscopes and macro-satellites have allowed us to bear witness to nature's patterns from the "intercellular to the interstellar," and what we find should "burst our bubble": all our inventions are already present, but in a more elegant form and at no cost to ecosystem health. Struts and beams are featured in lily pads and bamboo stems; termite towers keep a steady temperature, besting our central heating and air conditioning; a bat's multifrequency transmissions make our radar seem as if it is "hard of hearing"; smart materials are not as smart as a dolphin's skin or butterfly's proboscis. The list of heroics is as long as humanity's litany of eco-misdeeds: species navigate without maps, dive without gear, freeze themselves and come back to life. The hair of polar bears is like panes of a greenhouse; chameleons and cuttlefish change the patterns and color of their skin, hiding without moving; hummingbirds cross the Gulf of Mexico with almost no fuel; ants carry hundreds of times their weight in high temperatures; dragonflies outmaneuver helicopters. Notice that so-called "lower" life forms are included in this list—birds, fish, insects—since, similar to plants, recent research has revealed them as having intelligence, degrees of sentience, and communicative skill. And yet, these individual heroics pale in comparison to the intelligence of species within whole ecosystems, which juggle resources—and communicate via exchange of information—without producing waste.[56]

Benyus argues that failures are fossils, as evolution has weeded out what does not work, and thus nature knows best, a principle that Huaorani Indians, and most Native cultures "who have survived without fouling their nest," display by asking nonhuman species for guidance. For her, notions of superiority reflect ideology and narrative, not reality, leading us to expand beyond ecosystem limits.[57] In western Montana, where she has lived, the desire for no limits is played out in debates over the reintroduction of grizzlies into their natural habitat, with many not wanting to worry about being another species' meal. But what is truly at stake is the narrative that we are "top banana" rather than a life form within an intelligent Gaian self-organizing system that is itself a life-form.[58]

Benyus certainly seems to favor biocentric egalitarianism, arguing that we should think of ourselves as having one vote among thirty million (or much more) species, while realizing we may not be "the best survivors over the long haul," proven by the fall of many city-states due to over-consuming ecological capital.[59] As Derrick Jensen, another critic of the myth of supremacy, reminded, all that ultimately matters is the health of the landbase, and the humans-at-the-apex "Great Chain of Being" narrative has led us to "murder the planet."[60] For Benyus and Jensen, then, fitting responses will only emerge from new narratives, and thus changed epistemology, in which survival is marked by fitting-in, not standing out.[61]

All this is, well, mostly good, but a ladder expressed as an actualizing holarchy, or the Great Nest of transcending and including rather than a Great Chain of transcending without including, remains despite the many shadows of dominating hierarchies. Wilber, for example, points out that Arne Naess, rather than making claims for biocentric egalitarianism, like deep ecologists following his lead have done, acknowledges a "vast hierarchy of lower- and higher-order gestalts," or holons within holons, which relieves us from the problematic dualism of wholism, with everything having equal standing, versus atomism, with everything completely disconnected.[62] And Roszak, who Wilber calls a "sophisticated reenchanter," also acknowledges that "evolution progresses as a hierarchy of dynamic systems"; and, in a rare admission for an ecopsychologist or eco-philosopher, he further states that what embodies the full potential of what has come before—expressed as "finer orders of complexity"—stands at the crest and has a crowning position: namely, human imagination and the human world, despite our disastrous disassociations.[63] Wilber, of course, praises this admission, while criticizing him for returning to pre-rational animism, falling prey to pre-trans and dignity-disaster fallacies, instead of theorizing transrational realms. For Roszak, our crowning achievement would seem to be integrating the best of mythic-animism with ecological science, while for Wilber our crowning achievement is yet to be determined but evolves beyond first tier to second tier Integral and third tier Super-Integral awareness of Spirit unfolding.

Both Wilber and Roszak agree that repression of the biosphere by the noosphere, and thus hyper-separation and dominating hierarchy, is the root of ecocrisis—after all, without a biosphere, and physiosphere, the noosphere would collapse—but the way out is not the rejection of all hierarchy but embrace of holarchy. We must realize that we need a ladder of some type for ethics: on the Absolute plane, everything has equal Ground value, but on the relative plane a mosquito is not equal in value to a human child, a carrot is not equal to a cow. But, again, what Bateson, Benyus, and TEK disclose is that the noosphere is not the exclusive home of humans, and while Bateson argued that only humankind is capable of attaining Learning III, although we often fail to do so, we are genetically similar to nonhumans, and nonhumans oftentimes display superior mental characteristics, and evolution as a whole, of which humans and nonhumans are parts, displays mental characteristics.[64] All of this retains holarchic difference while evening things up quite a bit; but going further by throwing Wilber and Spirit unfolding into the mix, or the theosphere, the highest expression of human intelligence is to live in sympathy with a higher, wider, and deeper intelligence characterized by the dialogic interplay of order and randomness needed for creative emergence.

There is, however, one more thing: we express this intelligence by practicing an ecology of communication, listening within and without to better enact fitting responses.

This pretty much sums things up, but there is also much nuance, and thus still much to unpack, to better enact fitting responses. For example, while I fully support Benyus' goals, especially since biomimicry integrates multiple modes of knowing with aesthetic responsiveness to patterns, operating under the premise that "nature knows best" is problematic, as it leads to a series of related questions. Doesn't this phrase suggest another, "biology is destiny," which reduces human intelligence to selfish genes? Should we mimic everything in nature? Are not humans also nature? What is natural? And how can we make sense of the occasional example of evolutionary forms with no purpose, like the human appendix? "Nature knows best" does not answer these questions because it reduces intelligence to the biosphere, while a developmental holarchy and levels of learning respects evolving complexity, and thus nuance that aids critical unpacking of these questions.

Thus, instead of reversing the direction of the "humans know best" monologue practiced by conventional agriculture by stating that "nature knows best," or that nature should tell us what to do, we must realize that while nature and culture reflect different logical types, they are integrated within the higher logical type, or context, of the pattern that connects. Nature and culture are in dialogue when the subject matter leading the conversation—creating a spiraling flow of meaning—is the intelligence that resides within humans and nonhuman nature, or larger mind for Bateson and Spirit unfolding for Wilber. Otherwise, "nature knows best" leads to the dysfunctions of biocentric egalitarianism, especially the failure to perceive differences that make a difference.

The crowning human achievement for Bateson would seem to be integrating the best of mythic-animism with ecological systems science and the grace of aesthetic responsiveness; yet, he cites William Blake's famous lines—"To see the World in a Grain of Sand, And a Heaven in a Wild Flower, Hold Infinity in the palm of your hand, And Eternity in an hour"—to describe the highest potential of Learning III, and these lines are not only poetic, but mystical with reference to a transcendent dimension, which he typically rejects. Bateson further writes that some may experience their personal self merging with "all the processes of relationship in some vast ecology or aesthetics of cosmic interaction" and that the self may no longer be in charge of organizing behavior. This sure sounds like the nondual grace of "coming through" or effortless effort. And so, he also posited the possibility of Learning IV, suggesting that evolution as a whole does not stop evolving, although he doubted that such awareness could be achieved by "any adult living organism on this earth."[65]

It would seem that LSD and Esalen at times got the best of him, while he remained skeptical of Eco-Sage-like transformations. What is clear is that he saw the revolutionary potential of ecological systems science, which he explored in cybernetic terms—loops, codes, circuits, exchange of information—to greatly widen our awareness when conjoined with aesthetics and possibilities of grace. And, like Wilber, he knew that taking steps in human learning only occurred when we are "free from the bondage of habit" of earlier steps and thus free to change our self-sense.[66]

Bateson's focus was mostly on changing a modernist self-sense beholden to Cartesian epistemology or the need to practice distanciation rather than participation. Digital knowledge, of course, is beholden to Cartesian dualism, monologically striving for certainty and dominating over the analogic. Again, while purposive digital knowledge is needed for stability, it must be integrated with analogic flexibility and the context of not-knowing. Otherwise, the inability to recognize higher contexts leads to double binds and, ultimately, psychological illness, which we have taken as reality. The cure of metacommunication, or communication about how we are not practicing an ecology of communication, is a first step toward living sanely and sustainably.

But the cure is often undermined by our habitual participation in a Cartesian world view. Thus, it is instructive to learn how Descartes came to his radically dualistic theory of knowledge and self, a theory that deadened matter and muted the voices of nature. While the witchery of Descartes has made him a favorite, and often deserving, whipping boy for Bateson and many environmentalists, as well as feminists critical of "objective" science, Griffin showed that abstracting ourselves from nature goes all the way back to the Greeks and Christianity. And poor Descartes was responding to questions of his time, making him a case study for how what we deem to be fitting is dependent on history and circumstance.

That, at least, is the argument of Martin Lee Mueller in *Being Salmon, Being Human*, an Abram, and salmon, inspired treatise that is another wholesale rebuke of Cartesian dualism—he writes that the book, at heart, reflects a dialogue between author and salmon—and yet Descartes' intelligence is treated with sympathy despite, and because of, his legacy. In other words, Mueller practices transversal rationality, distancing from Descartes philosophy, while participating with, and thus embracing, aspects of his personal story and revelations.[67]

Mueller's defense of Descartes centers on the obvious but easily overlooked truth that he was responding to his information system, which could not supply the certainty, the grounding, he sought during the Thirty Years' War (1618–1648) in Europe, which, along with plagues and famine, led to tremendous suffering and death, with a third—a third!—of human population lost. Descartes, searching for a fitting response, found no claims that were not disputed, including the narrative of the Catholic church, which had previously provided stability. Out of this chaos, this crisis, and three vivid dreams in 1619 at twenty-three years old, he found what he interpreted as a next step forward. The first two were nightmares, likely mirroring his crisis-ridden age, but the third, with a dictionary and a book containing the question "what path in life shall I follow?" as symbolic images, led him to an answer that must have seemed like a graced "coming through": science, rather than being disparate, may lead to universal wisdom. All he needed was a method, and he found one 18 years later, by focusing on the only thing he could not doubt: his own thinking. Cogito, ergo sum. The rest is more history: only human thinking, divorced from the body, senses, and earth, could know reality.[68]

Of course, what actually came next, instead of certainty, are the ironies, like centering human identity in the individual thinking mind separating us from wider mind and the pattern that connects, which undermined sources of thought; or hyper-separation leading to a new crisis of narrative—the myths of bigger, better, and more—driving us to ecocide; or postmodernism celebrating uncertainty as liberation from dominant narratives rooted in Descartes' dualisms; or that all this emerged from a dream, which reflects the kind of analogic knowledge that Cartesian digital knowledge thoroughly ignores.

Perhaps the greatest irony, however, is that the discoveries of separate-yourself-from-the-object-under-study science reveals that we are not separate, that everything is hitched together, and thus what we can know of reality is not only disclosed from situational attempts at divorcing ourselves from nature but from also embracing preexisting patterns, or marriage that honors subjectivity and intersubjectivity. Or, more simply and bluntly, Descartes' dictum and the method that followed led to ecological science, which alerted us to a key force that is killing us: Descartes' dictum and method. Again, such ironies and paradoxes are perceived and unraveled via metacommunication that acknowledges holarchy and levels of learning, providing guidance for the perplexed.

Descartes is a man of two legacies, to be sure, with the positive legacy being a first step toward an ecology of mind. It is easy to see, for example, how objective ecological science has led to further steps within larger mind, like investigations into plant neurobiology, with forests and fungi revealed in all their brainy and communicating glory. Darwin's early experiments disclosed the sensitivity and signaling of plants, along with our evolutionary linkage to all life, and the field has steadily grown. More recently, books like Peter Wohllenben's *The Hidden Life of Trees: What They Feel, How They Communicate* are making mainstream the intelligence of plants, who act as social beings aiding each other in times of need via underground fungal networks and root systems, and above ground via electrical signals and olfactory messages carried by wind. Forests, or rather, undisturbed forests, which include shrubs, grasses, and other plant species, exchange information, acting like superorganisms.[69] Suzanne Simard, a Professor of Forest Ecology at the University of British Columbia who studies large "mother trees" that supply nutrients to smaller ones, calls this symbiotic superorganism the "wood wide web."[70] Other academic fields have also emerged, like biosemiotics, influenced by Bateson, which explores how signs are not just cultural, but biological, while multispecies ethnography explores specific places where species biologically, culturally, and politically interact. Going further, current science is showing support for "the songs of trees," "the language of plants," and "plants as persons," with vegetal life showing intentionality.[71] All these metaphors, of course, reflect the reality of another one: the intelligence and beauty of the pattern that connects.

Despite his witchery and whipping boy status, Descartes' dream led to rational steps up the developmental holarchy, or Learning I, then II, then III. It is also worth noting that the Western enlightenment project led to the end of burning witches and to rights for women, Times change. What seems fitting at one time may be unfit at another, bless his doing-his-best-within-the-context-of-his-time heart. Yet, the damage done, and still being done, in the name of Cartesian dualism is immense. We have taken steps up the holarchic ladder but are not fully awake to our eco-existential plight: more of the same processes of thought produce more of the same results, a runaway system of destruction due to epistemological error. Thus, with most humans residing at first tier, and many of them stuck at the Mythic level or content at Rational, we desperately need scholarship that discloses the unfit of Descartes' legacy.

Bateson did just that, but absent from his work is the link between androcentrism and dualism. This may be expecting too much, as feminist and ecofeminist criticism of the hegemony of dualism and I-It science were not fully articulated during his time. But as we saw with Griffin, the feminist voice—which has been historically silenced due to dualism—is a powerful and much-needed one. Interestingly, like Bateson, a prominent critical feminist voice, Donna J. Haraway, is inspired by cybernetics. In her "cyborg manifesto," she writes that "the cyborg is a cybernetic organism" that holds "incompatible things together because both or all are necessary and true."[72] Like the pattern that connects, the cyborg, as metaphor, counters Cartesian dualism by integrating nature and culture, but also goes further by adding technology, which reflects the fact that we are tied to our machines. While most eco-critics see this as problematic, for Haraway the nature-culture-technology nexus is not only necessary and true, it supplies possibilities for fitting responses in a post-gender world where we are not defined by essences, and technology, rather than always being foe or idealized as savior, may at least be friend.[73]

Haraway's original, and playful, manifesto fully embraces the postmodern turn, and thus breaches disconnections by deconstructing dominating hierarchies. Val Plumwood, in *Feminism and the Mastery of Nature*, is another influential ecofeminist deconstructing Cartesian dualism, although without the nod to cybernetics and techno-philia. She does, however, thoroughly take on the stupidity of hierarchal arrangements when intelligence is not shared by dualistically matched pairs, especially culture/nature, man/woman, and civilized/primitive. From an androcentric perspective, nature, women, and indigenous lifeways "includes everything that reason excludes": emotions, body, passions, animality, matter, physicality, sense experience, and the entire nonhuman world. For Griffin, the problem with rigid dualisms is that the dialogue among differences is distorted, and this distortion supports systems of domination, but Plumwood fully discloses how oppressive communicative habits—backgrounding, radical exclusion, incorporation, instrumentalism, and homogenization or stereotyping—are internalized such that they become "just the way things are."[74]

Backgrounding refers to the tendency of the oppressor to foreground their interests and needs while backgrounding the fact that the actualizing of these interests and needs is wholly dependent on the master/slave dynamic. Backgrounding, then, is a form of repression in which the master denies that their identity formation is dependent upon oppressive relationships. Radical exclusion refers to the tendency of the master to hyper-separate differences, and then characterizes these differences as lower and inferior. Incorporation refers to the tendency of the master to define the other only in terms of their self-concept; the other is incorporated into the master's perceptions such that they do not have independent needs or desires. Instrumentalism, as we've explored via Bateson and Buber, refers to the master's tendency to objectify the other, and thus relationships lack the ethics associated with intrinsic value, including empathy, respect, and care. And homogenization refers to the master's tendency to stereotype the other in accord with their "inferior" nature.[75]

All this sounds way too familiar, especially if you are an activist responding to a world of wounds, or protesting systemic sexism, racism, and class disparities, or reading headlines about the litany of eco-social crises. Yet, as Ramsey argued, relationships of subordination are built into social systems and ideology—like the I-It ideology of Cartesian dualism—leading us to fall prey to problematic myths and participate in our own oppression. Bateson, while recognizing barriers to learning, and that we must learn how to learn in response to crises, did not extend his analysis much to how capitalistic economic systems incentivize and reward oppression, or how "armoring" causes us to incorporate oppression into our bodies habitual tendencies. However, by making us aware of larger mind and continual exchange of information, we may compare information systems, like ecological and economic systems, to discern the unfit.

Plumwood's *Feminism and the Mastery of Nature*, like Bateson's vision, and Griffin's and Haraway's treatises, attempts to rekindle relationship by deconstructing Cartesianism as theory, yet her short personal essay, "Being Prey," returns to experience, dramatically revealing how dualisms and dominance are interwoven into our habit-bodies. Plumwood learned the hard way what we should already know: humans have instrumental value, especially when encountering crocodiles in the Australian bush while paddling a canoe. To cut to the chase, she was attacked, repeatedly, by a large crocodile who saw her as a food source, yet survived, barely, several underwater death rolls by grabbing a branch and dragging herself to shore. Plumwood, who has since passed away after years of healing, gleaned many lessons from this encounter, including the dangers of not listening within the information system of the bush.[76]

Plumwood failed to listen to outward signs and inner feelings—she received, but decided to ignore, a variety of warnings, due to a desire to find an Aboriginal art rock site—yet during and after her near-death experience, new insights were revealed. Her "narrative self," as in the story she was telling herself about who she was, was silenced when she could no longer deny that she, and we, are prey, not just top-of-the-food-chain predators. This reorientation led her to consider how easily we withhold compassion from animals, especially in factory farming: we don't empathize with prey since we don't think of ourselves as prey. Our skewed masculinist master-slave information system, to which we do habitually listen, too often rules our responses, making them unfitting. This was true before her attack, and especially afterward while she was hospitalized: the first reaction by most was to kill the crocodile. But Plumwood knew that she was the intruder and that the crocodile was just doing what crocodiles do, and what we all do: look for our next meal. From the perspective of the masculinist master narrative, the crocodile was a monster, not a species within an ecosystem, and despite having plenty of experience in the bush, and many years working to preserve habitat for crocodiles, she was positioned as a foolish woman who did not belong in wilderness.[77]

Plumwood, of course, had a different interpretation, in which the dualistic separation between humans and nature, and humans and animals, is overcome. We exist within the pattern that connects, and that includes the food web—but what should we eat? All carrot, no cow? Michael Pollan calls this the omnivore's dilemma, questioning what we ought to eat in support of personal and planetary health.[78] The three types of value help us to think through this dilemma, but cultural considerations, as well as individual metabolism, also come into play, as does the complexity of our postmodern world, as anyone who attempts to consciously engage with our food information system knows.

Some feel they need meat to be healthy due to their body type, and seek ethical ways to get it, like buying locally from thoughtful producers. Raising animals and killing them yourself dissipates dualism, especially the hyperseparation of supermarket meat wrapped in plastic and Styrofoam; that is, if we are mindful of backgrounding and radical exclusion. Hunters should be similarly mindful, hunting after having read Shepard and honoring predator-prey reciprocity, while realizing that we have instrumental value for top predators, and thus some humility and empathy would be a very good thing. Of course, vegetarianism or veganism, if they work for our bodies, are other ways to go—I have been mostly vegetarian for thirty years—but where we get our veggies is another ethical dilemma, with rainforests not only destroyed for raising cattle but growing soybeans. Local farmers markets where you can talk to the farmer and learn about their farming practices, as well as gardening our own veggies, are on the rise as people widen their a-whereness and concern.

Whatever our choices, Ground, intrinsic, and instrumental value and the pattern that connects may act as guide. As would the logical next step: an ethic of least harm that recognizes the language, the song, the voice, and thus the intelligence of nonhumans. No harm is not possible, even though everything has Ground value. We need to eat, and despite the wonders of plant intelligence, plants inevitably have more instrumental value as food for other species, and animals, including humans, have less instrumental and more intrinsic value, unless, of course, you are a crocodile. Still, for humans, least harm changes the how of eating, with the aesthetics of grace and the gift event further guiding our actions. This debate ultimately centers on degrees of complexity and sentience, yet research has revealed what TEK has long known, and what Bateson's focus on mind and exchange of information supports: the difference in degrees is not nearly as large as Cartesian dualism has led us to believe. And we do believe and pay homage via institutionalized economic exchange devoid of grace and gift.

On the other hand, sometimes differences that make a difference are larger, and our decisions are less complex. I kill plenty of mosquitos in summertime despite the beauty of Jain ahimsa or St. Francis' agape love, or aesthetic responsiveness generally, because I don't like being eaten. But mosquitos, and more dramatically, crocodiles and grizzlies, remind that we are also part of the food web—we are also food—and mosquitos play a role in the food web for some species, like swooping bats feeding in dim light at dusk, and bats play a role in the food web for other species, and on and on.

But when applying Ground, intrinsic, and instrumental value to agribusiness and the blood-letting of industrial slaughterhouses, where differences among species—human animals and domesticated animals—are not so large, one quickly discovers that the first two are nonexistent.

When economic information systems override ecological information systems, the few profit while the rest of us receive messages that we are not going to like. So many missives to choose from in an age of climate emergency, but we managed to add a pandemic to the list, likely due to wet market selling and eating of wild species, including bats, after encroaching on their habitat despite all species having the rights to be and become. There are so many warnings to choose from because so much has come around to kick us in the ass: mistaken epistemology, mistaken notions of value, a snowball's chance in hell.

Bateson's vision is sometimes dark, even without analyzing dualism and dominance associated with gender and economics, yet he lights a way out of the shadows of our Cartesian legacy via aesthetic communication, and thus listening within eco-aesthetic information systems that disclose the insanity of despoiling our home. However, it is important to point out that along with incessant bad news—a half-century later, Lake Erie, after many clean-up efforts, is still prone to oxygen-depleting algae blooms and dead zones due to agricultural and sewage run-off—there are plenty of sustainability solutions, including the citizens of Ohio voting to give Lake Erie rights in 2019.[79]

Paul Hawken's Project Drawdown provides many more hopeful examples, as he lists the top 100 responses to climate crisis, with one of the most significant being the education of girls leading to more social, economic, and political equality, more family planning and a reduction of population, and more community activism, adding much needed voices to the over two million organizations working on eco-social justice and restoration.[80] This is all good because it reflects the good and beautiful. However, many of Project Drawdown's solutions focus on technological innovations like solar, wind, electric cars, and high-speed rail, along with low tech approaches like regenerative agriculture. Biomimicry also leads to a plethora of tech solutions, but Benyus wisely noted that biomimicry is not equivalent with morality. The Wright brothers learned how to design the first airplane through close observation of birds, and within four years, humans were using them to drop bombs.[81]

Despite the wonders of technological innovation, the myth of technology as savior remains deeply problematic. Still, Haraway has a point: we are like cyborgs, living within a nature-culture-technology nexus. She finds inspiration and possibility in this nexus, as do many others, but abstract cybernetic language is rather I-It. Bateson's cybernetic self is not emotionally inspiring, nor is identifying as a cyborg—we sound like bots rather than embodied beings—and perhaps that is why Bateson was not quite sure of how to live this "self," even though he studied indigenous peoples, art, aesthetics, and was critical of education marked by discipline division rather than integration and wholeness. Perhaps that is why he turned to the religious language of sacredness and grace: for others to be inspired by his vision, but also for him to better live an embodied existence open to beauty. Technology, while not our savior, may be friend by playing a role in saving what deserves to be saved, at least when guided by aesthetic-ethical-spiritual values and insights.

Haraway, like Griffin, Plumwood, and Bateson, also takes on I-It science. Citing Sandra Harding's *The Science Question in Feminism*, she calls for "successor science" marked by "irreducible difference and radical multiplicity of local knowledges," or TEK, which will also play a part in saving what deserves to be saved, allowing for a better accounting of the world and therefore a better world.[82] While acknowledging the value of objectivity as a practice, she adds her voice to the feminist critique of objectivity in service of dualisms and hierarchical orderings, arguing for an embodied objectivity characterized by "contestation, deconstruction, passionate construction, webbed connections, and hope for transformation of systems of knowledge and ways of seeing."[83] Haraway, then, would seem to agree with Bateson that scientific practice should include the dialogic interplay of rigor and imagination. She further writes that "outcomes are not guaranteed" in our encounters with nonhuman others, whether "socially, ecologically, or scientifically," and the best we can often do is "getting on together with some grace."[84]

As we saw, a seminal graced scientific "getting on together" comes from Jane Goodall's practice of I-Thou science, yet her methods were questioned before being celebrated. Over the years, many more women have gone on to practice embodied science, rejecting the Cartesian dualism between mind and body by listening within and without, and making our information systems infinitely richer. For Birute Galdikas, who studied orangutans for twelve years, becoming like an adopted grandmother, and Denise Herzing, who studied a pod of wild dolphins for over a decade, this means that "observation is intimate."[85] The same is true for Katy Payne, an acoustic scientist who discovered the language of elephants and who also studied humpback whale song, which she called "fluid and ever-evolving, using structures and devices we find in human song and poetry," and "beautiful and moving to listen to," in her testimony to stop ocean sound testing.[86]

As far as I know, none of these scientists mention Bateson as an influence, yet they all understand that we exist within larger communicative systems and perceive the intelligence of nonhuman others, and, I would venture, they all find subjects under study "beautiful and moving to listen to." As a result, they have learned how to learn, work, and live in sympathy with the pattern that connects, rejecting the "witchery" of rigid Cartesian dualism and refining aesthetic responsiveness by meeting the other with recognition and empathy.

ROBIN WALL KIMMERER AS EXEMPLAR

Robin Wall Kimmerer may be the best exemplar of living Bateson's vision, especially since she turns to a grammar of animism and aesthetics rather than cybernetics, braiding together scientific knowledge, TEK, and "the story of an Anishinabekwe scientist . . . in service of what matters most."[87] Kimmerer's first memory of science was as a five-year-old kindergartener encountering first snow. Her wise teacher took a magnifying glass from her pocket, showing her the complexity and detail of a snowflake. She was surprised by how something so "small and ordinary" could be "perfectly beautiful." There was more to the world than what met the unaided eye: dazzling, secret, and mysterious knowledge. The already "gorgeous world" became "even more beautiful the closer you look."[88]

While Kimmerer's initial magnified gaze at a snowflake was surprising, the fact that she became a botanist is not. She grew up with seeds in shoeboxes and pressed leaves under her bed, stopped her bike to identify species, had plants color her dreams. When she applied to a college forestry school that had few women, especially those who looked like her, an advisor asked why she wanted to major in botany. Her most honest answer was that the plants chose her, but she responded that she "wanted to learn why asters and goldenrod look so beautiful together." This, apparently, was an unfit response, as she was advised that such concerns were not science, and then enrolled in General Biology so she could find out what was. And so began an education that taught her much but denied vibrant, living aesthetics, which, she was told, she could study in art school.[89]

Like her grandfather, who was forced to leave culture, language, and family behind in his first day of school, Kimmerer was instructed to forget her heritage, and that there was only one way to think. But questions persisted: why do visually stunning pairs of purple and gold "end up side by side." "What is the source of this pattern?" "Why is the world so beautiful?"[90] These seemed like good questions to her, especially since asters and goldenrod could grow alone. She had wanted to study botany *and* poetry, but was further instructed that she could only study one, as if they were not only separate disciplines but separate worlds; or, C. P. Snow's two cultures that hinder fitting responses to global problems. Kimmerer chose plants, or plants chose her, because in her Native upbringing, plants were "teachers and companions" to whom she was "linked in mutual responsibility." But the "who" of the plant was replaced by "what is it?" and other valid yet reductionist questions. Kimmerer struggled in school at first, but came to embrace scientific naming as a form of poetry and was fascinated by photosynthesis and the intricacy of leaves. She also found kind professors who did "heart-driven science," even if they could not admit it.[91]

Kimmerer's natural tendency was to perceive pattern and relationship, to "seek the threads that connect the world," but her education continued to hammer home rigid separation between observer and observed. And she learned her lessons well and excelled, privileged to know the "powerful tools of science" and eventually earn a Ph.D.; yet, "to walk the science path" she had "stepped off the path of indigenous knowledge," learning the Latin names of plants but no longer listening to their song. But then she heard a Navajo woman, with no university botany training, speak for hours about the plants in her valley, knowing names, where they lived, when they bloomed, who ate them, who lined their nests with what fibers, as well as their stories and relationships and the medicine they gifted. And she spoke of beauty. Kimmerer writes that this talk was like "smelling salts" waking her up to the limits of her new Ph.D., which offered one lens while this so-called uneducated woman provided knowledge that was "deeper and wider and engaged all the human ways of understanding."[92]

Kimmerer circled back to her original questions concerning beauty, questions that were "bigger than science." Science does disclose information on rods and cones in the retina and color perception, with specialized receptor cells absorbing light wavelengths attuned to red, blue, and purple and yellow, sending signals to the brain. This explains why the combination of Asters and Godenrod got her attention, but Kimmerer writes this does not explain why she perceives it as beautiful. Artist friends, however, pointed her to the color wheel, where purple and gold are complementary colors, making each more vivid. In Batesonian terms, they express reciprocity via exchange of information, and this attracts humans, as well as bees "bent on pollination," their "striking contrast" making them "the most attractive target in the whole meadow." Asters and Goldenrod growing together receive more pollinator visitors, a testable hypothesis informed by science, art, and beauty.[93]

For Kimmerer, Asters and Goldenrod are like science and TEK—complementary parts of a larger pattern—and it is a "whole human being who finds the beautiful path." Her questions, then, were scientific, yet ultimately about relationships and the "shimmering threads that hold it all together." Her questions were about love and caring for the whole world, not separate worlds of the sciences and humanities, and yearning to understand and make "something beautiful in response."[94]

Kimmerer, inspired by her first snowflake memory, eventually became a Professor of Environmental and Forest Biology at the State University of New York, her studies mostly focusing on mosses; a hand lens around her neck, its cord tangled with her medicine bag, she explored levels in the "hierarchy of beauty" beyond ordinary perception. Kimmerer writes that her knowledge of mosses comes from her scientific training, as well as the "plants themselves" and "intuitive affinity" for the knowledge of her Native heritage. Like the tangled cords around her neck, her science education twines with teachings from the old stories, in which "thrushes, trees, mosses, and humans" share a common language. Botany refines our ability to see, to observe, but despite science's power something is missing: listening as if "we are audience to conversations in a language not our own." When Kimmerer first heard the Anishinaabe word "Puhpowee," which means "the force that causes mushrooms to push up from the earth overnight," she discovered a grammar that respects unseen animate energies.[95]

In Kimmerer's award-winning *Gathering Moss*, her mission is to give voice to mosses, from whom we have much to learn. And listening to their voices inevitably leads to interactions among species, including mosses and rocks. Like any good scientist, or attentive observer, she once again begins with a question: Why do some rocks host ten or more moss species while similar nearby rocks host only one? Working in the Adirondacks, this question led to another as she and fellow researchers called out the Latin names of mosses as they discovered them: Why do most mosses only have scientific names and not common ones, like many rock outcroppings do? Kimmerer's daughters had named several rocks based on their relation to them—Reading Rock, Diving Rock, Bread Rock—but mosses tend to lie outside our circle of relations, Its rather than Thous, and thus without "sweet secret names . . . for the ones we love."[96]

At one point in her studies, she discovered a cave-like opening among the rocks leading to a grassy meadow and rock-filled room covered with mosses. All were Thous now, animate and alive, the mossy rocks "vibrating with energy exchanged at a very long wavelength." Rocks and mosses, she surmised, participate in ancient conversation; the rocks are "beyond slow, beyond strong and yet yield to a soft green breath as powerful as a glacier," with soft mosses wearing away hard rock, "grain by grain bringing them slowly back to sand."[97] Standing inside this conversation, this circle, losing track of time, she recalled that ancestors knew rocks held the Earth's stories, and mosses have names other than the Linnaean taxonomy. This recollection was a gift, and thus she did not feel willed to scientifically name mosses within this place, within this reciprocity of presences, only the grace of responsibility. She would carry forward the message that "mosses have their own names" and their "way of being cannot be told by data alone," even as she continued to explore the conditions that support diverse moss communities.[98]

Gathering Moss is filled with science—much more than I can share—but what resonates beyond the science is pattern, beauty, and reciprocal relations among mosses and other species and animate forests. In fact, Kimmerer writes that forests give thanks to the mosses, which offer their beauty along with many services, including capturing rainwater, which is nutrient-rich after running down the trunks of trees. Moisture-filled mosses also create humidity, which then aids the dispersion of mosses, which creates nursery-like moss-mats for seedlings, and decomposes fallen logs by supporting fungi that create soil. Moisture-filled mosses also keep insects alive that would otherwise dry out and die, supplying food for birds and other species, and material for nests. The entire food web also gives thanks to mosses.[99]

Kimmerer writes that rivers, clouds, mycorrhizae, algae, salamanders, and slugs also give thanks, as do indigenous peoples, who offered prayers. The only ones, it would seem, who do not give thanks are mistaken epistemology humans, but we could, and should, because gratitude creates openness for listening. Prayers may be answered, not in a petitionary ask-for-things way, but in an integrated mythic-animism, rationality, and aesthetic communication kind of way: ask good questions, with proper humility, and we may receive insights and guidance.

Kimmerer clearly mirrors Bateson: "The patterns of reciprocity by which mosses bind together a forest community offer us a vision of what could be."[100] However, she provides far more examples than he did of how to live this vision. From an indigenous perspective, we cannot begin to understand others unless we engage four aspects of our being: mind, body, emotion, and spirit. Science relies on the first two, at most: close observation and sense experience interpreted by individual mind—the information system is empirical—but she willfully integrates objective and subjective approaches, rigor and imagination, going beyond embodied objectivity to a dance of science, spirit, and story, receptive to grace, like the patterned beauty of a snowflake.[101]

When I encountered a sunlight-dazzled, jewel-like verdant green piece of moss when hiking the Appalachian trail, dropping to my knees to give thanks, as if to the Divine itself, I did not know why I was so moved by its beauty. I still do not know, in any complete, rational sense, but after reading Kimmerer I have a deeper rational, emotional, and spiritual understanding. And I feel like she would understand.

NOTES

1. Carol Wilder, "The Palo Alto Group: Difficulties and Directions of the Interactional View for Human Communication Research.," *Human Communication Research*, 5, 2 Winter, 1979), 171–185. See also Peter Harries-Jones, *A Recursive*

Vision: Ecological Understanding and Gregory Bateson (Toronto: University of Toronto, 1995).

2. Gregory Bateson, *Steps to an Ecology of Mind* (New York: Ballantine, 1972), 461–463.

3. Ibid.

4. Quoted in Morris Berman, *The Reenchantment of the World* (New York: Bantam, 1984), 227–228.

5. Bateson, *Steps to an Ecology of Mind*, 484.

6. Ibid., 462.

7. Ibid.

8. Gregory Bateson, *A Sacred Unity: Further Steps to an Ecology of Mind*, edited by Rodney E. Donaldson (New York: HarperCollins, 1991), 305.

9. Noel G. Charlton, *Understanding Gregory Bateson: Mind, Beauty, and the Sacred Earth* (Albany: SUNY press, 2008), 27.

10. Gregory Bateson, *Mind and Nature: A Necessary Unity* (New York: Bantam, 1979), 8.

11. Ibid., 8.

12. Ibid., 18–19.

13. Bateson, *A Sacred Unity*, 292.

14. Ibid., 267–268, 272.

15. Ibid., 269.

16. Ibid., 270.

17. Ibid., 296.

18. Ibid., 296.

19. Ibid., 298.

20. Ibid., 310–311.

21. Bateson, *Steps to an Ecology of Mind*, 462.

22. Ibid., 493.

23. Ibid., 462.

24. Buber, *I and Thou*, 69.

25. Peters, *Speaking into the Air*, 31.

26. Harries-Jones, *A Recursive Vision* (Toronto: University of Toronto, 1995), 29.

27. Quoted in Charlton, *Understanding Gregory Bateson*, 6.

28. Mary Catherine Bateson, foreword in Gregory Bateson, *Steps to an Ecology of Mind* (Chicago: University of Chicago Press, 2000), x–xi.

29. Ibid., xiv.

30. Quoted in Charlton, *Understanding Gregory Bateson*, 100.

31. Bateson, *Steps to an Ecology of Mind*, 75.

32. Bateson, *Mind and Nature*, 237.

33. Wilber, *Sex, Ecology, Spirituality*, 32–78. See also Wilber, *The Religion of Tomorrow*, 146–155.

34. Bateson, *Steps to an Ecology of Mind*, 287–300. See also Berman, *The Reenchantment of the World*, 212–214.

35. Bateson, *Steps to an Ecology of Mind*, 301–306.

36. Ibid., 306.

37. Berman, *The Reenchantment of the World*, 216.

38. Bateson, *Steps to an Ecology of Mind*, 304.

39. Ibid., 463.

40. Gregory Bateson and Mary Catherine Bateson, *Angels Fear: Towards an Epistemology of the Sacred* (New York, Bantam Books, 1988), 53.

41. Ibid., 50–56.

42. Wilber, *Sex, Ecology, Spirituality*, 517.

43. Thomas Berry, *The Great Work* (New York: Bell Tower, 1999).

44. Wilber, *Sex, Ecology, Spirituality*, 517–520. See also Wilber, *A Brief History of Everything*, 204–207.

45. Ibid.

46. Janine Benyus, *Biomimicry: Innovation Inspired by Nature* (New York: HarperCollins, 1997), 1–2.

47. Ibid., 16.

48. Berry, *What are People For?* 209.

49. This description is inspired by Hans George Gadamer's "genuine conversation," which is reviewed in Stanley A. Deetz, *Democracy in an Age of Corporate Colonization: Developments in Communication and the Politics of Everyday Life* (New York: SUNY press, 1992), 165–168.

50. Quoted in Roger Dunsmore, *Earth's Mind: Essays in Native Literature* (Albuquerque: University of New Mexico Press, 1997), 38.

51. Ibid., 41.

52. Ibid., 11.

53. Ibid.

54. Leslie Marmon Silko, *Ceremony* (New York: Penguin Books, 1977), 1.

55. Ibid., 135.

56. Benyus, *Biomimicry*, 6–7.

57. Ibid., 3.

58. Ibid., 8.

59. Ibid.

60. Derrick Jensen, *The Myth of Human Supremacy* (New York: Seven Stories Press, 2016), 22.

61. Benyus, *Biomimicry*, 9.

62. Wilber, *Sex, Ecology, Spirituality*, 50.

63. Ibid., 685.

64. Bateson, *Steps to an Ecology of Mind*, 306.

65. Ibid., 292.

66. Quoted in Charlton, *Understanding Gregory Bateson*, 58–59.

67. Martin Lee Mueller, *Being Salmon, Being Human: Encountering the Wild in Us and Us in the Wild* (Vermont: Chelsea Green, 2017), xxii.

68. Ibid., 2–5.

69. Peter Wohllenben, *The Hidden Life of Trees: What They Feel, How They Communicate* (Vancouver: Greystone Books, 2015), 1–13.

70. TED, Suzanne Simard, "How Trees Talk to Each Other," August 30, 2016, YouTube video, 18:25, https://www.youtube.com/watch?v=nH2BLg43QSQ.

71. See David George Haskell, *The Songs of Trees: Stories from Nature's Great Connectors* (New York: Penguin Books, 2017), Monica Gagliano, et al., *The Language of Plants: Science, Philosophy, Literature* (Minnesota: University of Minnesota Press, 2017), and Matthew Hall, *Plants as Persons: A Philosophical Botany* (New York: SUNY Press, 2011).

72. Donna J. Haraway, *Simians, Cyborgs, and Humans: The Reinvention of Nature* (New York: Routledge, 1991), 149.

73. Ibid., 150.

74. Plumwood, *Feminism and the Mastery of Nature*, 19–20.

75. Ibid., 48–55.

76. Val Plumwood, "Being Prey," Utne.com, July-August 2000, https://www.utne.com/arts/being-prey.

77. Ibid.

78. Michael Pollan, *The Omnivore's Dilemma: A Natural History of Four Meals* (New York: Penguin, 2006).

79. Vox, "Lake Erie Now has Legal Rights, Just Like You," April 29, 2019, YouTube video, 8:08, https://www.vox.com/future-perfect/2019/2/26/18241904/lake-erie-legal-rights-personhood-nature-environment-toledo-ohio.

80. Paul Hawken, *Blessed Unrest: How the Largest Social Movement in History is Restoring Grace, Justice, and Beauty to the* World (New York: Penguin Books, 2008). See also Project Drawdown, https://www.drawdown.org/.

81. Benyus, *Biomimicry*, 8.

82. Haraway, *Simians, Cyborgs, and Humans*, 187.

83. Ibid., 191–192.

84. Donna Haraway, *When Species Meet* (Minnesota: The University of Minnesota Press, 2008), 15.

85. Quoted in "Introduction," *Intimate Nature: The Bond Between Women and Animals*, edited by Linda Hogan, Deena Metzger, and Brenda Peterson (New York: Ballantine Publishing Group, 1998), xiii–xiv.

86. Ibid., xv.

87. Robin Wall Kimmerer, *Braiding Sweetgrass: Indigenous Wisdom, Scientific Knowledge, and the Teachings of Plants* (Minnesota: Milkweed Editions, 2013), x.

88. Robin Wall Kimmerer, *Gathering Moss: A Natural and Cultural History of Mosses* (Corvallis: Oregon State University Press, 2003), vi.

89. Kimmerer, *Braiding Sweetgrass*, 39–41.

90. Ibid., 41.

91. Ibid., 42.

92. Ibid., 43–44.

93. Ibid., 46.

94. Ibid., 47.

95. Ibid. 48–49.

96. Kimmerer, *Gathering Moss*, 3.
97. Ibid., 5.
98. Ibid., 6.
99. Ibid., 146–150.
100. Ibid., 149–150.
101. Kimmerer, *Braiding Sweetgrass,* 47.

Chapter 5

Discerning the Unfit

From New Age to Ascension

A NEW AGE FABLE

In the summer of 2000, I drove west from the Adirondacks through the traffic of Buffalo, Cleveland, and Chicago, and then along desolate roads, past way too many billboards for Wall Drugs in South Dakota, and past the Badlands, sacred land of the Lakota Sioux, on my way to the University of Montana for graduate school.

After a long day of driving, I began to set up my tent in a state park, noticing an older gentleman watching from the dirt road. I gave him a quizzical look and then an inviting smile. That was all the prompting he needed. He introduced himself as Bob, sat down at the picnic table, and proceeded to reveal his life story.

Bob was wide-eyed, giving me the impression that he was projecting oneness onto the diverse earth and universe. I soon discovered that he was a devotee of Jane Roberts' New Age books, which were supposedly written by Roberts as she channeled the angel, or the "energy essence personality," Seth. Bob began every sentence by stating "Seth says," prompting me to ask: What does Bob say?

Bob smiled, stating that his grown children asked the same question, but he desperately wanted to channel an angel and was trying to get rid of his personality in order to make room. He was on a mission to find a life-partner for his retirement years, and felt that an angel could lead him to find his mate via synchronicities. I liked Bob, especially after he dropped his "Seth says" sentences long enough to admit a painful past of alcoholism, divorce, and decades of work at a job he hated. He also claimed to have spiritual experiences of a larger self that woke him up to the shallowness of his identity, the hollowness of his life. But he interpreted his experiences via Jane Roberts and Seth, who were telling him that he could create his own reality, if only he believed.

Bob was surprised by my knowledge of differing spiritual traditions, and I tried to use this surprise as a lever to create some room for critical thought. I saw myself in him, not just in the spiritual sense, but because I had gone down a similar path, waking up to a spiritual life while struggling to make sense of the ongoing dialogue between knowing and not-knowing; still, for me, attempting to channel an angel was a bit much. And starting sentences with "Seth says" as a way to lessen ego and invite larger forces was likely to invite self-deception. But he was resistant to any criticism, as if doubt would keep the angels away. It was also clear that he was visiting because he expected to teach me, but we listened as best we could.

The Seth books assert that life either has meaning and reasons for everything, or life has no meaning and no reasons. The reader is instructed to choose a life of meaning and reasons, and then create their own reality by ridding themselves of all negative thoughts, focusing instead on the positive aspects of what they want and following synchronicities. I was familiar with New Age world views of this type, including New Thought and "the law of attraction," which assert that the right spiritual attitude will bring material supply. And, after study with Rafael, I focused on inner listening and was open to synchronicities, but also fell prey to thinking that all could be mine if I could just get into the right frame of mind.

Over time, however, I began to question whether such world views convey some truth within a larger lie. Negative thinking does drag us down, closing off receptivity to potentialities and possibilities, while positive thinking engages potentialities and possibilities within a deeply intertwined, gifting cosmos. But when life remains tangled and messy despite attempts to create our reality—due to the reality of randomness and material circumstances like patriarchy, racism, social and economic inequalities, and simply the life-narratives of others—we are left blaming ourselves. We are not spiritual enough, we don't believe enough, and we are not good enough. The long and ongoing historical interplay between freedom and determining forces, lived with the aid of critical rationality, is denied in favor of failure and guilt. Responsibility is a matter of degree: sometimes things happen and we are most responsible, sometimes things happen and circumstances or others are most responsible, sometimes things happen and nothing is most responsible, and most times it is a combustible combination.

By the time I met Bob I had long ago realized that we cannot create our own reality, but, along with others, we can influence outcomes by continually creating and living narratives that are responsive to eco-existential and spiritual realities. Positive thinking has its power, and we should remain open-minded, but, as Rafael once told me, we shouldn't be so open-minded that our brains fall out.

Bob and I talked for three hours. He and Seth kept trying to get in the last word, but I couldn't help making critical comments that would force another reply. It became dark, however, and I was ready to stop talking. I told him I was sleepy, offering to shine my flashlight along his path so he could see his way back to his campsite. He refused assistance, confident he could find his way. I assumed he planned on being guided by Seth or an angelic inner light, but heard him stumbling in the brush, and then a crash, so I flicked the switch and pointed in his direction.

Bob laughed sheepishly and thanked me, and I could not help thinking that most times what we all need is human help, even if just a simple gesture, like a mere mortal shining some light using a ten-dollar piece of technology.

QUESTIONING THE QUEST FOR A NEW AGE

Generally speaking, New Ageism may be defined as the mining of esoteric religious traditions in search of practices that will foster spiritual awakening, bringing forth a new age of enlightened consciousness. Such practices are often communicative, including channeling angels and following synchronicities, in the sense that information or messages are received. The focus, then, is mostly on various forms of spiritual communication, although we must question if received messages aid stage-steps up the developmental holarchy or act as impediments to growth. We must also question whether New Age spiritual practices challenge modernist assumptions that undergird ecocrisis or if they participate with monologic attitudes destructive of the earth.

Given such diametrically opposed possible New Age responses to modernity and ecocrisis, it is worth returning to Schrag's inquiry into what is fitting and unfitting before proceeding. In an interview, he states:

> The fitting response is a questioning. It is always a questioning of what is going on and then making hard decisions regarding the extent to which we can appropriate the tradition and the extent to which we have to intervene in it. There is a creative moment in the fitting response, a moment in which we have to invent something new, project something new, begin to enact something new. And it may call for a radical revision or indeed the overturning of traditional as well as current forms of thought and action.[1]

This quotation would seem to support New Ageism's desire to create a new reality, as well as Bob's enacting of spiritual practices new to him. However,

Schrag further states: "The fitting response has intelligibility only against the backdrop of a discernment and determination of what is not fitting. And it is out of this that action comes."[2] Thus, the new that New Ageism hopes to bring must be questioned based on its intervention into past traditions, interpretation of the needs of the present, and actions that are meant to call forth a future new age. I make no claim to be a final arbiter judging the spiritual experiences of New Agers or anyone; however, a transversal gathering of texts, insights, and experiences will better enable us to discern the unfit, and that is what I am setting out to do.

But what, exactly, is meant by "New Age?"

New Age is a slippery term, but based on personal experience and study of popular texts explicitly or implicitly labeled New Age, I can suggest three unfit tendencies: a projection of spiritual oneness that ignores material diversity; anti-rationality, or at least the privileging of nonrational insights over and against rationality rather than working to integrate them; and the belief that we can create our own reality, which dismisses direct action and the hard work of community building called for by Peters. Still, while I use the term "New Agers" to reference those who may fall prey to these tendencies, those interested in New Age literature and practices are not a cohesive group, and a review of critical texts discloses no consensus meaning to the complex phenomenon known as "New Age."

Michael Zimmerman argues that "countercultural dropouts and visionaries" initiated the New Age movement in the 1970s, which included "sometimes contradictory, sometimes flaky" practices: "To be sure, even though allegedly protesting against modern materialism, many hucksters have cashed in on the New Age interest in reincarnation, organically grown food, ethereal music, channeling, crystal power, astrology, and the like." However, despite criticisms he states that the focus on self-realization, or realizing a higher self that is one with the whole, has "influenced many modern institutions."[3]

Zimmerman sees positives in the New Age quest for self-transformation, or moving up the developmental holarchy. But how do we know when practices and experiences reflect information exchange that respects others and leads to growth versus projection of ego, which is another form of monologue? Wilber admits that it can be difficult to tell the difference between pre-rational and transrational experiences, but he calls the pre-trans fallacy of elevating "magical narcissism to transcendental awareness" the "single defining characteristic" of the New Age movement, despite good intentions.[4] Thus, the New Age focus on realizing a higher self, while laudatory, may devolve into hyper-individualism. In other words, in response to the age-old question, do we first change ourselves or society, New Ageism often focuses on the individual self and its "create your own reality" mantra, which degrades into a magical narcissism—ironic given New Age claims for oneness with the whole—that fails to critically apprehend that the question presents a false dichotomy, as self and society are dialogically related rather than monologically opposed.

Zimmerman is aware of this criticism. He explores New Ageism as a creative response to modernity's "dark consequences," including ecological destruction, while also acknowledging that it unthinkingly continues traits that produce these consequences, like hyper-individualism but also faith in progress and technology. In response, he argues that many New Agers turned to New Paradigm thought in the 1980s to counter Cartesian dualism, and materialism generally, whether expressed as overconsumption, nationalism, or militarism, by embracing the human potential, feminist, and environmental movements. Influenced by Bateson, New Paradigmers also turned to postmodern science, such as Fritjof Capra's synthesis of quantum physics, systems theory, and ecological thinking in his 1980 *The Turning Point*, to support their principles: we are in the midst of a global eco-social crisis that requires a paradigm shift; the needed change is tied to personal transformation marked by a relational self; and this relational self responds based on an ecological sensibility committed to cosmic holism, holistic health, and thinking globally while acting locally. This shift, this new way of thinking, perceiving, and acting, would also usher in a new politics.[5]

New Paradigm thought provides a hopeful list of principles with an emphasis on relationship and community, not just individualism, but Zimmerman, writing in 1994, questioned the optimism of assuming a new age or paradigm was on the horizon in regard to economics and politics, as both New Age and New Paradigm visions of transformation fail to take into account the power of societal structures: "Are New Paradigmers right in saying that we are witnessing the birth of a new era that leaves behind the violent binary of capitalism versus communism? Or are they politically naïve idealists who fail to see the social and ecological implications of a world run by multinational corporations?"[6] Over two decades later, we are still asking these and similar questions, as much has changed while much has stayed the same.

In contrast to Zimmerman, Stephen J. Sutcliffe argues that New Age thought and practice began as early as the late 1930s, with lay practitioners drawing on "occult, Eastern and neo-Christian practices and motifs" to prepare themselves for what they believed was an imminent "New Age." This new age would be "inherited by the chosen few" in response to "predicted social and economic collapse or—after 1945—possible nuclear holocaust." Sutcliffe contrasts this early expression of New Ageism, described as more ascetical and otherworldly, with post-1970s versions, which were more hedonistic and this-worldly.[7] This description is congruent with Zimmerman's analysis, especially the hyper-individualism and hedonism—which, again ironically, is consistent with the isolated, modernist self against which New Ageism is supposed to be a response—but I question his strong distinction between early otherworldly perspectives and later this-worldly orientations. As my Bob example suggests, there are plenty of otherworldly practices in post-1970s New Ageism, especially angel worship and channeling.

Sutcliffe further argues that New Ageism reflects heterogeneous alternative spiritualities and social networks dissenting against "established religious institutions and post-Enlightenment rationality alike." In other words, organized religion and science, with their contradictory world views, do not provide sufficient meaning and direction, and so New Agers turned to themselves, creating networks comprised of "self-taught practitioners, amateur thinkers and do-it-yourself seekers" influenced by popular culture and the freedom to choose, although often constrained by class, gender, and ethnicity. However, to Sutcliffe, these networks were never organized into a homogenous social movement.[8] Thus, he again highlights the individualism of New Ageism, or interest groups loosely bound together by common individualistic practices. Sutcliffe makes clear that individualism as response is fitting when Enlightenment rationality degrades into damaging instrumentalism and religion degrades into dogmatic fundamentalism, but problematic when defined over and against the co-creative solidarity and community organizing needed to create new societies, and thus a new age.

David Spangler also explores the history of New Ageism, as well as the promise and failures of New Age ideals. His background as a practitioner and leading thinker associated with New Ageism also provide unique insights. Spangler states that he first heard the term "New Age" in the 1950s in association with channelers and psychics, and in 1970 he joined Findhorn in Northern Scotland, a self-identified New Age community. Spangler soon published *Revelation: The Birth of the New Age*, giving him the reputation of founding New Age thought: "In that strange ahistorical immediacy that many Americans live in, I had the distinction of erasing everyone else who had ever thought about or taught about a New Age."[9] Spangler calls himself a "David-come-lately," arguing that envisioning a new age goes back much further than Zimmerman or Sutcliffe suggest, stating the idea is "at least two thousand years old, while I have only been talking about it for thirty years or so."[10]

Despite complicating the historical emergence of New Ageism, Spangler has specific insights on what should be New Age ideals: "To me, New Age represents each of us learning to embody more fully many of the essential qualities that lie within the soul of the planet itself, as well as within the deeper sacred dimensions of the cosmos itself."[11] This sounds like sympathy with intelligence, as he claims that "a living planet like ours is not just a place; it is a condition, a state of being" characterized by co-creativity, connectedness, intricacy, complexity, diversity, and synergy, with each individual offering a contribution and perspective "so that wholeness is manifested as well as the power of emergence."[12]

Spangler is a sophisticated New Age proponent focusing on emerging unity-in-diversity, sacredness, and communion. As such, he is critical of the "self-involvement, the narcissism, of some interiorly directed New Age elements that seems most irresponsible and open to criticism by an intelligent public."[13] He also admits that the term "New Age" may be hopelessly degraded, although for him it signifies an inevitable process. It is "not an image of a specific future" but "an image of the process by which futures emerge." This image includes "stepping from one pattern to another" and embracing mindfulness and conscious evolution, which sounds like Bateson's learning how to learn and Wilber's holarchic stage-steps. The focus on process rather than a predetermined endpoint is compelling, but the image of the New Age—thanks to too many practitioners relying solely on imagination, or magic-mythic thinking divorced from rigorous critical thought—may be beyond repair. Yet, "new age" is a powerful metaphor in times of crisis, one that counters the doom and gloomism associated with environmentalists; and, for Spangler, given that we are evolutionary participants more than victims, "the New Age is not optional."[14]

This "no other option but to transform" mantra is given full voice in Marilyn Ferguson's 1980 *The Aquarian Conspiracy*, which has been called the bible of the New Age. Ferguson was active in the human potential movement, publishing and editing the Mind/Brain Bulletin for nearly two decades, before writing the strangely titled manifesto, with "conspiracy" used in the positive sense of individuals conspiring together in small groups, and then larger ones, sharing their ideas and experiences on emerging new consciousness. And for Ferguson, there is no doubt that this transformation is happening in all aspects of life, including science, health, education, and politics. The appeal to science is central and not just focused on quantum physics, although she references Capra's *The Turning Point* claim that we should be living the lessons of quantum science, such as Werner Heisenberg's uncertainty principle—now a particle, now a wave, with the observer influencing the observed—which discloses a deeply intertwined reality.[15]

Ferguson focuses on what she calls "the science of transformation" via Ilya Prigogine, the Nobel Prize–winning physical chemist, and his theory of dissipative structures. Evolution is an irreversible process of more complex patterns and higher forms, and new patterns emerge from the dissipation of the old. Thus, instability is key to change, or order out of chaos. She writes that dissipative structures may be described as "flowing wholeness," another way to say holons, and as energy dissipates, shaking up old forms, there is potential for "sudden reordering."[16] Entropy is not our fate, restructuring is, and change may happen fast, especially when conspirators work together to transform individual habits and collective culture. For Ferguson, Margaret Mead's oft-quoted adage that a small group of thoughtful and committed citizens can change the world is borne out by science.

Ferguson points to many antecedents to this call for and expectation of change, which she calls "premonition of transformation." She first turns to famous mystics—Meister Eckhart, Jacob Boehme, Emanuel Swedenborg, William Blake—who implore us to go within and evolve our vision. She also mentions the spiritual rebellion of the transcendentalists, as well as Richard Bucke's 1901 *Cosmic Consciousness*, which later influenced 1960s hippie subculture, and William James, who wrote that changing inner attitudes can change outer lives. In the 1920s, Jan Smuts, the prime minister of South Africa, argued for holism and that both matter and mind evolves, and H. G. Wells wrote that "the world is heavy with the promise of greater things." She further mentions Arnold Toynbee, who as early as 1935 predicted that Eastern philosophy would significantly influence the West, Jung and the collective unconscious, Alfred North Whitehead and process philosophy, Buber and the hunger for I-Thou relationship, Ludwig von Bertalanffy and systems theory, and Aldous Huxley's views on the perennial philosophy and his novel, *Island*, which is as utopian as *Brave New World* is dystopian. She then regales Marshall McLuhan and the promise of communication within the global village, Abraham Maslow and peak experiences in support of self-actualization, Colin Wilson on the poetic outsider as the forefront of spiritual evolution, Barbara Marx Hubbard, who would write of conscious evolution, and Roszak on the promise of 1960s counterculture and 1970s human potential. The list continues with Lewis Mumford and Erich Fromm, who tackled the humanization of technology, while Carl Rogers saw an "Emerging Man," and Joseph Campbell honored the hero's journey. Ferguson also included a bit of Bateson, who, she writes, predicted the 1970s would be a decade of new ideas and thinking, similar to the creation of American democracy in the 1800s, although she left out his snowball in hell response to looming ecocrisis.[17]

For Ferguson, all these thinkers, and many more, argued that humans are ripe with potential and new ways of living were not only likely but on the way. Hope was not a false palliative but a response to knowing more. But the most optimistic, and the most influential to her fellow Aquarian conspirators, was Pierre Teilhard de Chardin, the Jesuit priest, philosopher, and paleontologist who wrote *The Phenomenon of Man* and *The Divine Milieu*, which detailed the evolution of the noosphere toward the "Omega Point." The Omega Point represents the pull of worldwide enlightenment that draws humanity forward, although, like Prigogine's dissipative structures, the new overtaking the old is disruptive and disorientating, but out of the chaos will come "not only survival but superlife."[18] Like Mead's inspiring words, Teilhard de Chardin is also known for oft-repeated quotations, including the claims that "we are not human beings having a spiritual experience" but "spiritual beings having a human experience" and "Someday, after mastering the winds, the waves, the tides and gravity, we shall harness for God the energies of love, and then, for a second time in the history of the world, man will have discovered fire."[19] Conspirators took these quotations to heart, despite Teilhard de Chardin being criticized for the certainty of his Omega Point teleology; still, his basic theology disclosed the Divine expressed as transcendent unity and immanent intelligence, life force energy, and fire of love.

Ferguson also spills a lot of ink writing about Thomas Kuhn and paradigm shifts, with *The Aquarian Conspiracy*'s uber-optimism expecting critical mass and tipping points soon; what followed for her was an international bestseller, which in itself suggests the beginning of transformation, or at least the wish for it, although critics pointed out the danger of false assurances that change was inevitable. Her post-book travels led her to find evidence for "the awakening of a more artful humanity," but what also followed was the election of Ronald Reagan and a decade of trickledown economics and greed is good profiteering. The seesaw of good news, bad news led her to be "skeptimistic," a hybrid of skepticism and optimism, while she continued to point toward uplifting storylines, including that "bad news can be good news" and "the evolution of revolution," in which crisis forces us to wake up and get active. She predicted a new world by 2000, but her analysis whiffed on the many blocks to change.[20]

Like Wilber, Ferguson saw a potential leap in consciousness marked by awareness of awareness and stages of growth. To her credit, she explored the influence of individual intention but did not fall for "creating your own reality," calling instead for the sharing of wisdom among conspirators who would collectively act. However, she fell prey to quadrant absolutism, focusing too much on the interior individual quadrant of the developmental holarchy, and thus she did not take into account the power of social structures to contain change. She was also overly optimistic in regard to the global village and emerging communications networks and did not anticipate the internet and the downsides of misinformation, including the endless propagation of conspiracy theories, the culture wars, and blocks to development at the Pluralistic level. She also did not foresee the specter of climate emergency, and that while order emerges out of chaos, too much chaos can lead to wholesale collapse.

The most academic attempt to capture the phenomenon known as New Age is likely Wouter J. Hanegraaff's *New Age Religion and Western Culture*, although he also admits to the difficulty of defining the term.[21] Hanegraaff focuses on twenty years of New Age influence from 1975 to 1995—a two-decade heyday that resembled a movement—while tracing its roots to various expressions of esotericism, including the early twentieth-century theosophy of Alice Bailey, who he claims was the first person to use the term in any significant sense. To Hannegraaff, New Ageism is mainly a gnostic response—listen to yourself, listen to intuition—to the limits of the twin poles: religious faith and scientific rationality.[22]

Whatever its historical roots, Zimmerman, Sutcliffe, and Spangler do not dismiss New Age thought and practice; instead, like Ferguson, they see the desire for a new age as a complex response to modernist monologue that traverses past traditions in the valid pursuit of alternative spirituality. However, unfit practices and interpretations must be discerned, especially in regard to solipsism, as well as "spiritual bypassing," a phrase coined by John Welwood, an existential and Buddhist psychologist, to describe focusing on the spiritual at the expense of the material, including our emotional lives.[23] Both lead to a failure to develop a rational and active response for overturning social structures and habits of mind that undermine planetary well-being. In this respect, *The Aquarian Conspiracy* is more New Paradigm than New Age, both in its focus on science and appeal to community, or co-conspirators working together to change the world, but Ferguson also supports research into parapsychology, arguing that the human brain, along with the larger noosphere, is ever-evolving, with telepathy and the like pointing to human potential.

Despite criticism, New Ageism marches on, often under the guise of "Mind, Body, Spirit" or "spiritual but not religious," as well as a new moniker, Ascension. I first heard the term from a mechanical engineer with technical left-brain abilities far beyond my own who also consults with intuitives and is deeply into astrology. Basic Ascension claims include that cosmic energies greatly influence earth-dwellers, and those awake to these energies are transforming themselves and collective consciousness, moving from third dimension, to fourth dimension, and then fifth dimension awareness. For Ascensioners, the state of the world reflects our consciousness, and we need not worry about ecocrisis because enough of us have shifted to a higher consciousness or frequency. The message is not just optimism, but certainty and comfort: don't worry, a shift to a new age is definitely underway, and when (not if) enough of us raise our vibrational level—since everything is energy—Gaia will be able to restore itself and a new world will be born.[24]

Depending on your point of view, Ascension takes the ascent tradition to new heights or new lows by valorizing the unfitting aspects of New Ageism, or some combination. My engineer friend recommended Vidya Frazier's *Ascension: Embracing the Transformation* for an overview of "the Shift," which is explored as if predetermined due to prophecies and the cosmic position of the earth, with otherworldly Ascended Masters and this-worldly

Lightworkers, Wayshowers, and Gatekeepers leading the way. To me, these terms sound like something out of a fantasy novel, but for Ascensioners they denote somewhat differing means for being conduits to higher realms. Frazier details her spiritual experiences and personal struggles, and finding her way from 3D to 4D and 5D consciousness via Ascension literature and practices. She acknowledges climate science and crisis times, stating that it is "natural to conclude that things are definitely getting worse" and that "humanity may be on its way to extinction," but exhorts readers to not only embrace alternative spiritual practices but to go beyond mainstream negative headlines to alternative media and websites, where one may find evidence for "the Shift."[25] However, Frazier does not cite a single source, as she, and others like her, are "called to enter the Unknown and live solely by trust and intuition" and thus are in contact with nonmaterial sources.[26]

One such alternative website comes from Sandra Walter, a self-proclaimed Ascension Guide, Gatekeeper, and Wayshower. Walter, like Frazier, hits on common spiritual themes—unity, higher purpose, and collective awakening—but they are interpreted via astrology, or stargate flows and sun energy (as of January 2020—new readings emerge each month), to claim that a new earth is on its way, causing disorientation or new openings depending on how awake you are. Unlike Wilber, no blocks to growth are considered other than 3D consciousness, but that will be overcome as Walter claims that a new age is accelerating, individually and collectively, leading to healing events on a global scale. To her, old forms are breaking down, while new forms of love, peace, compassion, and unity are emerging as spiritual realms are physicalized. She further claims that all previous Ascension work has led to a big heart shift that is not only changing awareness but DNA and Gaian systems. Partnering with Gaia via a multidimensional self, Walter states we have reached political, economic, and environmental choice-points of death and rebirth. She admits there will be physical death of plants, animals, and humans, but also many positive transformations: mass healing, rethinking of priorities, moving to other realms, intention of choice, Gaia shedding the old, and the breakdown of illusion. We should expect the unexpected, with stargate flows helping us to express new vibrations.[27]

Walter, and Frazier, document the challenges of these transformations for 3Der's and 4Der's as old habits and patterns are torn down. However, given their belief in reincarnation, they argue that we have chosen to be on the planet, and thus we have chosen to have a physical ascension experience with bodily, emotional, mental-egoic, and spiritual patterns transmuted. Walter argues that many will be mournful, weeping for seemingly no purpose and coming apart as slates are wiped clean.[28] In online research, I have seen posts with astrological charts and long lists of ailments—depression, aches and pains, restlessness, and on and on, often with reference to differing chakras—all used as evidence for the reality of emerging Ascension energies rather than this-worldly reasons and maladies, like leading unhealthy lives on an increasingly damaged planet. Walter further claims that awakening may include experiences of euphoria and bliss and visions, which corresponds with Wilber's articulation of subtle states of consciousness and deity mysticism, while surrendering to the surreality of a new awareness: we are one being guided by our individual souls with energy gateways providing support in the Ascension process. Not surprisingly, she states that we receive what we are prepared to receive, our reactions creating reality.[29]

Other prominent Ascension proponents include Esther Hicks, or rather, Abraham-Hicks, as she claims to reflect, not channel, Abraham, or "group consciousness from the non-physical dimension," and Matt Kahn, who claims to have direct experience with Ascended Masters and Archangels.[30] Both Hicks and Kahn have websites, write books, and give numerous talks, which can be found on YouTube, and both provide plenty of basic spiritual advice on love and joy and listening within and following your path.

Hicks is known for her create-your-own-reality "law of attraction" advice, or "anything you can imagine is yours to be or do or have," often going off on long Abraham soliloquies on how we are responsible for our thoughts and therefore our reality, and this includes collective thoughts when natural disasters strike, as she suggests that low vibration people may have to die to learn their lessons.[31] Kahn's mantra is "whatever arises, love that," which is good positive psychology advice for creating inner space to explore emotions rather than merely reacting emotionally, although, to me, he tends to speak with an "I know more than you" smile-smirk, while sometimes following up good advice with messages from the Pleaides, a cluster of stars where Pleaidians supposedly created an enlightened civilization. Kahn tells his audience to just feel, rather than think, while he conveys insights about and from these perfect beings.[32] The certainty of Kahn is startling—he also claims heaven is a real place, as he visited when he was six—although consistent with New Age channelers and message-bearers.[33]

Both Hicks and Kahn are popular—he has over sixteen million YouTube subscribers—although Kahn utilizes Ascension language more than Hicks, continuously advocating for advancing to 5D consciousness, while Hicks exhorts us to embrace the energy vortex that is reality. Both claim that "the Shift" is happening and we must raise our vibrations, to which the law of attraction responds.

As we've seen, New Age ideas have been around for a long time, and Wilber mines the deep history of mysticism, but the "Fifth Dimension" initially makes me think of the late 1960s musical group and their recording of "The Age of Aquarius/Let the Sunshine In" from the musical turned movie *Hair*, and then John Hick's *The Fifth Dimension: An Exploration of the*

Spiritual Realm, who, unlike Hicks and other Ascensioners, utilizes many critical sources—he is a renowned professor of religion—to explore the fifth dimension of our nature. Based on religious traditions, Hick writes that we may be "continuous with, or akin to or in tune with, the ultimate reality that underlies, interpenetrates, and transcends the physical universe"—in other words, the sympathy with intelligence advocated by Thoreau and the ascent tradition articulated by Wilber.[34] For Hick, then, 5D awareness would be living in harmony with this 5D reality, of which we are whole/parts or holons; for New Agers/Ascensioners, 5D means a wide range of practices, often astrologically based, including the 1987 "Harmonic Convergence," a global meditation event on the date when the planets were supposedly aligned, which was the brainchild of Jose Arguelles, a New Age author and artist who also had a Ph.D. in art history and taught at several colleges.[35] A more recent Harmonic Convergence occurred in July 2020, with ten days of global online meditations and workshops focused on helping humanity connect with all life and embrace our role as planetary stewards, which included a "global meditation to invite peaceful extraterrestrials to show themselves with thousands of lightships across the sky."[36]

Before continuing, and considering what is fitting and unfitting, I want to be clear that exploring spiritual communication, and thus transrational experiences, or at least being open to mystery and possibility, is part and parcel of taking stage-steps. However, if you are exploring and open, you are often stuck between a rock and a hard place: those at Magic or Mythic, or situationally interpreting from those levels, often too quickly embrace the nonrational, and those at Rational are unwilling to explore or be open. When working with Rafael, who was very open and who had experiences I did not, I came to my mantra: listen and hold rather than affirm or deny. I have continued with this attitude for decades. One of the appeals of transversal rationality is its embraces uncertainty: there is room within the in-between for dissonance and learning, and thus instead of falling back to rigid opinion, we may grow forward with unfolding positions. Yet, research on transrational experiences and meditative states has gone on for many decades, as has research into various pathologies. There is unfolding evidence for what is fitting and unfitting, all of which Wilber reviews in his many books.[37]

Wilber has written that nondual experience and having one's center of gravity at Super Integral reflects the most certain of human knowing. I cannot personally speak to the Super Integral part, but the clarity of nondual moments reveals unity-in-diversity as obvious and true, although one can also have that realization through contemplation. And from that realization, and others, some actions are clearly unfit, like creating social systems that treat others like Its. I also listen to what scientists say, who, along with studying brain states of meditators, posit that it is highly likely that there are intelligent civilizations on other planets. Does that mean UFOs are real? My first reaction is to be skeptical, mindful of self-deception, but I find it humbling that cosmologists, who understandably seek certainty, state that what we claim to know only deals with 4 percent of the universe.[38] The rest is dark matter and dark energy and beyond our current ability to explore. And I also love the scene in Tim Burton's *Mars Attack*, in which a large group of Harmonic Convergence-like devotees go to the desert to greet aliens with peace and love, and then get zapped by their ungracious visitors. To me, the scene is funny because it reveals the hubris of certainty informed by what is supposedly spiritual awareness.

As I attempt to discern the unfit in regard to New Ageism and Ascension, then, I do so without claims for certainty, and yet with deep concern that energy is too often directed at inner change rather than outer structural change, or that all we have to do is change our energy, our vibrations, and outer structures will change. Thus, I am also concerned that certainty, especially in regard to claims for spiritual communication and a coming new age, undermines our ability to fully respond to ecocrisis. Of course, what I ultimately argue is that we need to practice an ecology of communication, which, rather than a path to certainty, is an ongoing interpretative structure that allows us to know more and live more in sympathy with intelligence.

The spiritual path has long valued the unseen as much as the seen, and following one's heart, or as Pascal famously wrote, "The heart has its reasons of which reason knows nothing."[39] Yet, Hick, like Wilber, argues that while spiritual experiences may catalyze self-transcendence and aid our ability to live in sympathy, they also have a dark side when poorly interpreted, which, at worst, results in cults, and at best are merely harmless eccentricities.[40] The heart is home to our deepest desires, intuitions, and callings, but also the metaphoric place where multiple modes of knowing come together, including critical thought. To live in sympathy, then, it is wisest to think through the heart rather than only interpret heart through the mind or make heart and mind a false dichotomy. Kahn, and Frazier, and many other New Age and Ascension enthusiasts, advise us to live solely through the heart expressed as feelings—indeed, one cannot be an intuitive and receive messages without getting out of one's head—but despite this being a necessary condition for certain experiences, there is still room for critical interpretation post-experience: were you really in touch with the Pleiadians or was that pure imagination without rational rigor? Without asking such questions, we must wonder if enlightenment traditions have gone off the rails; or rather, if New Agers and Ascensioners have embraced the enlightenment of the east and esoteric west, but often misread them due to not honing practices from the western enlightenment project like critical and scientific thinking. In this respect, New

Paradigm thought at the Pluralistic level is a step forward, but its fittingness must also be questioned.

One of the byproducts of the "create your own reality" ideal is meditation as a form of activism. The Harmonic Convergence event was not only considered global meditation but also collective action meant to bring peace to the planet. Such "world work" is common, with Walter and others conducting weekly global unity meditations aimed at raising planetary vibrations, as well as a global meditation held in April 2020 as a response to coronavirus.[41] Such meditations may relieve individual stress related to illness but are also meant to collectively heal people and planet at a distance. The latter is basically self-fulfilling prophecy writ large, or, as Abraham-Hicks claims, we become what we think, whether negative or positive.[42] However, when I attended an outdoor ecopsychology conference in Northern Arizona in the late 1990s, Greenway illustrated the biases involved with meditation as activism, remarking that frontline activists, based on their frustrating experiences fighting the powers-that-be, often think that the planet is going to hell, while meditators sitting on cushions tend to think all is rosy and well. Balance is obviously needed, but believing we can change the world simply by thinking it, or wishing it, or meditating our collective troubles away is not the same as being active with the activists.

Still, New Age "create your own reality" rhetoric is appealing to many, not only because it counters doom and gloomism but because of the partial truths embedded within a muddy mix of fitting and unfitting responses. We can't magically change the world via magical thinking, but quantum physics has revealed that matter is bound energy, and there is an emerging science of resonance and vibrations, providing some clarity to the mind and matter conundrum. Tam Hunt, who works in a Psychological and Brain Sciences lab with Professor Jonathan Schooler at the University of California, Santa Barbara, writes that they have developed a "resonance theory of consciousness," with resonance another way to say "synchronized vibrations" that are at "the heart of not only human consciousness but also animal consciousness and physical reality more generally." Hunt admits that this "sounds like something the hippies might have dreamed up—it's all vibrations, man!" or, of course, New Agers and Ascensioners. Yet, quantum physics does disclose stationary objects "vibing" at various frequencies, and thus "all things in the universe are constantly in motion" and "all of nature vibrates." Resonance reflects this motion, or the "oscillation between two states." When differing vibrations "sync up," or vibrate at the same frequency, spontaneous self-organization may occur, or order out of chaos.[43]

Much of this exploratory science of resonance matches the rhetoric of New Agers and Ascensioners, although instead of turning to the energy vortex or Pleiadians for answers, Hunt and Schooler, as well as New Paradigmers, point to neuroscience, with differing types of brain waves producing differing experiences, and vice-versa, as well as neurons synchronizing via a kind of communication. Hunt and Schooler make the panpsychist claim that all things are "at least a little conscious," and this consciousness was always present in evolution, not just an evolutionary emergent, yet mind has evolved to be more complex. Like Bateson, Hunt writes of information exchange, but the focus is on resonance and the speed at which vibrations sync up: the speedier the sync up, the more complex and evolved the consciousness. Boulders and sand are slower, humans and other animals are faster.[44] Referencing the work of mathematician Steven Strogatz, who argues that "feats of synchrony" are everywhere occurring, Hunt and Schooler also explore vibrations of groups, like fireflies that flash their lights in sync, although this can be explained via internal biology and automatic response without turning to consciousness.[45] Still, they claim that biological structures with "the right kind of information pathways and processing power" may self-organize and produce "larger-scale conscious entities."[46]

Hunt and Schooler probe fundamental questions of consciousness by exploring the philosophy of mind, biophysics, and neuroscience; in doing so, Hunt remarks that consciousness is best explained via types of vibrations and shared vibrations, which, like quantum physics, seemingly provides scientific support for New Age and Ascension views. However, we need to take critical care here. In *Quantum Questions*, Wilber long ago countered claims that quantum physics proves mysticism by anthologizing the mystical writings of well-known twentieth-century physicists, including Heisenberg, Einstein, and Schrodinger, who knew the difference between their science and direct experiences of unity, with physics focusing on the physiosphere and mysticism on the theosphere. In fact, the physiosphere, despite the uncertainty principle and vibrating energy, is more deterministic than biological life, intersubjective mind, and unfolding Spirit, as much more freedom and creativity are displayed as we move up the holarchy. Quantum physics does not prove the Tao, or God, or Spirit—such a claim is the ultimate reduction—but reveals a cosmos of deeply intertwined vibrating energy. However, as the theory of resonance suggests, differing degrees of consciousness have differing exchange of information and sync up potential, which may be in sympathy with Spirit unfolding

The question, ultimately, centers on the subtlety or density of the communicative system. New Agers and Ascensioners argue that the system is exceedingly subtle, with every thought, every feeling sending vibrations out into the world, not just influencing but literally creating our reality—forces with more density, like matter and materiality, be damned—not to mention the subtle thoughts and feelings of others that have their own degree of co-creating power. Science once again seems to explore similar terrain, this time via chaos theory and the "butterfly effect," in which fluttering wings may influence events at a distance, including hurricanes, which became popularized as the ripple effect. But this hypothesis disclosed the possible dialogic interplay among complexity, chaos, and order, in which small changes in initial conditions may have large aftereffects, not scientific certainty.[47] In contrast, New Agers and Ascensioners claim that reality can be changed with ease due to the power of intention: mind rules over matter, with no action needed other than thought.

Lynne McTaggart, who is usually more New Paradigmer than New Ager, advocates for this power, doing plenty of science research and talking with experts studying paranormal activity, including from the Princeton Engineering Anomalies Research lab (PEAR) and the Institute of Noetic Science (IONS), while also doing experiments herself. In her 2007 *The Intention Experiment*, she begins with some caution, asking whether thought can "nudge things in a particular direction" when "highly motivated and targeted," which is quite different than claiming one can create reality. Her research concludes that we can, from influencing dice throws in the lab to healing at a distance.[48] But in March 2020, she started leading weekly guided meditations via social media in response to coronavirus, the intention directed at creating less fear, but also creating, not just nudging, reality, although toward specific, doable goals, which, she claims, provides focus and increases the likelihood of success. She opened with an intention statement on reducing deaths and transmissions by 10 percent, asking fellow meditators to take thought into their hearts while visualizing particular hot spots and all things being connected. At the end of the meditation, she mentioned the experiments of Transcendental Meditators to reduce crime in specific cities and then the fact that crime did decrease as proof of the power of intention. She did not mention that other factors may have been the real cause of the reduction, or if Covid-19 deaths and transmissions did decrease by 10 percent, this may be due to material rather than spiritual reasons, like wearing masks and physical distancing.[49]

Abraham-Hicks also weighed in on the coronavirus pandemic in mid-March 2020 from, of all places, a cruise ship in the Caribbean. In a workshop she stated that fear was the real virus, and that intention can provide immunity; what we attend to acts as a summons and thus we must focus on stability, optimism, and source energy. For Abraham-Hicks, the communicative energy system is so subtle that the energy vortex or Source immediately answers: "When you ask, it is given." However, she slips in a disclaimer: "Your vibration has to match the vibration of what you seek."[50] In other words, if you don't get what you ask for, it is because you are not asking right, not vibrating correctly. The tautological blame game should be obvious here, or obvious to anyone thinking critically, but other than taking thought to ask for what you want no other kinds of thinking are required; as always, one is instructed to feel instead of think. Interestingly, the claim that "thinking-feeling it equals being it" counters the whole point of meditation and mindfulness: being aware of something without giving our energy to it and thus *not* becoming it, whether bad habits, anger, or panglossian optimism.

Abraham-Hicks' response to coronavirus does not appear to be any different than the response of many evangelical Christians, who insist on going to church despite social distancing protocols, thinking they are safe because they are clothed in Jesus' blood or some other magic-mythic reason. But while rigid belief gets us into a lot of trouble, including wars, it is worth noting the potential positive power of belief, scientifically shown in the placebo effect and by reducing fear-caused stress that impacts our immune system. Mind may well rule over matter to various degrees, depending on circumstances, but there are no guarantees. And humans do indeed have amazing physical and mental potential, whether Win Hof's current ability to withstand extreme cold, although with much training (don't try this at home, as injury or death are also in the realm of the possible), or the "supernormal" feats of yogis of old.[51] Dean Radin, a lead scientist at IONs, explored these yogic abilities, which are documented in the *Yoga Sutras* of Patanjali and other religious sources, stating that "yoga" comes from "yoke," meaning to "combine, connect, or unify," and spiritual practices not only induce enlightened states but may lead to extraordinary abilities. Turning to scientific experimentation rather than "New Age twaddle," he attempts to find support, if not outright proof, for "mystical realities underlying most religions" and what he terms "real magic," including telepathy, clairvoyance, precognition, and psychokinesis.[52]

Despite scientific investigation of extraordinary abilities—Hof is being prodded and pricked—unfitting responses always abound, with pseudo-religious ones amped up during crisis times. Yet claims for spiritual communications are always problematic when interpreted at magic-mythic levels, or when reflecting a strange mix of insight and poor thinking, or not thinking. Also, useful spiritual insights, whether experienced or taught by others, may lull as much as awaken. In other words, one helpful comment by a teacher does not mean that the next one will be equally profound. In their YouTube videos, Abraham-Hicks and Kahn go from good advice to troubling comments in a matter of moments. Kahn claims that getting what we ask for depends on being aligned with God, and that ego gets in the way of living God's will. The second part is a common insight, but also his way of guarding against complaints that one is not getting results: not receiving what you ask for is due to too much ego. He then follows with the most egoic claim imaginable: "The depth to which I have surrendered to the universe has the power and potential to save the entire planet."[53] However, he later suggests that he might not be the one saying this, as he is channeling heaven and "did not know that this was going to come through." Audience members lulled by the good advice seem to raptly accept it all, with Kahn smirk-smilingly amazed by the profundity of the words that have come through, asking everyone to pause and soak it in.[54]

The appeals to health and extraordinary abilities are further reasons why New Age and Ascension claims find followers. It is no surprise that Abraham-Hicks and Kahn have both published with Hay House, started by Louise Hay, the bestselling author of *You Can Heal Your Life* who has been called the queen of the New Age. After studying Religious Science and New Thought literature, Hay claims to have rid herself of cancer through a mix of forgiving thoughts, nutrition, and alternate healing modalities. In her book, she made a list of ailments and probable causes, which are all rooted in thought, or the wrong thought: symptoms are metaphysical more than physical. Hay ended up working with AIDS sufferers in the 1980s, using affirmations to give them peace more than healing while adding to her fame, and then started her publishing empire, with Deepak Chopra, Wayne Dyer, and Marianne Williamson included as clients, becoming a master, as journalist Mark Oppenheimer puts it, of addressing "spiritual hunger, a symptom of modernity that, along with oil and war and sex, maybe one of the best business models of all."[55]

Wilber, who nonreaders of his books may think is the king of the New Age, went from early sympathy with the movement to being a longtime critic, not only because of pre-trans fallacies but because his wife Treya was diagnosed with cancer and then bombarded with New Age guilt trips: since thought creates reality, then she must have created her our own illness. But, for Wilber, this view has little to do with mystical traditions and much to do with narcissism and hyper-individualism.[56] Treya, after much research and experience, posited that cancer is likely 30 percent genetic and 55 percent environmental, which leaves 15 percent for the nonphysical.[57] She was open to and explored positive thinking and alternative healing modalities, recognizing that medical science, or science generally, is not enough—the same conclusion that the great physicists came to in their mystical writings—but Wilber argues that science, or critical rationality generally, should strip away pre-rational misinterpretation, creating space for transrational insights of the higher stages of development. Treya, who died in 1989, suffered with guilt before stating that those playing the blame game "don't know what they are talking about" and are basically trying to sell something.[58]

That mind and body are connected and influence healing is obvious, but health has many sources, and in a universe that is both spiritual and material, subtle and dense, we can't always get what we want. Still, we also exist within larger mind and shared life force energy; Reiki, for example, has become increasingly used in hospitals in conjunction with Western-style medicine, with some practitioners working at a distance. Larry Dossey, an internal medicine physician and former chief-of-staff of Medical City Dallas Hospital, has written numerous books exploring the interconnecting energy of nonlocal mind (including for Hay House), referencing Emerson's Over-Soul and Schroedinger's claim that "the overall number of minds is just one," to argue for the influence of prayer at a distance in the healing process.[59] This is, of course, the same as McTaggart's power of intention, but the message from Hay and Abraham-Hicks is that illness is your fault, and thus we get what we deserve. Hay did wonderful work during the AIDS crisis, creating understanding within families and even helping patients to let go of guilt and presiding over funerals when traditional religious leaders refused care and concern. But while she hedged about whether AIDS patients could create healing via thought, she claimed that wrong thought could cause it. And she also stated that the root cause of the Holocaust may have been karmic retribution for behavior in past lives.[60]

Those in need of healing are often desperate, and that makes them susceptible to self-blame, and thus self-deception, suggestion, and charlatans. John of God, the so-called Brazilian psychic surgeon who not only sells healing but an array of products, has also been accused of sexual assault by over 600 women from ages nine to sixty-seven and arrested for rape. To the degree that anyone was healed after being in John of God's presence—or rather, the Entity he supposedly channels—the actual source could easily be the placebo effect, with one's own mind influencing healing rather than John. This example once again attests to the power of belief, as well as the dangers of belief without critical thought. John of God's reputation as a healer was enhanced after appearing on *Oprah* and promoted by marketing advisors, and the economy of the small town in Brazil where he resides is largely based on travelers feeling they must believe to find some cure for their ailments.[61]

But people are not only desperate for health, they are desperate for wealth. New Thought has a long history of connecting positive thinking to money, including Napoleon Hill's *Think and Grow Rich* and Norman Vincent Peale's *The Power of Positive Thinking*. Rhonda Byrne took this claim to new extremes in her 2006 film turned into a book, *The Secret*. The secret, of course, is that we can create our reality, and that includes cash and cars or any material thing we want. As always, what is lacking is any connection between thought and action and that you need to follow up your thoughts by actually doing something. And, as always, if you don't get what you want, you are not doing it right. Another partial truth is abused here: we live in a giving universe because we live in an interdependent universe, but all that exchange of information does not guarantee material wealth, it guarantees learning and growth to the degree we embrace all levels, lines, and states of the developmental holarchy.

Like any *Music Man* huckster knows, a surefire way to get rich is selling promises and dreams. Byrnes' book has sold over 30 million copies, and the many talking heads in the film also have something to sell, whether books, or workshops or themselves, and when they get rich they use that as proof that the dream they are selling is true. One of my students, a Business major, fell for the dream, disappearing for a month and only handing in one assignment all semester long, and then showing up on the last day without his final paper. After class I asked him why he bothered to come, since he didn't have his paper. He struggled to answer at first, and then admitted that he watched *The Secret* the previous night. I knew where he was heading—he thought he could think himself an A without actually thinking about course content, or acting by handing in work. Like Abraham-Hicks and Kahn, *The Secret* has an answer for everything. When bad things happen, you need to stop blocking your good vibrations with critical thinking: just express your heart, just feel, and all will be yours. No questions, no mystery, all certainty, all the time.

For Dossey, the point is not to chase "One Mind" experiences but to use them to better enact fitting responses, as the overall message of One Mind awareness is a hopeful golden rule—"Be kind to others, because in some sense they are you"—which supports collective thinking and action in response to crisis, but also unanticipated epiphanies emerging from the "informational dimension" of nonlocal Mind.[62] Dossey's version of mind is different than Bateson's, which is far more grounded and material, although both argue that solutions arise when we open ourselves to the intelligence of the larger communicative system. Dossey further argues that we can dip into this intelligence as needed, or when compelled by crisis to do so, and it may respond to intention, like stem cells transforming into specialized cells depending on the body's needs.[63] The keyword here is "may," yet he writes that after instances of wishful thinking and fraud are dismissed there remains "a substantial number of really puzzling and carefully researched cases," including NDEs, reincarnation, remote viewing, telepathy, and communication with the dead, which suggest the existence of a One Mind repository.[64]

Put it all together—an optimistic salve to doom and gloomism (which may degrade into panglossianism), the allure of partial truths (which typically results in treating the partial as the whole truth, leaving us with falsehoods), the superpower of self-healing (which may not work, leading us to play the blame game), and "show me the money" zeal (which also likely won't work, and even if it did, or appears to, we probably have our priorities wrong), and you've got a persuasive stew of New Age and Ascension create your own reality appeals. A more hopeful takeaway, however, is that many are not only seeking health, wealth, and superpowers, but respite from a screwed up, eco-social crisis world, which both fundamentalist religion and science do not provide. And the turn to spirituality does provide promise, especially if partial truths are listened to, and woven, not into a set-in-stone map leading to certain treasure, but an unfolding stepping-stone tapestry of human potential and possibility that transcends but includes, always includes, critical rationality.

Fortunately, there are many transcend and include spiritual voices out there, including Elizabeth Lesser in *The New American Spirituality*, a guidebook similar to *The Aquarian Conspiracy* in its comprehensiveness. *The New American Spirituality* was published in 1999, when Ferguson believed that we would all be partying, not, like Prince, because everyone has a bomb and its judgment day, but because we had finally woken up and implemented the shift. Turns out that Prince's lyrics were closer to the truth, and yet instead of end times we saunter on, in need of more help, which Lesser, a co-founder of the Omega Institute (notice the nod to Teilhard de Chardin), a spirituality and wellness learning center, attempts to provide. Along with a clear articulation of many manifestations of the spiritual search, including some history, a bit of Wilber, and the landscapes of body, mind, heart, and soul, what makes this book stand out is a list of shadows to be wary of. She even admits to Omega's role in sometimes providing access to charlatans, or the unenlightened acting as if they are enlightened, advising readers to not trust teachers who claim perfection.[65]

Such claims for spiritual perfection is an all too common guru trap, leading Lesser to share Chogyam Trungpa's criticisms of what he called "spiritual materialism." Trungpa, a Tibetan Buddhist who taught for many years in the West, warned students about the self-deception that blocks progress toward enlightenment, or the tendency to play an ego-driven game, spiritual materialism, in which they project a spiritual self rather than living "the self" as transitory and fluid. The ego-self imitates being spiritual, creating dense armor preventing true transformation. Ironically, the distortion of spiritual techniques leads to the strengthening of egocentricity.[66] An important insight, yet Trungpa's teaching methods earned him the title, "the roaring tiger of crazy wisdom." No student would mistake him for a saintly guru, as he claimed to come from a "long line of eccentric Buddhists." Such eccentricities may disclose lessons on lightheartedness, and they likely did for Trungpa. But these eccentricities also included giving teachings reeking of alcohol, surrounding himself with bodyguards, and dressing in military garb of his own design.[67] And despite stating that the guru is like a spiritual friend, he also argued for hierarchical guru-student relationships because it taught devotion.[68] Such devotion greatly enhanced his sex life—being a guru is a super way to get laid—but he died in 1987 from complications associated with alcoholism.[69]

Trungpa's accomplishments are many. Along with writing numerous books introducing the east to the west, he also established over 100 meditation centers in Europe, Canada, and the United States, and was the founder of the Naropa Institute in Colorado, a spiritually and ecologically based university.[70] The irony here is immense, especially since east meets west is supposed to be a harbinger of a new age, but perhaps the real problem is the tendency for students to expect a teacher not to be human. Alan Watts, another heralded spiritual advocate who demonstrated his awareness through many accomplishments, also died from alcohol-related causes. Watts became addicted to vodka because it made him feel less alone, which is astounding considering that he was a meditation teacher, and his many books and talks point toward transcending the separate self-sense and its attendant loneliness and alienation. Watts at least recognized his folly, joking that he was an "authentic fake."[71]

Lesser argues that lurking amid the spiritual search is the problem of when and how to listen to spiritual teachers, who, traditionally, are needed for enlightenment; but, wisely, this tradition has been critically questioned to discern the unfit.[72] Unfortunately, such questioning did not keep the Omega Institute from hosting visits from John of God, although they have put out a statement that no abuse was ever reported at their site, and that the organization stands with the victims.[73]

New Ageism would seem to be a fitting response to the guru trap, since it is rooted in the direct path of listening to spiritual experiences, intuitions, and our higher self; however, New Agers often give authority to nonphysical others like angels and ascended masters. Also, New Agers are not immune to following flesh and blood huckster-gurus. Near the New Age mecca of Sedona, Arizona, about an hour down the road from where Greenway spoke of the problematic division between meditators and activists, James Arthur Ray, a New Age seminar leader and author of the New York Times bestseller, *Harmonic Wealth: The Secret of Attracting the Life You Want,* who was featured in *The Secret*, held a "Spiritual Warrior" retreat in 2009 where three people died and eighteen were hospitalized during an utterly inauthentic and appropriated sweat lodge event. Participants in the workshop paid $10,000 to attend, allowing Ray to attract the life he wanted, although he fled the scene, because he was scared and ended up in jail for negligent homicide.[74]

Ray's folly is another New Age fable filled with lessons, yet those who claim to be on the path to 5D consciousness, or to have attained it, may not have ears to hear. We do not need faux-spiritual warriors attempting to prove themselves inside makeshift plastic sweat lodges; we do need Eco-Sage-inspired warriors who integrate insights from meditation with on-the-ground activism. But some New Age and much Ascension literature simply admonish us to raise our vibration level as the fitting response. For Ramsey, the "coming through" of what is most fitting is thoroughly material, in which we respond to a range of circumstances, from the mundane to the world-historical, while I have added the spiritual via Taoist wu-wei or effortless effort. New Agers and Ascensioners go still further into intention as action and channeling but we can't solve eco-social crises solely with our thoughts or poorly interpreted experiences of what seems to be spiritual communication.

Synching up sounds pretty good, though, pretty radical, especially if it reflects a spiritual and material coming through, and thus a four-quadrant approach, integrating transcendent and immanent, and embracing an ecology of communication. Maybe, just maybe, right intention plus action will lead to tipping points and transformation of minds and societies. Coronavirus, like other world-historical events like Pearl Harbor, showed that change can happen fast in response to crisis—reordering out of chaos—but we must ask if changes will be positive, negative, or a mix of both, and whether they will they last. They will likely only last if social structures change. Order out of chaos in the physiosphere and biosphere is quite different than order out of chaos within the noosphere, within the realm of socio-cultural systems where the well-being of the earth meets the promise of intersubjective communication and interpretation.

THE DREAM TEAM: SPIRITUALISM, TELEPATHY, ANGELS, NDEs, PRECOGNITION, CHANNELING, AND SYNCHRONICITIES

Transformation expressed as stage-steps demands that we question the quest for a new age, discerning the unfit by separating the wheat from the chaff. Despite a lack of collective direct action—unless you count group and global meditation—the New Age and Ascension desire for individual direct experience has merit. The "wheat" is self-work that challenges organized religion's hierarchy of official intermediaries communicating to the faithful messages of proper belief and behavior. The "chaff" is taking it way too far. Peters, however, cautions against the assumption that any experience can be unmediated, such as the dream desire for oneness. In doing so, he explores the age-old roots of the New Age fascination with angels and communication with the dead.

In Peters' history of the "dream of communication" Socrates' coupling of consciousness via eros was followed by Saint Augustine's privileging of transcendent communicative ideals over immanent realities like language, writing, and the body. Augustine suggested that God speaks directly to the angels, and if we are open, God will speak to us as well, transcending "the mediation of physical signs" such that messages may be "perceived (not heard) within."[75] But for Peters, angels are much more that immaterial entities, they represent the mistaken desire for perfect communication, providing "a lasting vision of the ideal speech situation, one without distortion or interference."[76] Angels are considered messengers without limitations, whether due to body or distance, and thus they "effortlessly couple the psychical and the physical, the signified and the signifier, the divine and the human."[77]

Peters further catalogues the dream by exploring nineteenth-century spiritualism, which began with Dr. Franz Anton Mesmer, a "grandiose if sincere quack" who taught that everything is connected via "universally diffused fluid." Sickness occurred when the invisible fluid, acting like a magnetic or gravitational force keeping "souls in love and health and communication," became blocked. For health to be restored, people needed to be "en rapport," or in dialogic contact with others via the universal fluid. Despite the extremely odd focus on fluids, his theory suggests respect for and responsiveness to a universal life force coursing through all things. But Peters points out the dangers of being too open to such suggestion. We take from Mesmer the terms mesmerism and mesmerize, which leads to a "nightmare loss of self to another's will."[78] We don't become like angels in rapt rapport with each other and the cosmos; we become like zombies or Nazis.

Spiritualism, however, did not focus on perfect communication with alive others but on communicating with the dead. And instead of direct contact, a new intermediary emerged: the professional medium. Like New Age channelers, mediums claim to be conduits conveying messages from far off places, spaces, and times. Interestingly, this communication was not always understandable, with channeled utterances and table rappings needing to be interpreted. Peters writes that the spiritualist movement "has always explored the troubles and utopias of communication across the gaps."[79] He also writes that Emerson discounted Dr. Mesmer's theory as an authentic expression of the Over-Soul, and that Thoreau quipped that "if spirit rappers gave a reliable picture of eternity, he'd gladly trade in his prospects for immortality for a glass of cold beer."[80] Thus, Emerson wisely questions methods of contact and connection, while Thoreau humorously questions if the dead have anything worth saying.

Peters' brief turn to the history of angels and communication with the dead discloses vast potential for misinterpretation. Yet, despite criticism in the nineteenth century, the desire for perfect communication increased in the twentieth along with the advance of New Ageism, and then into the twenty-first with the addition of Ascension. Peters' skepticism is much needed, then, to discern when differing forms of spiritual communication are a distracting dream or if they may be deemed spiritual communication at all, and when they gift insights that counter modernist monologue and eco-destruction.

Rupert Sheldrake, a Ph.D. biochemist from Cambridge University and prominent New Paradigmer, has a lot of experience taking on skeptics of the paranormal, which he calls psi, short for psychic investigation, because he considers such experience to be natural rather than supernatural and worthy of scientific inquiry.[81] Sheldrake explores telepathy in *Dogs That Know When Their Owners Are Coming Home* and his follow-up *The Sense of Being Stared At*, citing numerous examples and arguing that personal experiences are a valid measure. Peters, of course, considers telepathic claims a product of the dream of communication, and dangerous due to their impossible expectations, but Sheldrake's examples suggest its possibility, as well as benefits, since, for him, such dialogic experiences undermine the illusion of rigid separation and thus the downsides of modernist monologue.

One example includes a specific word in rural Norway, "vardoger," used when someone believes they heard someone arrive, and then soon after they actually do.[82] However, he also cites a long history of lab experimentation, including dream telepathy, in which "senders" attempted to influence the dream content of subjects by concentrating on images at a distance, and conducts his own experiments, adding rational support to a voluminous amount of anecdotal stories.[83] Lab experiments, despite artificiality, consistently show statistically significant results beyond what one would randomly expect, with participants having an emotional bond displaying even higher results.[84] This is also true for presentiment, with friends and family having higher instances of thinking of someone just before receiving a phone call, or these days an email or text, which, like feeling you are being stared at and then turning around and finding you are right, is common for most everyone.[85]

Skeptics point out the fallibility of experiments and the lack of 100 percent certainty, claiming that anything extrasensory demands extra evidence, which, for Sheldrake, is just a way to "move the scientific goalposts."[86] In *Science Set Free*, he argues there is a larger issue here, mechanism, or the Cartesian world view, which still predominates and keeps so-called orthodox scientists from accepting new evidence. For Jeffrey J. Kripal, a Professor of Comparative Religion, the evidence is overwhelming, and thus scientific materialism must be transcended and included, with larger mind or consciousness, rather than the brain, being the basis of reality, and thus the basis of experimentation.

Kripal is active with the Esalen Institute—Esalen marches on—and has written broadly on the history of parapsychology, as well as specifically on near-death experiences. In *Changed in a Flash*, which is coauthored with Elizabeth G. Krohn, who was struck by lightning and experienced an NDE that caused her to leave her body for what felt like a two-week conversation with larger forces but was only a couple minutes in "normal" linear time, he does an extensive interpretation of this event and others like it. This interpretation includes the necessary admission that Krohn's experience was colored by personal history and culture, including her Jewish upbringing, and is ultimately ineffable; and yet, he argues NDEs should not be dismissed as mere imagination but, like dreams, reflects the use of symbolic communication to deliver messages and information, in this case about the afterlife, and thus reality as a whole. In other words, instead of NDEs reflecting the dream of communication, they reflect dreams and the productive use of imagination as forms of communication. Krohn's NDE occurred within what she described as "the Garden," and included a "beckoning glow," a presence or guide, unconditional love, linear time as an illusion, reincarnation as fact, and instant communication—any question she asked was immediately answered.[87]

For Kripal, Elizabeth's NDE, and so many others, which are being studied at the Noetic Institute but also the Department of Perceptual Studies (DOPs) at the University of Virginia, does not fit into our dominant materialist world view, and thus this world view is clearly unfitting and must be expanded to include the primacy of consciousness, and the in-between of objective and subjective inquiry.[88] Such openness leads to considering the possibility of time loops and precognition, which Krohn experienced after recovering from the lighting strike, most often in dreams, which she documents by sending dated emails in the middle of the night. It also leads to the possibility of guiding spirits, whether interpreted as angels or the Greek daemon or some other presence, as aspects of our larger self that act as an intermediary between the divine and mortal, or between worlds. For Kripal, the soul likely has a double nature, or what he calls the "Human as Two," which enhances the ability to "know things outside or before the ordinary senses."[89]

Kripal celebrates the religious imagination, arguing that it is not counter to science but adds to it—after all, NDEs and psi phenomenon, while not reliant on science for legitimacy, are being legitimized through scientific studies—while pointing to the long history of religious traditions and experiences that corroborate Elizabeth's narrative. He also has nice things to say about Mesmer, who apparently conducted "astonishing cures and healings," and he celebrates the imagination of science fiction novels and movies, which point toward our human potential, and yes, the possibility of a new age of new possibilities. The X-Men movies, for example, disclose the future evolution of consciousness in the form of mutants with super normal, not supernatural, abilities.[90] Again, Kripal foregrounds the reality of differing interpretations; he just does not want to close them down such that we do not include NDEs, precognition, and other psi phenomenon.

Not surprisingly, Kripal's writings do not endear him to more traditional scholars of religion, just like Sheldrake's writings do not endear him to many of his scientific colleagues. Yet they both march on, with Kripal arguing that more scientists are becoming converts, having "flipped" due to having psi or mystical experiences that cannot be easily explained away.[91] And with Sheldrake creating his theory of "morphogenetic fields" or the "invisible blueprints that underlie the form of the growing organism."[92]

Sheldrake argues that morphogenetic fields act like collective patterns or habits that evolve and influence the responses of species. These nonlocal, intercommunicating fields, which provide structure to Dossey's amorphous conception of "One Mind," may account for exchange of information and learning beyond what can be explained via materialist means. Sheldrake gives an example of horses being cut and mutilated when first introduced to barbed wire, but he states that later generations learned to avoid the barbs via the silent communication of morphic resonance. In other words, he speculates that it was easier for later horses to learn what earlier horses had learned the hard way because their habits changed and then were communicated via invisible morphogenetic fields.[93] Sheldrake's theory is hardly accepted but criticism does not deter, since, to him, the Cartesian paradigm prevents critics from considering the possibility of morphic resonance, or telepathy and other psi phenomenon, and such outright dismissal goes against the spirit of science.

Sheldrake is a sophisticated proponent of telepathic communication, which, for him, is rooted in our biological nature and survival reaching back to hunter-gatherer lifeways. He mentions Laurens van der Post, who studied the Bushmen of the Kalahari Desert in southern Africa and stated that telepathy was an accepted part of their culture, with success in the hunt often known in advance of returning home: celebrations were prepared and songs were already sung for the specific species killed. Unfortunately, similar to Bateson arguing that most anthropologists did not acknowledge right-brain phenomenon when studying indigenous societies, Sheldrake writes that most did not pay attention to experiences of telepathy.[94] However, Bateson made a distinction between hunters who performed ritual imitations of an animal to better understand them and improve their skill and those that were thought to telepathically draw prey to their net, which he considered misguided magical thinking.[95]

Like all New Paradigmers, Sheldrake is influenced by quantum physics, arguing that "fields and energy are now more fundamental than matter" and that telepathy may be similar to quantum entanglement, in which particles that move apart remain connected and change with each other regardless of distance.[96] Einstein famously called such entanglement "spooky action at a distance," but, spooky or not, interconnectedness is a proven scientific reality. Sheldrake, of course, goes further, adding his voice to the argument that mind is not confined to the brain, which is necessary to explain phenomenon like telepathy, precognition, and the sense of being stared at, as well as the power of intention, with morphic fields "linking us to the objects of our perception, and capable of affecting them through our intention and attention."[97] New Agers read this as proof of our ability to create reality, but Sheldrake is merely attempting to theorize reasons for why people have psi experiences. And he states we are capable of *affecting* the objects of our perception, not creating our reality, which, once again, is a difference that makes a considerable difference.

Sheldrake also delves into the experience of angels, arguing that New Paradigm science reveals that we are living in an energized, entangled, and alive quantum world that is open to the possibility of intelligent entities rather than a dead mechanized one that is not. In *The Physics of Angels,* Sheldrake dialogues with the Creation Spirituality theologian Matthew Fox, who was kicked out of the Catholic Church because of non-traditional views, on the existence of angels, and they speculate that there are startling similarities between angelology and quantum physics, particularly Einstein's descriptions of photons of light as massless phenomenon that can only be detected by their action.[98] For them, the theory of morphogenetic fields also suggests that intelligence extends throughout the universe, lending support to the possibility of angels as objective realities.[99]

Like Kripal, Sheldrake and Fox argue that there is a long tradition of angels, existing in nearly all cultures, as telepathically communicating intermediaries between the Divine and the mortal in a holarchically ordered cosmos. It is only during the scientific revolution that angels have been excommunicated, considered an embarrassment to the rational mind. But, for Sheldrake and Fox, this privileges an impoverished scientific imagination over an enriched religious one, while also neglecting a possible response to ecocrisis: "Angels . . . may well prove to be indispensable allies, truly guardian angels, instructing us in safeguarding our inheritance of a once healthy but today endangered planet."[100] This is quite a claim, but once they start down the path of legitimizing message-bearing angels via quantum physics, they assume they have something worthwhile to say. It makes sense to them that angels, as intelligent spiritual beings or entities, would be awake to the problems of ecocrisis. Of course, another interpretation is Kripal's "Human as Two" theory, in which there is not a rigid separation between our mortal and angelic selves, and when we listen to our higher nature, we will be more protective of living nature.

Sheldrake and Fox also argue that angels are similar to spirits in the shamanic tradition. Roger Walsh, a scholar and psychologist, takes up this theme in *The Spirit of Shamanism*, using Wilber's descriptions of subtle state experiences to make sense of shamanistic practices, including journeying to the spirit world. Walsh writes that shamanic journeying is characterized by visions of "archetypal spirit figures" and that shamans were likely "the earliest masters of this subtle realm," soul traveling beyond the body and "bringing back information and power from this realm to their earthbound inhabitants."[101] Walsh also explores journeying as a form of channeling, as the shaman typically speaks from the perspective of a power animal. He is well aware of the possibilities for dissociation and self-deception, but remarks that a number of positive "psychological mechanisms may be involved," such as acting as if one is empowered or has a desired quality, role-playing, belief, and identification.[102]

Like the history of angels, Walsh argues there is a long history of channeling from differing traditions, including the Greek Oracle of Delphi and Tibetan deity yoga in which the yogi merges with a "godlike figure who embodies virtue upon virtue" and then "attempts to move, speak, and act as the deity."[103] Walsh, and Wilber, also testify to the mystical profundity of *The Course in Miracles*, a 1200-page, 3-volume text channeled by Helen Schucman, a "reluctant and astounded Jewish professor of psychology at Columbia Medical School."[104] Schucman experienced clairaudient dictation from within, in which "the Voice," which she claimed was Christ, would not leave her alone. However, some, including Wilber, came to doubt whether the source was anyone other than Helen, or her higher Christ consciousness self, as she was "mystically inclined" as a child, asking for miracles, grew up in a household in which she was exposed to metaphysical ideas, and her own poetry is very similar to the *Course*.[105]

Walsh attempts to make sense of channeling by questioning whether such communication emerges from subpersonalities within the unconscious or from nonmaterial entities separate from the self. A psychological view would suggest subpersonalities, or a higher self, as there is much that lies in the unconscious beyond conscious awareness. Jung, for example, had a wise old man dream in which Philemon appeared, who he considered to be a fantasy figure yet real enough to converse with and learn from, as he represented superior knowledge. The religious view would suggest the possibility of transcendent beings, whether experienced within ourselves or as outer entities, including "the shaman's 'helping spirits,' the Hindu's 'inner guru,' the Quaker's 'still small voice within,' the Naskapi Indian's 'great man,' or the Christian's 'Holy Spirit.'"[106] To this list we must add the Greek daemon, or guiding intelligence, to which Socrates turned when making decisions. Both the psychological and religious views may be true, along with immense potential for self-deception, pathologies, pre-trans fallacies, anti-rationality, and a faulty, rather than enriched, religious imagination.

Wilber argues that subtle state experiences, when dissociated from the immanent earth, may result in "exaggerated visionary awareness" and "contact with ultraimaginative" so-called "other worlds," "alien beings," "masters," and "guides," leading to rejection of conventional life and embrace of "my kingdom is not of this world."[107] As we saw, Wilber considers such beings to be experiences of archetypal involutionary first forms that are then interpreted as ontologically separate and real from magic-mythic levels. Experiences of such luminous, creative, and vital forms is often so overwhelming that one may consider them higher beings, and get lost in higher worlds, when one is actually in contact with their own higher self.[108] Channeling, then, may get the ego out of the way, allowing higher or unconscious intelligence to come through. When studying with Rafael, for example, I meditated on the "I" that is all things, placing my personal sense of "I" within the larger context of the All. But, of course, channeling may not get the ego out of the way at all, but put it front and center. The quality of channeled material, then, depends on what level of the developmental holarchy one is expressing, or "channeling," when we have subtle state experiences. Which is why Walsh remarks that the "range of quality of channeled material is enormous," while quoting Wilber's summation: "Higher intelligences have got to be smarter than the drivel most of these channels bring through."[109]

Jon Klimo, a longtime Psychology professor, writes that channeling has a long religious history as revelation, including Moses on Mount Sinai and many other biblical prophets, and Mohammad, who channeled the Koran as transmitted from Allah through the angel Gabriel. Klimo breaks down the claims of possible channeled sources into eight categories: our higher self (or intelligence); the universal mind and the collective unconscious; group beings; Jesus and other ascended masters; nonhumans including angels, devas, elementals, plants, and animals; extraterrestrials; discarnate human spirits; and God, or gods and goddesses.[110] That's quite a range, with our higher self and the unconscious being the most easily believed, and more skepticism arising as the source moves outward, beyond the self to transcendent sources. Again, Wilber states that intelligence pervades the Kosmos and may be personalized in productive ways, although dysfunctions always lurk. Of course, many worshippers, while not channeling, claim to use prayer to converse with God, but, to be productive, such converse should put us in sympathy with intelligence within and without. Neale Donald Walsh claims to have recorded such converse in his bestselling *Conversations with God* books; like Schulman, he states that the dictation happened to him, although he is sure of the source, God, who responded to his questions in words that he and everyone could understand. To me, that sounds like a bit much, but regardless of the source we may focus on the message, discerning what is fitting and unfitting.[111]

Klimo credits Shirley MacLaine with mainstreaming channeling, and channelers, in her popular 1980's New Age books *Out on a Limb* and *Dancing in the Light*, including J. Z. Knight, who supposedly channels Ramtha, or "the Enlightened One," who in turn claims to have been incarnated 35,000 years ago and whose main message—you guessed it—is that we can create our reality and are headed toward a new age. Knight has made beaucoup bucks as a channeler, conducting well-attended seminars, selling recordings, and starting Ramtha's School of Enlightenment.[112] Three of her students made the 2004 film, *What the Bleep Do We Know?!*, which, like *The Secret*, which appeared two years later, has lot of talking head "experts" telling audiences we can create reality due to the reality of quantum physics. The film wisely questions the degree to which we are slaves to conditioning and habit, but the appeal to quantum physics once again confuses levels of reality, not to mention what the majority of physicists say. Which raises a series of questions: Why do New Agers want the ability to create reality to be true and for physics and metaphysics to not only play nicely together but be the same, and why do they keep turning to science rather than the long history of contemplation and spiritual practices that lessens the influence of the isolated ego and increases compassion and care?

Quantum physics is indeed suggestive, challenging dead matter world views and observer and observed separation, but New Agers are clearly seeking legitimacy, which shows up in these films but also New Ageism generally. Science remains a dominant narrative, as it should, and if New Agers can hook their wagon, well, then there is proof for their claims. But harmony among holarchic levels of reality is one thing, false equivalency another. Yet there also appears to be other needs: certainty and control. How can we question a channeled source outside ourselves? They are so ascended, so advanced, what they say certainly must be true. And the desire to create your own reality, which has got to be the most egoic endeavor imaginable, promises control. A kinder interpretation, and a genuinely spiritual one, is that New Agers and Ascensioners desire to be empowered to make an individual difference in a crisis-driven world where we so often feel systematically disempowered. The larger question, then, is how can we take effective action in a post-metaphysical world while discerning unfitting distractions—like the illusions of certainty and control—which just become more blocks.

John Hick writes that the basic criterion for the authenticity of religious experience is transformation: we cannot describe experiences of ultimate Reality, or ultimate Reality itself—both are beyond words and thoughts—but we can describe responses, the effects such experience can have on lives, including being more centered in Spirit and thus more loving, compassionate, and responsible. Great mystics have often been "spiritual warriors" at work in the world. Meister Eckhart took on a variety of organizing and teaching roles as Vicar, Provincial, and Professor, Plotinus was in demand as a guardian and trustee, St. Bernard was a gifted organizer, Catherine of Genoa was a nurse and administered large sums of money, Ignatius Loyola co-founded and directed the Jesuit order, St. Teresa, as we saw, is not only known for visions of angels but her reformation of convents and founding many others despite operating in a male world, St. Juan of the Cross displayed similar practical abilities, and on and on.[113] The Buddha, of course, did not just sit under the Bodhi tree after enlightenment but got up and traveled and taught, inspiring the Bodhisattva vow of listening to the suffering of the world and responding. Jesus traveled and taught that the kingdom of God was within, and also kicked butt when he wasn't seeing it among money-changers in the temple. Within spiritual traditions, you know the degree of enlightenment by the fruits, by what is brought forth, by the fittingness of one's responses. Trungpa and Watts still measure up well by this criteria, given their many contributions, while also falling short in many ways; not surprisingly, their lives, like all of us, reflect an unfolding mix of levels, lines, and states. However, genuine spiritual warriors are consciously moving up the developmental holarchy, still human and fallible but with their center of gravity at second tier, or third, which is then disclosed in their responses and actions.

Hick also explores the dark side of religious experience when poorly interpreted and when delusions are thought to be religious experiences. Crazed cults are many: Reverend Jim Jones and the mass suicide at Guyana, the Order of the Solar Temple's murder-suicides in France and Canada, the Heaven's Gate members who committed suicide thinking they would be beamed up to a spaceship, leaving the rest of us behind to perish rather than create a new age, to name some relatively recent examples.[114] Of course, the great religions also have their cultic dark side, leading to untold violence and making "God is not great" atheists out of many. Again, what is unfitting is not always complex: if you treat others like "its," living a life of most harm rather than least harm, you should reconsider your ethics and actions. The same is true for channeled texts, which may be judged by their validity and what they lead us to do. Some may provide good spiritual advice, which may also be found in non-channeled sources, some may reflect harmless eccentricities, and some may lead to harm. At the least, critical rationality should make us wary of wasting infinite potential quantum energy on fakery and foolishness.

We are all capable of more fitting "coming through" responses to material circumstances, as well as "channeling" the wisdom of the unconscious and our higher self, but the desire for more may end up being another block. If you desire meaning and growth, and have begun to consciously take stage-steps up the developmental holarchy, ascending by transcending and *including* rationality, you are on your way; if you are at Ramtha's School of Enlightenment and find yourself with a head injury after running blindfolded among other blindfolded students because you did not intuit how to avoid each other, perhaps it is time to say "What the BLEEP" and question the quest.

Telepathy, message-bearing angels, and channeling are common forms of communication within the New Age orbit. Such communication is often exceedingly dreamy, but for New Agers, they provide more evidence for an emerging critical mass of transformed consciousness. Sheldrake's morphogenetic fields would obviously aid such transformation, as would synchronicities, which, for Jung, are messages disclosed by connections that do not have a discernible cause and effect relation. His famous example was of a beetle that tapped at his window at the same time a client described a dream featuring a scarab jewel, which, he noticed, mirrored the beetle's colors and patterns. Jung opened the window and the bug flew in, leading him to remark, "Here is your scarab." This synchronicity supported the significance of the dream, which, as always, was open to interpretation, while helping the client overcome rational disbelief and become receptive to the wisdom of the unconscious.[115] The lesson behind this story is also open to interpretation but a possible reading is that hyper-rationality may impede growth, and openness to synchronicities, displayed by Jung's need to help his client and noticing the confluence outside his window, may aid our growth.

Wilber writes that synchronistic experiences are real and rare, but in the hands of New Agers they are considered common, which psychiatrists call "delusion of reference."[116] Synchronicities may well bring unconscious messages to conscious awareness—the inner synching up with the outer—but for those who believe they can create their own reality, habitual misperception may be the actual common occurrence. The inflated ego sees what it wants to see: uncommon synchronicities transfigured into happening-all-the-time meaningful coincidences, making us the grand interpreters of all events, and also the center of all events. Bateson's vision of the pattern that connects and information exchange would seem to support New Age notions of meaningful coincidences—mind is everywhere—but, as we saw, despite living among human potential movement advocates at Esalen, he argued that "receipt of information" occurs via identifiable material pathways, which "rules out such variants of extrasensory perception as telepathy, distance perception, second sight, etc.," as well as "that superstition called 'the inheritance of acquired characteristics,'" or the basic idea behind morphogenetic fields.[117] Bateson also argued that coincidences are natural occurrences that are much more probable than people expect.[118]

Both Bateson and counterculturalists turned New Agers and New Paradigmers seek to re-sacralize the world, providing a more magical and aesthetically alive communicative experience, but, for him, such magic does not include ESP and the like, which work on the user, in the sense that they are being used, and impeded, by irrationality.[119] Characteristics of mind are embedded within matter, but this does not lead to mind over matter abilities or supernatural explanations for natural phenomenon.[120] Bateson is most definitely onboard with experiencing a different reality by changing our perception—after all, his core claim is that ecocrisis is a crisis of perception—but a different reality is not the same as creating reality. To him, we need saner perceptions that recognize ecocrisis, and solutions, by understanding ourselves as co-creating communicating subjects within the unfolding material reality of the pattern that connects.

Interestingly, Bateson describes his reservations in *Angels Fear,* which references the phrase "where angels fear to tread" to acknowledge his trepidation when exploring religious themes. This is obviously a quite different interpretation of angels compared to the embrace of angelic presences by New Agers, Ascensioners, and some New Paradigmers, as well as some religious scholars, although, again, interpretations vary. Thus, despite inspiring New Paradigm science like Sheldrake's morphogenetic fields, Bateson would likely be critical and unconvinced, practicing distanciation more than participation, although remaining open to new evidence, including experiments on the sense of being stared at and dogs who know when their owners are coming home. Wilber, on the other hand, interprets morphogenetic fields as ordering processes that influence upward causation, which may result in new Kosmic grooves or habits, like second and third tier emergent structures that provide guidance while always remaining open to change. However, as Wilber well knows, distanciation is always necessary to evaluate new habits.[121] Morphogenetic fields, for example, have been linked to the 100th monkey story popularized in the late 1960s by Ken Keyes' *The Hundredth Monkey*, in which he conveys a critical mass myth where monkeys on differing islands telepathically learned to wash their sweet potatoes after a hundred learned the trick. There was supposed to be scientific backing to this story, but it turns out that the monkeys learned independently—they didn't like sandy potatoes—and were further taught by their mothers on the two islands.[122]

Many 1960s and 1970s counterculturalists found hope in the 100th monkey myth for the possibility of sudden social change. We may still derive hope from our potential for spiritual growth, although given the potential self-deception associated with nonlocal extended mind communication within the noosphere, like telepathy and synchronicity, as well as subtle state spiritual communication, like converse with angels, perhaps we are better off sticking with the grounded dialogic interplay among mythic-animistic, rational, and aesthetic communication. Bateson's vision may be all we need to take next steps.

That may sound good at first blush, but all holarchic levels and forms of communication have their shadow-side, and there is much evidence for psi phenomenon and nondual spiritual communication. To return to Heisenberg, "Not only is the Universe stranger than we think, it is stranger than we can think."[123] Or, as William James famously wrote long ago: "Our normal waking consciousness, rational consciousness, is but one special type of consciousness, whilst all about it, parted from it by the filmiest of screens, there lie potential forms of consciousness entirely different. . . . No account of the universe in its totality can be final which leaves these other forms of consciousness quite disregarded."[124]

ALWAYS AND FOREVER: FALLING IN LOVE WITH INTERPRETATION AND ACTION

I am no stranger to non-ordinary states and seemingly extraordinary experiences. My time on the Appalachian trail, which included experience of timelessness, was only one period in my life when the screen disappeared.

In second and third grade classes, I had no-thought, no-separation extended moments, which even then made me wonder about "normal" consciousness and what we call "reality." If I were taking a true or false test, I would have marked those moments true. Also while younger, after spending hours searching for my lost dog, Casey, who slept near me every night, I stopped, prayed with all my heart for him to return, and then he ran around the corner. And when playing a baseball dice-rolling game with my friend, I threw double-sixes, which equated to a grand slam, and then did it again, and again, with both of us expecting it, and it continuing to happen; later, I reflected on how it seemed like our expectations, which is another way to say intentions, influenced the dice.

On the actual baseball diamond, I had many "in the zone" effortless effort playground moments, although full of effort thoughts often got in the way when playing organized Little League; but, during one at bat, all the worry washed away and a silent mind emerged, leading my twelve-year-old self to hit an over-the-fence home run despite being small for my age. I knew I was going to hit it out before the ball was pitched; it became clear to me at that time that we have potential beyond what we would expect, and belief is a big reason why.

Later in life, I had plenty of moments with partners in which we were so in-synch that reading thoughts was relatively common, and once had a shared dream, both of us suddenly waking up and describing the same details. What seemed like synchronicities, or at least happy coincidences, has led me to meet people I needed to meet, including partners, but also Rafael when going through an existential crisis after graduating college, not knowing what path to take; and so many times in bookstores, including accidently knocking over a pile of books and finding Krishnamurti's *Think on These Things*, which was exactly what I was looking for.

I've also had empathetic, if not telepathic, silent communication with dogs, including Mac, a nearly 200-pound Newfoundland with whom I walked in the woods in Maine and who "saved" me, due to acquired instinct, when we swam together in the Penobscot river, with me holding onto his tail while he swam for shore. Once, he suddenly ran off trail, making me nervous that he might get lost, made worse by the fact that he was my roommate's dog. I stood there worrying, but then a few minutes later, Mac burst through the brush about fifty yards ahead of me. I immediately realized that if I had kept walking at our usual pace, I would have been at that spot. It was a lesson in trusting the bond between us, which remained when he was out of sight.

When living and studying with Rafael in New Mexico, who was an Emerson-like mentor rather than a crazy wisdom guru, I walked amid the expansive night sky and saw a shooting star at the exact moment of intense emotion thinking about a deceased friend. I did not care if it was a projection; what mattered, at that moment, was the feeling. Amid the trance of endless miles driving cross-country, I once had a time loop-like vision of myself as an older man laughing at my youthful worries, the message being we are always already enlightened, but we push it away through incessant striving. In an extreme moment, a car accident, I displayed extreme strength, squeezing and slightly bending a steering wheel in anticipation of the crash, and in many dreams I have had extreme smarts, doing outlandish math problems and writings, which I cannot do when awake.

I have also experienced the sense of being stared at, although perhaps Bateson is right and such experience, and others mentioned here, are likely mere coincidence with me adding meaning through interpretation.

Of course, most of my life I have bumbled and stumbled, more lightly when mindful. And the less extreme has been more significant: years of meditation and contemplation aided bliss currents and buoyancy when doing simple things, like landscaping or building rock walls. But my most significant moments did occur on the Appalachian trail with moss, moose, and expansive vistas, because I was ready for them after years of struggle to find vocation, with unity-in-diversity experiences leading me to feel called to read and write and eventually teach environmental ethics and the power of listening.

A couple of experiences, occurring a few years after the trail, opened me up to what is possible, but also to the stupidity of misinterpretation. I once played blackjack with friends in Atlantic City, winning a nice sum by combining logic with intuitive hunches, our big score happening when I saw a card in my mind before it was dealt, leading me to hit instead of stand. We left with money that night, but I went back in a future night by myself, convinced I could do it again. We all know the end of that story. But what if I had continued to win? Should I spend my infinite potential quantum energy life gambling in casinos?

The high of winning at gambling is similar to the high of drugs, which many turn to for non-ordinary experience and insight. But there are limits to such pursuits, with life mostly lived in mundane moments of getting things done, although non-ordinary experiences can open us to spirituality, and vice-versa: the yogis of old sometimes displayed "supernormal" abilities due to spiritual practices and experiences. However, all spiritual traditions focus on the path, not the cloud nine experiences, as Rafael used to say, and Teilhard de Chardin is right: it is a path of love, and if we rediscover that fire, we will be on our way. And that's the point of other forms of consciousness aiding our accounting of the universe. The drug of spiritual experiences, or drugs that induce spiritual experiences, may be the kick in the pants we need, but chasing them may be more distraction, especially in age of ecocrisis.

Spiritual communication, then, as receiving insights from spiritual experiences, may lead to both fitting and unfitting responses. However, when guided by the fire of love, we will likely be graced with spiritual experiences and feel motivated to move up the developmental holarchy. Or, a spiritual experience may communicate the fire of love. Wilber writes substantially about a variety of unity-in-diversity spiritual experiences and sparingly about telepathy and the like, and that says a lot about what he considers most important to growth. He also questions the New Age and New Paradigm claim that a new age is imminent, as he recognizes the hard work needed for rational advance let alone spiritual ascent; yet, similar to Teilhard de Chardin's omega point, he argues that Spirit unfolding, or eros, is pulling us forward, and Kosmic grooves, which may be disclosed via spiritual practices and experiences, show the way.

Our responses will make or break us, and the mistaken assumption that we can easily change the world by changing our thoughts is not going to get it done. We need to be motivated by the fire of love, but we also need to fall in love with critical interpretation and action. While Wilber admits that it is not always easy to differentiate between pre-rational and transrational experiences, we must take care not to confuse magical narcissism with transcendental awareness. And wise discernment does not include the constant appeal to quantum physics; it does demand comparing methods and discoveries with competent and experienced spiritual practitioners, both past and present and preferably outside of New Age circles.[125]

There is clearly way more subtlety to the communicative system than Cartesian science allows, and freeing science from materialism is a step toward eco-sanity, but the complexity of material constraints and limits do not disappear. The lesson of post-metaphysical thought is that before metaphysical speculation, there exists the physical, and material events have a density that are collectively interpreted rather than individually created. Thus, if we experience subtle state energies, or what appears to be higher beings, we also need to excel at subtle thinking, discerning nuance and making distinctions.

Like Schrag and Wilber, Peters argues that our ultimate destiny is to interpret, stating that the fact that interpretation always involves conflicting desires does not "signal a fall from the supposed grace of immediacy" but marks the possibility of interaction, in which there are "no sure signs in communication, only hints and guesses." Instead of a meeting of minds, at best we "dance" and "sometimes touch." Communicative praxis, then, compensates for "the fact that we can never be each other."[126] In other words, communication and interpretation exist because we exist in relationship, not in an unmediated oneness. But while interpretation is inevitable, we still must ask whether experience of a nondual relationship of oneness is one of the ways we sometimes touch.

Peters continually celebrates the existential realities of communication, and in doing so warns against perceiving the transcendent as the only path to wisdom: "Once we are stung by miscommunication, it is tempting to imagine communication as an escape from mortal modes."[127] While he considers transcendence a natural response to communicative breakdown, he also finds it dangerous, as the "immanent work of love and justice will be disdained as nothing but wreckage and refuse."[128] Peters misses the point here, at least if we are to learn lessons from traversing religious traditions and forms of spiritual communication, which is to integrate transcendent and immanent, not pit them against each other; still, he continues by referencing New Age thought and practice: "Short of some redeemed state of angels or porpoises, there is no release from the discipline of the object in our mutual dealings."[129] For Peters, interaction, not oneness, is clearly the beginning of wisdom, in which we care less about whether the "self is authentically represented" and more about whether "the other is caringly served. Such service "beats anything the angels might offer."[130]

And that returns us to Bob and his desire to have angels speak through him. New Age rhetoric, at its best, challenges instrumental forms of rationality, fundamentalist religious dogma, and modernist separate self-understanding that contribute to ecocrisis, celebrating the power of the individual and providing a hopeful narrative for the future. Spangler's depiction of the New Age image of stepping away from problematic habits to embrace transformation is particularly compelling. But New Age rhetoric, at its worst, uncritically gathers practices from exoteric and esoteric religious traditions that fit its vision of self-transformation, claiming that we merely need to change our personal reality to transform the world. Right thought is powerful, as the Buddha taught, but thought must be integrated with the other seven principles of the eightfold path, including right view, speech and action. Or, to return to Schrag's tripartite of critique, articulation, and disclosure: New Ageism tends toward anti-rationality, avoiding critique of its assumptions; fails to acknowledge the articulation of differing interpretations, privileging unmediated experience and projection of oneness; and discloses individualistic experience rather than relationship and the hard work of social change.

Spangler, who claims to channel a "spiritual force" named John, defends New Ageism despite its "narcissistic and silly stages," arguing that what motivates most New Agers is the "quest for a new relationship with the sacred and an accompanying sense of power and value—a sense that now I can make a meaningful contribution in service to my world."[131] However, too often this degrades into the desire to create reality via the subtle interplay of quantum energies, which, interestingly, has much in common with postmodern relativism: in the absence of post-metaphysical guarantees, we are free to create our own truth. This helps account for New Ageism's heyday in the 1980s and 1990s when postmodernism was abuzz, and post-truth Trumpism today, but subjective desire must be informed by intersubjective interpretations and discernment of partial truths, especially if we are to fittingly respond to eco-social crises.

New Agers and Ascensioners likely think they are Eco-Sages, but we would have to create a new term, Eco-Individualists, which fits because it is a contradiction in terms. A new consciousness may well churn within a growing number of individuals, but, ultimately, we create new habits together, co-creating perspectives and responses, which must be turned into action if we are to live into a new age. Is this process influenced by morphogenetic fields? I don't know. It's nice to think so. We can say that everything is subtly related, yet there is density to our interactions and communications rarely reflect direct spiritual connections and influence.

Within the context of this project, New Ageism, while too diffuse to fully explicate, attempts to embrace and express spiritual communication, including nonlocal extended mind communication expressed as parapsychology, while appealing to New Paradigm science. Mythic-animistic and aesthetic communication are mostly absent, other than the call for a new narrative and an embrace of Gaia theory and thus an alive planet. However, these attempts are limited by the failure to fully practice rational communication. Thus, perhaps we should follow Emerson's example and rationally question how a message is sent, and then grab a beer along with Thoreau, locally and organically brewed is more fitting, and simplify things still further by rationally attending to what the message says. Evaluating New Age discourse demands that we critically question the how and the what of communication.

Whatever New Age's merits and demerits may be, it is not an integrated response to ecocrisis, especially modernist realities of duality, domination, and control supported by the myths of unlimited growth, linear progress, and technology as savior. Spangler's insightfulness and my fondness for Bob notwithstanding, as popularly understood, New Ageism, followed by

Ascension thought and practice, are mostly unfitting responses to these damaging dualisms and narratives. However, New Paradigm science, like Sheldrake's investigations into telepathy and theory of morphogenetic fields, does challenge the damaging dualism between mind and nature. And perhaps meeting Bob was a meaningful coincidence, providing thinking and writing fodder for me, and, I can't say what, for Bob. Or perhaps the deeper meaning was our willingness to engage and have a conversation at all, and a three hour one at that.

Despite the possibility of time loops and precognition, everything does not happen for a predestined reason; but our spiritual paths, when open to new possibilities that stimulate our growth and the growth of others, may lead us to find what we need to find. What we find, and the quality of our interpretations, call forth the future. Reasons can be drawn from everything that happens; better reasons, and better responses, can be drawn from practicing an ecology of communication.

NOTES

1. Calvin O. Schrag interviewed in *Experiences Between Philosophy and Communication,* edited by Ramsey Eric Ramsey and David James Miller (New York: SUNY Press, 2003), 28–29.
2. Ibid., 30.
3. Zimmerman, *Contesting Earth's Future*, 69–70.
4. Wilber, *One Taste*, 117.
5. Ibid., 81–85.
6. Ibid., 87.
7. Steven J. Sutcliffe, *Children of the New Age: A History of Spiritual Practices* (New York: Routledge, 2003). 3.
8. Ibid., 196.
9. David Spangler and William Irwin Thompson, *Reimagination of the World: A Critique of the New Age, Science, and Popular Culture* (Sante Fe: Bear & Co., 1991), 35.
10. Ibid.
11. Ibid., 47.
12. Ibid., 48.
13. Ibid., 32.
14. Ibid., 44.
15. Marilyn Ferguson, *The Aquarian Conspiracy: Personal and Social Transformation in Our Time* (New York: Penguin/Tarcher, 2007), 143, 148.
16. Ibid., 164–166
17. Ibid., 26–48.
18. Ibid., 32–33.

19. Goodreads.com, https://www.goodreads.com/author/quotes/5387.Pierre_Teilhard_de_Chardin

20. Ibid., 466–488.

21. Wouter J. Hanegraaff, *New Age Religion and Western Culture: Esotericism in the Mirror of Secular Thought* (New York: SUNY Press, 1998), vii.

22. Ibid., 95–96.

23. John Welwood, "Human Nature, Buddha Nature: An Interview by Tina Fosella," *Tricycle*, spring 2011, https://tricycle.org/magazine/human-nature-buddha-nature/.

24. Vidya Frazier, *Ascension: Embracing the Transformation* (Sarasota, FL: First Edition Design Publishing, 2015).

25. Ibid., 8.

26. Ibid., 10.

27. Sandra Walter, Sandrawalter.com, retrieved January 2020, https://sandrawalter.com/.

28. Ibid.

29. Ibid.

30. See Abraham-Hicks.com and MattKahn.org.

31. See Abraham-Hicks.com. See also Abraham-Hicks Publications, "Abraham: How Source Sees Natural Disasters," April 13, 2012, YouTube video, 10:10, https://www.youtube.com/watch?v=aupLxCokSZI.

32. Matt Kahn All For Love, "The Pleidian Prophecy," June 2013, YouTube video, 1:03:44, https://www.youtube.com/watch?v=-5WElscYSFw&t=785s.

33. Matt Kahn All For Love, "Thy Will Be Done," March 2020, YouTube video, 1:37:55, https://www.youtube.com/watch?v=Ct2vWHz-DM0&t=3757s.

34. John Hick, *The Fifth Dimension: An Exploration of the Spiritual Realm* (London: OneWorld Pub., 2013), 2.

35. Margalit Fox, "Jose Arguelles, New Ager Focused on Time, Dies at 72," *New York Times*, April 2, 2011, https://www.nytimes.com/2011/04/03/us/03arguelles.html.

36. Theharmonicconvergence2020.org: https://www.theharmonicconvergence2020.org/.

37. See Wilber, *The Religion of Tomorrow*. See also Ken Wilber, Jack Engler, and Daniel P. Brown, *Transformations of Consciousness*.

38. Richard Panek, *The 4 Percent Universe: Dark Matter, Dark Energy, and the Race to Discover the Rest of Reality* (New York: Mariner Books, 2011).

39. TheSolitaryWriter.com, https://thesolitarywriter.com/2009/03/the-heart-has-its-reasons-of-which-reason-knows-nothing.html/.

40. Hick, *The Fifth Dimension*, 163.

41. Sandrawalter.com, retrieved April 2020.

42. Abraham-Hicks.com.

43. Tam Hunt, "Could Consciousness All Come Down to the Way Things Vibrate?" *The Conversation*, November, 2018: https://theconversation.com/could-consciousness-all-come-down-to-the-way-things-vibrate-103070.

44. Ibid.

45. Steven H. Strogatz, *Sync: How Order Emerges from Chaos in the Universe, Nature, and Daily Life* (New York: Hyperion, 2003), 1.

46. Hunt, "Could Consciousness All Come Down to the Way Things Vibrate?"

47. See James Gleick, *Chaos: The Making of a New Science* (New York: Penguin Books, 1987).

48. Lynne McTaggart, *The Intention Experiment: Using Your Thoughts to Change Your Life and the World* (New York: Simon & Schuster, 2007), xxv.

49. Lynne McTaggart, "Lynne McTaggart: A Live Intention to Limit the Spread of Coronavirus," March 19, 2020, YouTube, 24:07: https://www.youtube.com/watch?v=aTDJ15t5PPo.

50. Fun and Budget with Tinesha Davis, "Abraham Hicks' First Talk on the Virus," March 2020, YouTube, 13:05 https://www.youtube.com/watch?v=wUYPa5H3Pnk.

51. See Scott Carney, *What Doesn't Kill Us* (New York: Penguin/Rodale, 2017).

52. Dean Radin, *Supernormal: Science, Yoga, and the Evidence for Extraordinary Psychic Abilities* (New York: Random House/Crown Pub., 2013), xx-xxi.

53. Matt Kahn, "Thy Will Be Done," March 2020, YouTube, 1:37:55: https://www.youtube.com/watch?v=Ct2vWHz-DM0&t=3757s.

54. Ibid.

55. Mark Oppenheimer, "The Queen of the New Age," *New York Times*, May 4, 2008: https://www.nytimes.com/2008/05/04/magazine/04Hay-t.html.

56. Ken Wilber, *Grace and Grit: Spirituality and Healing in the Life and Death of Treya Killam* Wilber (Boston: Shambhala, 1991), 266.

57. Ibid., 50.

58. Ibid., 51.

59. Larry Dossey, *One Mind: How our Individual Mind is Part of a Greater Consciousness and Why It Matters* (Colorado: Hay House, 2013), 11.

60. Oppenheimer, "The Queen of the New Age," *New York Times*.

61. Laura Premack, "John of God: My Encounter with Brazil's Accused Faith Healer," *The Conversation*, January 4, 2019: https://theconversation.com/john-of-god.

62. Dossey, *One Mind*, xxviii–xxx.

63. Ibid., 28.

64. Ibid., 122.

65. Elizabeth Lesser, *The New American Spirituality: A Seeker's Guide* (New York: Random House, 1999), 54, 61–62.

66. Ibid., 53–58.

67. Georg Feuerstein, *Holy Madness: The Shock Tactics and Radical Teachings of Crazy-Wise Adepts, Holy Fools, and Rascal Gurus* (New York: Penguin Books, 1992).

68. Chogyam Trungpa, *Cutting Through Spiritual Materialism* (Boston: Shambhala, 1987), 31–50.

69. Lesser, *The New American Spirituality*, 372.

70. See the Shambhala Publications website, Shambhala.org.

71. Monica Furlong, *Zen Effects: The Life of Alan Watts* (Woodstock, VT: Skylight Paths Publishing, 2001).

72. Lesser, *The New American Spirituality*, 61–62.

73. See the Omega Institute website, eomega.org: https://www.eomega.org/about-omega/press-center/statement-from-omega-regarding-john-of-god-allegations.

74. Miguel Sancho and Kimberly Brown, "Arizona Sweat Lodge: The Inside Story of James Ray's Fatal Retreat," abcnews.com: https://abcnews.go.com/Primetime/james-arthur-ray-arizona-sweat-lodge/story?id=11016900.

75. Peters, *Speaking into the Air*, 70.

76. Ibid., 74.

77. Ibid. 75.

78. Ibid., 89–91.

79. Ibid., 101.

80. Ibid., 93, 97.

81. Rupert Sheldrake, *Science Set Free: 10 Paths to New Discovery* (New York: Crown Pub., 2012), 233.

82. Ibid., 242.

83. Ibid., 237.

84. Ibid., 236–238.

85. Ibid., 244.

86. Ibid., 253.

87. Elizabeth G. Krohn and Jeffrey J. Kripal, *Changed in a Flash: One Woman's Near-Death Experience and Why a Scholar Thinks it Empowers Us All* (Berkeley, CA: North Atlantic Books, 2018), 23–32.

88. See the Tom Tom Foundation panel discussion, "Is there life after death," at the University of Virginia with scholars from the Department of Perceptual Studies, May 30, 2018, 1:41, YouTube video, https://www.youtube.com/watch?v=4RGizqsLumo&t=11s.

89. Krohn and Kripal, 261.

90. Ibid., 227, 143–148.

91. Jeffrey J. Kripal, *The Flip: Epiphanies of Mind and the Future of Knowledge* (New York: Bellevue Literary Press, 2019), 54–57.

92. Rupert Sheldrake, *Dogs that Know When Their Owners Are Coming Home: And Other Unexplained Powers of Animals* (New York: Three Rivers Press, 1999), 302.

93. Ibid., 309–310.

94. Sheldrake, *Science Set Free*, 242.

95. Bateson, *Angels Fear*, 36.

96. Rupert Sheldrake, *The Sense of Being Stared At: And Other Aspects of the Extended Mind* (New York: Crown Pub., 2003), 4.

97. Sheldrake, *Dogs that Know When Their Owners Are Coming Home*, 308–309.

98. Matthew Fox and Rupert Sheldrake, *The Physics of Angels: Exploring the Realm Where Science and Spirit Meet* (New York: HarperCollins, 1996), 21.

99. Ibid., 10–11.

100. Ibid., xii.

101. Roger Walsh, *The Spirit of Shamanism* (New York: Tarcher/Putnam, 1990), 239.

102. Ibid., 122.

103. Ibid., 123.

104. Ibid., 125–126.

105. Jon Klimo, *Channeling: Investigations on Receiving Information from Paranormal Sources* (Los Angeles: Jeremy P. Tarcher Inc., 1987), 42.

106. Walsh, *The Spirit of Shamanism*, 130–131.

107. Wilber, *The Religion of Tomorrow*, 376.

108. Ibid., 379.

109. Walsh, *The Spirit of Shamanism, 125.*

110. Klimo, *Channeling,* 168–184.

111. Neale Donald Walsh, "Introduction," *Conversations with God: An Uncommon Dialogue, Book 1* (New York: J. P. Putnam & Sons, 1995.).

112. Klimo, *Channeling*, 42–45.

113. Hick, *The Fifth Dimension*, 170–171.

114. Ibid., 161–162.

115. Bernard D. Beitman, "Jung's Scarab as a Psychotherapeutic Technique," *Psychology Today*, August 12, 2015, https://www.psychologytoday.com/us/blog/connecting-coincidence/201508/jung-s-scarab-psychotherapeutic-technique.

116. Wilber, *The Religion of Tomorrow*, 416.

117. Bateson, *Angels Fear*, 54.

118. Ibid., 57.

119. Ibid., 55.

120. Ibid., 51.

121. Wilber, *The Religion of Tomorrow*, 217.

122. Alan AtKisson, *Believing Casandra: An Optimist Looks at a Pessimist's World* (White River Junction, VT: Chelsea Green, 1999), 190-191.

123. This quotation can be found in Werner Heisenberg's *Across the Frontiers.*

124. This quotation can be found in William James' *The Varieties of Religious Experiences.*

125. Wilber, *The Eye of Spirit*, 85.

126. Peters, *Speaking into the Air*, 268.

127. Ibid.

128. Ibid.

129. Ibid.

130. Ibid.

131. Spangler, *Reimagination of the World*, 33.

Chapter 6

Discerning the Unfit

Interspecies Communication

GRIZZLIES, DOLPHINS, AND SECRET LIVES

The "wilderness effect" occurs when we cross the boundary from physically inhabiting wilderness to the psychological experience of wildness marked by a primal sense of community with fellow nature that is already within us. Encountering a fortunately docile Moose enhanced my experiences of wildness, but, like all things fitting, there exists unfitting flipsides; the communal bonding and healing potential of the wilderness effect may potentially degrade into solipsism, or just plain foolish behavior, especially when dialogic relations with nature is sought sans rational communication. This point continues to be crucial to this project, and there is perhaps no better illustration than Werner Herzog's film *Grizzly Man*.[1]

The grizzly man in question is Timothy Treadwell, a former actor and waiter who spent thirteen summers among the grizzlies of Alaska. Treadwell filmed the grizzlies and himself, amassing amazing footage of wilderness, wildness, and human foible. The grizzlies do what grizzlies do, including fighting to be dominant, searching for food, and occasionally eating their young when the river and salmon run dry. Treadwell speaks to the bears of his love for them, with his video recordings meant to teach humans of their awesome qualities, despite the fact that he may well end up being their dinner. In his most revealing moments, he speaks of how the grizzlies gave him a meaningful life after years of alcoholism and searching, frequently stating he is there to save them, but his videos disclose a man who needs saving.

Treadwell is a sympathetic character, but as a Native American and Alaskan states in the film, he fails to respect the grizzlies by not honoring the boundaries between species. While kin from an evolutionary standpoint, grizzlies are radically other, deserving of space without human interference or romanticized attempts at dialogue. Treadwell does commune with diverse nature, especially red foxes with whom he playfully frolics, finding a joy rarely experienced elsewhere. And his ability to survive thirteen summers with the grizzlies suggests an empathetic connection, or some type of nature wisdom, or perhaps just luck. Respectful dialogue with grizzlies would elicit humility, and, as Plumwood learned the hard way, the realization that we are not always atop the food chain. Treadwell continually acknowledges that he is food for the bears, but this assertion is repeated to highlight his survival abilities, including his ability to dialogue with nature, rather than to elicit humility in the form of respectful distance.

Treadwell's experiences give new meaning to the notion of communicating at a distance, with keeping grizzlies close in our imaginations much preferred to physical closeness. Such imaginative closeness liberates our ecological unconscious to speak to us of wildness, limiting both romanticized converse and monologic arrogance. But Treadwell has clearly gone to nature for therapy and to escape from a human world within which he does not feel at home. The end of the story is predictable: Treadwell, along with a girlfriend, who fails to listen to her fears, are eaten by a hungry bear.

It is easy to see Treadwell's need for healing, and easy to see his desire to experience, and keep experiencing, the wilderness effect as cure. Near the end of his fateful last summer, he leaves the grizzles but gets so aggravated by dealings with airport personnel—or the messy miscommunication Peters embraces as part and parcel of being human—that he returns to the wonder of wild nature. Greenway supplies insight here, as he states that mourning often occurs when reentering our cultural lives after we have psychologically attuned ourselves to rhythms of nature. Treadwell, and others like him, seek to escape the insanity of dominant dualistic forms of psychological processing, especially when split from all other forms of processing like mythic-animistic communication, but sanity lies in being able to integrate back into culture with the dialogic wisdom discovered in nature rather than losing ourselves in romanticized converse or searching. We do lose habitual notions of self from the experiencing the wilderness effect, but the next step is to embody a healthy relational self so that fitting responses may come through, not externalize unconscious personal issues. Treadwell reveals a soul wisely sensitive to mind monologically separated from nature. It is his response that is unfitting.

Treadwell's life and death provides more fable-like lessons, especially in regard to the potential for interspecies communication. The twentieth century offers a stream of books testifying to the human ability to communicate with animals and plants, as well as communication among and within differing species. In the early 1950s, Bateson researched octopi communication, arguing that they display a range of rule-like nonverbal behaviors, and monkeys at a San Francisco zoo, who used metacommunication to signal that their bites were play rather than hostile attacks, as well as dogs trained to help the blind to explore levels of learning. In the 1960s, he turned to dolphin communication, working for a time with John Lilly, who is not only famous for dolphin research but his creation and use of isolation tanks. Dolphins presented a unique case for study, as they have no external ears or facial expressions, nor can they easily signal with their bodies like land-based mammals. Instead, they signal with a range of sounds, leading Bateson to conclude that humans and dolphins have differing coding systems and that it a mistake to judge one by the other. Differing coding made interspecies communication difficult, but researchers could read patterns of behavior, like between mother and child, and participate in those patterns. As always, perceiving relationship, difference, and exchange of information was highlighted.[2]

Lilly went on to write books in the 1960s and 1970s, including *Man and Dolphin*, a 1961 bestseller, exploring the potential for communication with dolphins, who he called the humans of the sea. For Lilly, the first step toward exploring such communicative potential, despite differing coding systems, is respect for their intelligence, which he considered higher than humans given the size of cetacean brains. He envisioned a day when oceanaria would progress from long-term prisons to temporary-stay interspecies schools where humans and dolphins learn from each other. However, his research was marked by strange experiments, with one of the strangest involving Margaret Howe Lovatt, who, in her early twenties, lived on the Caribbean island of St. Thomas where Bateson was directing a lab and working with Lilly. She visited and became a research assistant thanks to being an intuitive observer; however, after leaving each night and imagining the big-brained dolphins all alone, which felt wrong to her, she proposed living with them full-time. Lilly, a scientific yet radical spirit, went for it, and a joint living area was constructed for her and Peter, a young male dolphin, with a goal of teaching him human-like sounds. Progress was made from daily lessons, but young Peter began to have distracting sexual urges. He was occasionally moved to another area with female dolphins, but that was laborious, and so Lovatt respectfully quelled the itch, so to speak, so that Peter could focus, giving new meaning to "all observation is intimate." Not surprisingly, when the story got out to the public, including in *Hustler* magazine, it overshadowed their research.[3]

Lilly, who was a neuroscientist and medical doctor, further overshadowed the research by taking his enthusiasm for LSD to the lab, giving doses to two dolphins, Sissy and Pamela, but not Peter, who was protected by Lovatt. Lilly's LSD research provided no experimental results—the dolphins seemed unaffected—but did eventually result in the closing of the lab, with Bateson moving on to do further research at the Oceanic Institute in Hawaii due to Lilly's disregard for the welfare of the dolphins. Lovatt ended up being in charge of decommissioning the lab while Peter was sent to Lilly's other lab in Miami, which had more confined spaces and no Lovatt, whom he had lived with for the previous six months. Peter committed suicide soon after, refusing to take another breath and sinking to the bottom of the tank. Andy Williamson, a veterinarian, argues that Peter died of a broken heart.[4]

Lilly continued to spend years studying and writing about dolphin communication, including attempts at telepathy, but the attempt to teach dolphins English wisely went away, with more focus on their unique coding system, which is what Bateson advised. Despite Lilly's surprising disregard, and forays into isolation tanks and LSD, and later, communication with aliens, he was a champion for cetacean intelligence and communicative abilities, influencing popular culture in films, such as *Day of the Dolphin* and *Altered States*, as well as other media.[5] But the tale of Lovatt and Peter, rather than being sordid, suggests more than intelligence: the emotional bonding among differing species.

Another influential cultural phenomenon was Peter Tompkins and Christopher Bird's 1973 bestselling *The Secret Life of Plants,* which drew from a provocative range of sources. *Secret Life* celebrated the work of George Washington Carver, who began life as a slave and yet received a master's degree in agricultural science and made numerous contributions, including crop rotation to preserve soil health and increase productivity, and many new peanut products, among other innovations.[6] Carver worked within whole systems and had a biomimicry mentality, while stating that God spoke through the beauty of nature.[7] But Tompkins and Bird also document the more controversial experiments of Cleve Backster, an expert on polygraphs, who argued that plants are sentient, responding to emotions and states of mind.[8]

The Secret Life of Plants was criticized for being pseudoscientific—it also includes arguments for the reality of fairies and other unseen yet intelligent entities—and Backster was lampooned, yet he claimed to have proof that plants are telepathic, first discovered when a dracaena hooked up to a polygraph responded to a threatening fire image in his mind. Backster has spent decades experimenting on plants, and they have all led him to the same conclusion: characteristics of mind go beyond the brain. Scientists have mostly ignored his research into the "Backster Effect," or biocommunication, which he claims is due to difficulty of replication, as plants bond with experimenters and anticipate planned experiments, and also because you can't pretend: a plant will not respond to the threat of fire if the intent is not real. They need to feel it. However, Sheldrake has been receptive, and Backster has even done experiments in which he takes a walk, returning when a randomized timer goes off in his pocket, and then finding that the plant responded at the moment he turned around: in other words, plants that know when their owners are coming home.[9]

Today, the secret of sentience is not so secret, with one of the most intriguing books exploring plant intelligence being Monica Gagliano's *Thus Spoke the Plant*, in which she discloses scientific and personal encounters with a variety of vegetal life. Gagliano, who is a research associate professor of evolutionary ecology, among other impressive-sounding research titles, admits that some of her stories may "appear weird, out of the ordinary, or totally unbelievable," yet she feels called by plants, leading her to integrate scientific objectivity with the "subjectivity of transcendental experiences and intuition."[10] Her sense of being called is so strong that she claims her book is coauthored with plants, similar to Mueller's assertion that his book reflects a dialogue between author and salmon. Gagliano also makes claims for dialogue, rather than ventriloquizing, or channeling for that matter; the human, as listener, "filters out personal noise to hear plants speak." For her, such practice reminds of our long history of connection with nonhuman others, no doubt awakening our ecological unconscious, and provides a "truly inspiring future for the whole."[11]

Of course, as Roszak pointed out, there is a long history of animal and plant sentience in fairy tales, and modern-day children's books and movies are filled with nonhuman speakers. Roszak is critical of the goofiness of Disney characters, but some language-using nonhumans are powerful. A personal favorite is the ass-kicking ancient tree Ents from *The Lord of the Rings*, who go on the march to defend their land. A perusal of recent books also reveals scores of titles on interspecies communication by horse whisperers and animal communicators. And, as we've seen, Abram argues for preverbal and bodily based converse with the other-than-human world, which mostly reflects mythic-animistic communication, while Kimmerer draws upon her Native American and scientific background to argue for a new grammar of animacy and aesthetics, reflecting an integration of mythic-animistic, rational, and aesthetic communication. Abram and Kimmerer, along with Shepard and TEK advocates, reveal that what has been considered secret was actually repressed; traversing our hunter-gatherer past, and indigenous past and present, discloses the potential to know more, and learn more, by overcoming species separation, and dualistic separation generally, which is a root cause of ecocrisis.

Emily Plec, in the anthology *Perspectives on Human-Animal Communication*, has taken the conversation further by coining the term "internatural communication" as a kind of missing other half of intercultural communication. We are not just cultural animals communicating, we are also nature communicating with other expressions of nature, which, for her, has empathy at its core.[12] Boundaries should be respected, but imagining ourselves as fellow nature, or fellow animals, rather than separate species who sometimes interact, opens a door to the kind of listening needed to have deeper experiences. Gagliano, for example, has pioneered the field of bioacoustics, devising experiments to show that plants have "voices" and respond to sounds, a project borne of crossing boundaries.[13] For Plec, internatural investigation is a radical response that overcomes radical duality and long-held anthropocentric biases.[14]

Treadwell's fateful decisions remind, however, that we must question the doors we open and how we walk through, discerning the unfit to better enact fitting responses. And that means critically interpreting claims for interspecies or internatural communication, whether centered in mythic-animistic, aesthetic, or spiritual communication, or some combination, from the standpoint of rational communication. New Agers, for example, tend to cast interspecies communication as solely spiritual in character, failing to make distinctions between differing modes of communication, or, surprise, surprise, to question via rational communication; after all, numerous New Age channelers claim to channel dolphins.[15]

THE NOT-SO-AMAZING CASE OF STRONGHEART AND FREDDY THE FLY

One of the most read twentieth-century books on interspecies communication is J. Allen Boone's 1954 *Kinship with All Life*, in which he details his close relationship with Strongheart, a German Shepard for whom he was caretaker. It is impossible to measure *Kinship*'s influence on the plethora of today's animal communicators; however, Sheldrake mentions Boone's book as an example of the telepathic capabilities of animals, and its popularity suggests it has been very influential. Boone claims to write within the mystical tradition and to have established an easy communicative rapport with nonhumans: all life is an expression of Intelligence, or the Mind of the universe, and thus communication among differing forms of life is not only possible but the true reality. *Kinship with All Life*'s turn to mysticism and telepathy gives it a New Agey vibe, and thus it provides another example of spiritual communication, or nonlocal mind communication, in need of critical interrogation.

Strongheart is a Hollywood film star, as well as a highly trained former military and police dog—he is clearly quite special—but according to Boone, he is so special that they are in "rational correspondence." His first example is when he verbally insists that Strongheart sleep on the bed with his head toward the top. Strongheart refuses, communicating why by tugging on his pajamas, forcing him to look at "insecure old French windows covered by curtains." Strongheart pulls a curtain open a few times, lets it drop, and then barks back and forth between Boone and the windows. It becomes clear that Strongheart, through "a language made up of simple sounds and pantomime," must sleep with his head toward the bottom of the bed so that he can protect Boone from an intruder that might enter through the windows.[16]

Boone then gives further examples of Strongheart's intelligence and perception—he acknowledges his training, but this seems secondary—providing two instances in which he barks vehemently at strangers that falsely represent themselves, fooling humans but not the dog. Boone also details how Strongheart knew when he changed his mind about spending the day writing, deciding to take him for a long walk instead. The moment Boone made the decision inwardly, Strongheart rushed excitedly to him, bringing him his walking sweater, blue jeans, boots, and stick: "I stared at him in stunned amazement. How did the dog know that I had changed my plans and was going to take him on an outing? There was no outward communication between us at all."[17]

These experiences lead Boone to assert that Strongheart reads minds, neglecting the possibility that he may read body language and emotions. But for him the evidence keeps growing, although at this point, the communication, or at least the telepathic communication, is one way. Strongheart can read Boone's mind but Boone cannot read Strongheart's. Boone sets out to remedy this situation, attempting to learn from the dog how to communicate silently and directly. After studying Strongheart's character and realizing that he is superior to most humans, he further realizes that his unconscious sense of superiority over the dog, and all animals, was placing a communicative block between them. Boone sets out to perceive Strongheart as an equal, projecting humility and openness: "Each of us, I know, was an individual expression of life and intelligence. And that plainly being so, I assured myself, there had to be a point of contact where he and I could meet with perfect understanding."[18]

Boone's appreciation reaches new heights as he recognizes Strongheart's zest for life, powerful agility, sense of wonder, and "complete interest in the immediate thing he was doing," among other qualities. However, he then has a mystical realization that he is not merely witnessing "a dog expressing great qualities" but "great qualities expressing a dog."[19] This description brings us back to Wilber's claim that everything has equal Ground value in Spirit, as well as Taoist effortless effort, with Strongheart, like an enlightened adept, seeming to express the Self that is all things via nondual awareness. In fact, Boone describes Strongheart's ability to silently sit on mountain ledges and watch the sun set on the tiny towns and city and the immense Pacific Ocean as a meditative state: "His gaze was focused on a point in the sky considerably above the horizon line. He was staring off into fathomless space . . . giving him great satisfaction, great contentment, great peace of mind. That fact was not only written all over him, it was permeating the atmosphere like a perfume . . . I had watched human pilgrims in such meditative poses on sacred mountains in the Orient."[20]

As a dog lover, I am both amused by and fond of the idea of Strongheart having qualities superior to humans, and even the possibility of an enlightened pooch. Still, the obvious critical concern is anthropomorphism. Is Boone projecting biases and poor interpretations, or does his description reflect fitting responses to his experiences? We project all the time, usually our ego desires, but when done consciously, projection can be a spiritual practice, like seeing God or Spirit or the Divine in all things, especially since we so often habitually live in a world of "Its." Recall, for example, Archbishop Oscar Romero interpreting God's will as seeing with God's eyes, asking if God would want the poor to suffer, or the Council of All Beings ritual, in which participants project themselves as differing species, seeing from their perspective to better understand and respond to ecocrisis. Such practices are a wise first step to dialogue, but, for Boone, seeing Intelligence expressed as Strongheart opens the door to two-way telepathic communication.

One day while sitting on a mountain ledge, Boone mentally asked Strongheart numerous questions. Strongheart kept staring into space, but suddenly turned toward him, staring "right through," before returning to the view and his meditative pose. Boone writes: "I knew that Strongheart had been silently talking back to me. And I had actually been able to understand what he said to me! The proof was that I had answers to practically every question I had asked, answers that were subsequently verified in every detail."[21]

Boone does not provide specific telepathic conversations, but does state that he went into a "blank state of mind" that made him receptive to Strongheart's responses, creating a "mental bridge" carrying "thought traffic" between them. However, the bridge closed down whenever Boone fell back into a superior attitude, or if his thoughts were not high-minded enough. In fact, he claims that they were not the true source of their silent dialogues, but "were being communicated through by the Mind of the Universe": "We were being used as living instruments for its good pleasure. That primal, illimitable and eternal Mind was moving through me to Strongheart, and through Strongheart to me."[22]

This passage reflects the claim that spiritual Oneness is reality, as well as channeling, although instead of a person being an instrument, a person and dog become vehicles of One Mind expressed as telepathy. Potential dysfunctions here, other than self-deception, include dissociation, in which one identifies with the transcendent at the expense of the immanent, which can degrade into differences not making a difference between communicative participants; or rather, the New Age tendency to project Oneness as ego desire rather than a spiritual practice that lessens the influence of ego desire. Boone goes so far as to state that we are all "mental nudists": "Neither my inner life nor his inner life nor the inner life of any other living thing is private or concealable. We are all mental nudists, always on public display for all freely to observe and evaluate."[23] Mental nudity is another aspect of Boone's conception of Oneness, and a far cry from Peters' claim that we can't read our own minds, let alone the minds of others.

Wilber sometimes uses the language of oneness, but always reminds that reality is not-one, not-two when describing nondualism. Perhaps moments of nondualism and telepathy are what Boone is experiencing with Strongheart, but there is evidence in the text of solipsism—of an enlarged ego rather than the absence of ego—and thus the projection of Oneness supported by the belief that he can create his own reality. For example, he speaks of his power to create his relationship with Strongheart, although this claim is a bit tempered: "I saw that I was primarily responsible for whatever happened in our relationship, and that the responsibility lay not so much in what I said or did but in what I was really up to mentally."[24] Perhaps he is stating that he is responsible for elevating himself to Strongheart's level, but this also suggests that he alone creates their relationship, with Strongheart backgrounded and incorporated.

In *Kinship*, the how of communication is addressed, but not critically, and the what is not disclosed, as he gives no details about what was telepathically shared and so we have no way to interpret its validity. However, there is obviously much subtlety to the communicative system in Boone's view of reality. So much that his telepathic abilities extend to Freddy the Fly. Freddy landed on Boone's mirror while shaving one day, and instead of squashing him he admired his ability to fly and walk on mirrors and walls. Freddy proceeded to follow him into another room, landing on his finger, leading Boone to anthropomorphize:

> For a few minutes he was very quiet; perhaps he was planning what to do next. Then with the quick step so characteristic of flies, he began parading up and down the full length of my finger as if he were marching to the music of an invisible brass band. Now and then he would pause and then resume his parading. He gave the impression that he was having a perfectly wonderful time and hoped I was too.[25]

Freddy ends up living with Boone, although in constant risk of being killed by a non-telepathic guest. But Freddy does not live so much as respond to Boone's mental whims or his creation of reality: "I simply was compelled to realize that as I identified Freddie as either intelligent or unintelligent, good or bad, friendly or unfriendly, co-operative or unco-operative—that is precisely how he behaved. For Freddy was nothing more nor less than the state of my own consciousness about him being made manifest in our outward experience."[26]

Humans respond to others based on how they are treated, as do dogs, and bugs, but Freddy does not have an iota of independent existence in this admission, which provides a means for interpreting what happens next. Boone, pondering the human-fly relationship, wonders why Freddy was so different from other fly experiences. While pondering, Boone noticed that Freddy kept landing on typewriter keys, causing him to think that he was trying to get his attention. Boone waited for a message, and suddenly received a mental impression: Job 22. He went to a Bible to read the twenty-second chapter of Job, finding his answer in a verse that basically states that when you "decree" something, "it shall be established onto thee."[27] In other words, we create our own reality. Boone had decreed differently with Freddy, and so he established different results.

The book of Job is actually a fine example of respect for earth and animals, one that Christians and non-Christians too often overlook. Job, after complaining to God, is given a biting response, told to consider the lions, vultures, donkeys, giants of the deep, rain and snow—all of wild nature—to learn that life is not all about him. God challenges Job's anthropocentrism and selfishness, demanding that he change his perspective by becoming more humble, and thus more joyful in response to creation. God does not just love humans, but all wild things.[28] Given Boone's experiences with Strongheart, he is likely on board with this, but apparently Freddy has taken him to the book of Job to glorify his telepathic talents and ability to create his own reality. The fact that he mentioned the book of Job earlier in *Kinship*, suggesting that he is familiar with its contents, does not register as a possible interpretation of why his unconscious conjured up this particular passage.

Kinship with All Life provides much to applaud, but also much to critically question. I am particularly appreciative of Boone's sympathy with intelligence and reverence for all forms of life, which are genuine mystical attitudes. However, his version of sympathy does not seem to revere differences that make a difference—he celebrates "All is One" rather than "One in many, many in One," or unity-in-diversity—and thus what he interprets as impartations of One Mind may be projections from his individual mind. Some of those projections are pretty darn good, it's just that the how of communication may be misinterpreted. The thoughts and feelings that come through Boone when he is in I-Thou relationship, attuned and in sympathy with Strongheart, are definitely worth listening to, but they are not definitely telepathy.

Some distinctions clearly need to be made among silent dialogue, in which thoughts and feeling emerge from within when in relationship, telepathy, in which thoughts are transferred from mind to mind, and channeling, in which one is a vehicle for imparting messages from some transcendent other, although the more psychological view seems more accurate to me, in which we are actually channeling the unconscious or our higher self and intelligence. Boone has discovered a fine spiritual practice: Sit quietly on a majestic mountain with a majestic view and a majestic dog. Insights that feel like messages emerge from within when connected without—typically the reality of unity-in-diversity, joy, wonder, and potential for transcendence—but the more long-winded the message, the more likely that ego desires have infiltrated and made themselves heard. Self-deception and misinterpretations occur when we don't even consider this possibility.

We may well experience moments of mirror mind, nondual awareness, and eternal now and we may well find answers to our questions emerging from our unconscious or higher self. However, along with concerns about self-deception, we are also left with Peters' lamentation of expecting too much from communication. Animals most commonly speak through nonverbal bodily comportment and emotions, often through touch and their eyes, as Goodall experienced with the chimpanzee she named David Greybeard, or, as she reports, when animals suffer, their eyes say "Won't *anybody* help me?"[29] They also speak via verbal sounds, perhaps of joy, but also of suffering, especially in ghastly slaughterhouses or animal experimentation; before seeking to communicate via telepathy and channeling, we should make sure we are not disregarding these "lesser" material forms of communication.

While Boone's narrative leads critical interpreters to wonder if he is dialoguing with nature or mostly monologuing within his head, it also causes readers to consider the possibilities for subtle communications, including at a distance. A couple of times a week he ate lunch in Los Angeles, having a friend sit with the Strongheart. He had no routine, returning at different times, but Strongheart knew when he would arrive, moving to his favorite observation spot to wait for him to "turn the bend in the road and head up the hill."[30] Not surprisingly, Boone once again fails to explore interpretations other than One Mind telepathy. Perhaps Strongheart's acute sense of hearing allowed him to hear his car at a distance, leading him to move to his observation spot. On the other hand, Boone may provide more evidence to support Sheldrake's hypothesis that dogs often know when their owners are coming home, and that experiences of telepathy, while significant in themselves, are most significant because they challenge us to rethink our Cartesian legacy of mind and extension, and thus dualities and dominance at the heart of our ecological despair.

NATURE AS SELF: MICHAEL ROADS' TALKING WITH NATURE

In the *Tao Te Ching*, Lao Tzu essentially wrote that nature does not speak in long speeches, while Thoreau stated in his journals: "To the highest communication I can make no reply; I lend only a silent ear."[31] These sentiments are in direct contradiction to chatty displays of telepathy, in which nature, and animals, and even flies, are quite noisy with a lot to say. Brief moments of telepathy, and brief moments of influence from intention, may well disclose insights, but when taken to extremes of long-winded mind-melding converse marked by claims of creating our own reality, they deserve more scrutiny, especially when deemed as a virulent response to ecocrisis. Nowhere is this more true than in Michael Roads' *Talking with Nature*.

H J Kramer, an acknowledged distributor of New Age titles, published Road's *Talking with Nature* in 1985. Like *Kinship*, *Talking* is supposed to be a nonfiction account of dialogic relations with animals and nature. Unlike *Kinship*, *Talking* is firmly placed within the New Age heyday of the mid-1980s along with Shirley MacLaine's *Dancing in the Light* and Gary Zukav's *The Dancing Wu Li Masters*, another book equating mysticism with quantum physics. However, Roads' dance of telepathy and channeling may be more monologue than dialogue.

The publishers, Harold and Linda Kramer, extol the wisdom of Roads' ability to channel nature's voice, which speaks to him "of her deep concern with the role of man in determining the fate of the earth." Roads, they say, does not only communicate but merges his consciousness and becomes one "with water, plant, mineral, animal, and even the terrifying fury of a raging storm." They continue in their joint preface: He writes of these experiences with chilling veracity, telling us what Nature is longing for us to hear. . . . The message of this book is one of hope. A solution is in sight. Like Michael Roads, humanity can learn to accept and listen to the voices of Nature. . . . Nature can be our friend rather than our enemy. She can instruct us in finding our way back to the true path. The birds, plants, rivers, and winds are praying for us. Will we listen to them before it is too late? Can we learn to hear in time?"[32]

Given that this description emphasizes the core argument of this project—we need to practice deeper and wider listening as a response to ecocrisis—*Talking with Nature* needs critical unpacking. One might think that I would love Boone's and Roads' writings; after all, the voice of nature is my thing, and I have been teaching a course with that title for twelve years. Also, as with New Age literature and spiritual experiences, I am not the final arbiter on interspecies communication, especially for experiences that I have not had myself. However, it would seem that New Age dysfunctions have infected some of the interspecies communication literature, with anti-rationality, the projection of Oneness, and the belief that we can create our own reality readily present. To me, *Talking* falls into this category, with Road's version of spiritual communication and deep listening too often jettisoning insights from rational communication. Thus, if interspecies and internatural communication are to provide fitting responses, the wheat must once again be distinguished from the chaff.

Roads, an Englishmen who moved to Australia to become a farmer after marrying his beloved Treenie, begins his narrative by describing a natural setting replete with a Platypus disappearing underwater, wind whistling through the needles of a River Oak, and leaves moving near his seat on the river bank. He then makes a request of his surroundings: "Help me?" Roads feels nature smiling, inside himself, and nature responds: "Help yourself. If you wish to tell the story of our connection, then write from the point of contact which you are." He complains to Nature that this exchange is not going to be easy: "What I need is help to explain in a concise and simple way the human connection with Nature. I need something that people can relate to. Something which is easy to understand and accept." Nature chastises him for wanting something easy, because the "slumbering Self" within humans must be disturbed: "We have a choice, my friend. Either you write it as it happens, as it is revealed, or forget the whole project. I can offer no compromise. Accept it. This will be written as a synthesis of man and nature."[33]

This exchange sets the tone for the narrative and mirrors both Mueller's and Gagliano's assertion that their books are coauthored with nature, although they don't have extensive literal conversations and Roads does. Also, he claims that nature is speaking for itself rather than via his perceptual biases, which means we are back to channeling. Soon after his encounter with the voice of Nature as a whole, he converses with a Heron that swoops down and fixes him with a stare: "Okay," I said. "If you have an opinion we might as well hear it." He then experienced a "powerful feeling of bird energy moving into the more subtle areas of awareness," and the Heron speaks: "We would also like to be involved. If you are going to write down our connection, then we will represent our own point of view." Roads asks whether the Heron represents all birds or just Herons, and the Heron responds: "We represent our own species only. We will contact you when the time is right."[34]

The remainder of *Talking with Nature* catalogues such contact, with Roads calling forth literal conversations with differing species and differing species calling to him. He writes that his first exposure to telepathic communication occurred when he was a farmer and cattle rancher. One day Treenie suddenly exclaimed while sitting in their living room that the "cows want moving." Roads took this statement as a challenge and drove to the pasture. To his surprise, most of the cows were "standing impatiently at the gate, waiting to be let out." Apparently Treenie continued to hear the cows desire to move "regardless of the amount of pasture available." This led Roads and his wife to combine their thoughts to try to contact the cows, telling them that they will move them the next day at nine o'clock. Roads writes: "Smile if you like, but next morning at nine o'clock the cattle were all waiting to be moved.

Now, I am not suggesting that the cattle can tell the time, because we eventually proved that they would begin to congregate well before the chosen hour; but the method worked, consistently, showing a stunning disregard for logic and reason."[35]

I did smile when reading this, wondering if Roads and his wife are displaying a stunning disregard for logic, reason, and alternative interpretations, while also remaining open to the possibility of subtle connections and communications. Roads next contacts wallabies that, not "unnaturally," are eating pasture on a new parcel of property meant for cattle. The cleared land is located between wilderness areas, and the "forest wildlife" extracts "a heavy toll." Roads decides to defend his pasture by shooting the wallabies, finding this solution distasteful but necessary. However, one night while raising his rifle, his prey suddenly turned toward him, fixing him with its gaze: "Transformed into glowing red jewels, the eyes met mine, and I gazed spellbound into the soul of a wild and wonderful Nature. For long moments our eyes held, locked . . . I stood silent, shocked to the core. Compassion, a comparative stranger to farmers saturated in death, surged powerfully from somewhere deep inside."[36]

This passage resonates with Goodall's soulful eye connection experiences with David Greybeard and many other animals, and especially Aldo Leopold's description when hunting in "Thinking Like a Mountain." However, Leopold shoots the mother wolf because fewer wolves meant more deer for hunting, but her dying green fire eyes made him realize that neither the wolf nor the mountain that is being decimated by too many deer, agree with this view.[37] Leopold learned a profound lesson concerning life force, empathetic connections with nonhumans, and ecological interdependency. Roads, on the other hand, while also realizing his error, turns to telepathy as response rather than thinking like a mountain: "I talked the problem over with Treenie, and together we reached the only solution possible. If we could 'think'-communicate with our cattle, why not try to 'think'-communicate with the wallabies."[38]

Roads telepathically communicates an agreement to the wallabies, suggesting that he would stop shooting them if they only ate twenty yards along the edge of the pasture, leaving the rest for the cattle. He writes: "Within only a few weeks, the pasture was thickening so rapidly that I was able to introduce an extra ten cows and calves. . . . For three years we maintained this tenuous agreement, the pastures continuing to thrive and flourish." Similar to the admission about the cattle gathering well before the telepathically sent chosen hour, Roads admits that the wallabies ate closer to forty yards, but they stayed to the boundary, leading him to declare: "One fact which emerged was obvious. We were able to communicate our wishes to the wildlife and reach an agreement for mutual benefit."[39]

What is obvious to Roads, of course, is likely not obvious to others, since, once again, other interpretations are possible. Still, Roads is not the only one to make claims for similar communications. In *A Language Older Than Words*, Derrick Jensen tells a story of coyotes repeatedly killing his chickens until he respectfully, and verbally, asked them not to, mentioning his work in defense of the wild and adding a promise to feed them the head, feet, and guts after he killed them for his food. Jensen states that the coyotes kept the bargain, although he considers whether it was just coincidence, or if he was crazy. However, he tried the experiment again when his dogs started eating eggs, and it once again worked, making him think it might be more than coincidence. He also consulted his friend, Jeannette Armstrong, an author, teacher, and Okanagan, who stated that attitudes about the possibility for such interspecies communication are the "primary difference between western and indigenous philosophies."[40] Of course, Roads claims that his experience with the wallabies occurs telepathically from a distance and Jensen verbally speaks with the coyotes when they approach the chickens, although he is verbalizing what is in his mind and that may resonate nonverbally, or telepathically.

To be fair, then, just like mystical experiences are ineffable, and thus impossible to make fully understandable to those who have not had similar experiences, one can imagine the same can be true for long-winded telepathic converse. And, Roads adds a kicker to the story: After selling the land, he revisits and discovers from the new owner that the pasture is overrun with wallabies, despite 6,000 shot in two years. Roads sums up: "Considering the huge amount of forest, I had suspected there were large numbers in the area when we made our agreement, but I had no idea I was dealing with such numbers. It was only then I fully understood the extent to which our agreement had been kept."[41]

Roads offers these stories as rational proof of his, and nature's, communicative abilities. However, he later argues that rational proof is unnecessary, making the ineffable experience argument, or rather, a pine tree tells him so: "Proof has become mankind's obsession. It is an intellectual attachment which, rather than expanding your horizons, is rapidly becoming a restriction. Proof can easily become a denial of that which 'IS,' and the seeds of tragedy are contained in this. Those who feel these words resonate in their hearts will need no proof, while those who do not could accept none."[42] That which "IS" is Oneness, and Roads argues throughout *Talking with Nature* that Oneness is the true reality. Roads, or the pine, also make the New Age listen to your heart, not your head, argument. Of course, this dismissal of "proof' treats rational discernment as a problem to be solved, or disregarded, rather than a mode of knowing to be integrated.

Talking with Nature includes other passages that question the worth of logic and rational inquiry, but his conversation with the pine tree, which becomes his Christmas tree, is especially void of critical thought. Roads seeks out a small pine amid the large Hoop Pines on his property, intending to transplant the tree into a pot. He finds a worthy specimen, and an inquiry finds the Pine "very willing." However, the transplanting does not go well, with Roads realizing he will have to chop through a main root, which will likely kill it. He apologizes, and then suggests the Pine withdraw its energy, telling it he will return the next day to finish the deed. Roads returned and severed the root, leaving the Pine "rather pathetic with only one short root stump and a single wisp of side root." The Pine is decorated, admired, and enjoyed by his family, but after New Year's Day he placed the Pine outdoors in a shady spot, giving it plenty of water but expecting it to die. Instead, the Pine flourished, becoming their Christmas tree again the next year. Roads asks the tree for an explanation: "Tell me, did the other Hoop Pines refuse to become our Christmas tree because you needed moving, or did you as a particular tree wish to be more involved with us? Also, how did you survive the transplanting, which was little short of brutal?"[43]

The Pine gives a lengthy response:

> To answer your question I would ask you to discard the physical aspects of transplanting a tree. I am not separate in consciousness from each separate tree you wished to take. Each tree can be likened to an aspect of "one" consciousness. Within this is your challenge. Your science has made great inroads in the study of plants and physical forms, but in the more subtle regions of life-energy-consciousness so very little has been realized or recognized. . . . My long root was destroyed, my connection with the earth and its elements erased. Yet in the time lapse when you suggested I withdraw my energy, it was withdrawn on nonphysical levels of which your science knows nothing.[44]

The guru-like Pine continues to explain that if it had been not received energy from Roads and his family, then death would have come, but since they became like a "socket" in which it could "plug for energy," it was able to grow a new root system.[45] The tree then gives the critique of proof already discussed. It's interesting that this passage dismisses scientific knowing and proof because, as we've seen, New Agers often appeal to quantum physics as proof for all kinds of subtle communications and energy exchange. Regardless, perhaps we are in "Backster effect" telepathic territory here, although Backster claims the polygraph as proof rather than long conversations that apparently need no proof. Also, Backster's main claim was for sentience and the emotional lives of plants, which included short moments of telepathy. Roads may have reasonably claimed that his family's energized, emotional response influenced the well-being of the plant, along with material care, allowing biological needs like rich soil, water, and shade to be met.

Instead, he turns to telepathic consciousness, literal conversations with nonhuman others, and the claim that Oneness is the true reality.

There is much in *Talking with Nature* that deserves rational respect. Roads explains how his growing spiritual awareness called him to sell his cattle farm, becoming involved in the organic farming movement instead: "A different truth was calling to us, for we were no longer the same people who had commenced farming a decade earlier in Australia."[46] He also details a difficult, yet transformative 110-mile wilderness effect hike that exposed him to vibrant dreams and mystical insights; displays impressive knowledge of nonhuman species despite remarking that he is not a botanist; and provides thoughtful advice on modes of engagement with the natural world, including humility, asking for guidance, loving attention, appreciation of beauty, and the recognition of energy flow. And his remarks on rational proof and intellectual attachment, or the Pine's remarks which he then takes as gospel, disclose a partial truth. Some wish to stay at the Rational level of development and disregard transrational experience. However, he falls for the head versus heart false dichotomy, and thus the obvious criticism of Roads is our old friend the pre-trans fallacy, resulting in a failure to differentiate and regressive interpretations. Roads also does not make a distinction between logic and logos. His criticisms of a narrowly defined logic, which are potentially valid when directed at scientism, are undermined when unthinkingly extended to a more integrated logos, which Roads, and New Agers generally, desperately need to practice.

Roads refers often to the Intelligence of Nature, which he interprets as Oneness, but, again, claiming that Oneness is the true reality may lead to causal-witnessing dissociation, in which the immanent dimension is not integrated into nondual awareness. This dysfunction, at it's extreme, leads to the view that ecocrisis is perfect or illusory because all is Spirit. All *is* Spirit from a transcendent perspective, and thus perfect in that sense. However, recall Zimmerman and Esbjorn-Hargens' dictum: "Things are getting worse, better, and are perfect." Roads tends to skip the first two, focusing on "all is One" perfect communication. Missing is Spirit unfolding as diverse matter, life, and mind; or, to return to Ramana Maharshi: "The world is illusory; Brahman alone is Real; Brahman is the world." "Brahman is the world" adds the finite realities of intrinsic and instrumental value, as well as plenty of opportunities for imperfect responses. As we saw, instead of recognizing the potential for denial embedded in "All is One" or "All is Spirit" assertions, New Agers turn to their "create your own reality" response; in other words, if we could all just see Spirit in pollution and eco-destruction generally, then this destruction would magically disappear.

Road continually falls prey to this "create your own reality" view. A Jade advises him: "Believe, believe. Whatever you believe—it is so."[47] Roads does believe, especially after his cold disappears after he converses with the Jade's "magnificent plant energy": "The sore throat, headache, and cold had gone. Were they ever a reality, or were they an illusion I created to hide behind? Whichever, I controlled them. I felt excited because I had never done this before . . . Expansion—love—creativity—Nature—knowing. Does anyone have a better offer?"[48] Not surprisingly, critical rationality is not included in this list. Once again, the claim that body and mind are connected, and that health is influenced by mental states, is logical, but to then leap to the assertion that we can control physical events is not.

Wilber, in response to Treya's cancer, provided eleven possible interpretations of its cause according to varying world views, from fundamentalist Christianity, which may claim that illness is punishment for sin, to science's claim that while there are physical causes, there is no meaning to illness, only chance or necessity due to genetics or unhealthy living, to the New Age assertion that illness is controlled by thought and the meaning of physical illness is to receive spiritual lessons.[49] Many do receive lessons from illness, just like they receive lessons from other hardships, but the New Age response is one of many, and, as Treya argued, deeply problematic, especially in regard to guilt; yet the "create our own reality" mantra is treated as if it is the only response, which is always a clue to likely unfittingness. Transversal rationality, on the other hand, would traverse all eleven interpretations, critically considering insights and integrating or dismissing them while remaining open to the possibility of new interpretations, new knowledge, and new traditions.

Rational communication of this type is absent in both Boone's and Roads' books, as they stick to one interpretation, Oneness, and thus force the same question: Are they talking with nature or mostly with themselves, creating reality but in the sense of seeing what they want to see? Unlike Boone, Roads does provide specific telepathic conversations, but literal evidence, if contrived, is not much better than no evidence. Is he projecting wishful desires onto the natural world, and then cataloguing his pathologies as spiritual communication? Instead of lessening the influence of the isolated ego and embracing more of Spirit beyond and within all things, Roads may be guilty of psychic inflation. Is *Talking with Nature* about nature, or is it really about Roads?

At best, Roads' conversations are inter-spiritual rather than interspecies or internatural. Even though the book is called "Talking with Nature," there is no silent dialogue reflecting Abram's always already bodily converse and grounded animistic perceptions, or insights emerging from the ecological unconscious, or learning the songs of other species, which come as a gift when in mythic-animistic rapport. Nor is there aesthetic rapport reflecting Kimmerer's integration of animism, scientific observation, and heightened awareness of beauty. A nondual rapport would reveal the inter-spiritual—that all is not-one, not-two, typically experienced in silence, induced by a quiet, receptive mind—but, for Roads, is expressed as lengthy telepathic converse or channeling. Thus, when Roads claims that the heron is speaking, we must wonder if the heron is backgrounded and incorporated into his perceptions, and unconscious, such that it does not have independent existence, just like Freddy the Fly.

I have no problem with imagined conversations, if that is what is mostly happening with Roads when feeling connected to others. I do it often, including with imagined students when walking after teaching, and with a deceased, yet living in my personal world, friend and mentor, and my father who passed away at eighty-seven, his ashes spread in the lake he dearly loved. And, again, the Council of All Beings discloses the worth of imagining ourselves as animal and plant others, when aware of what we are doing. Productive inner conversations of this type are aspects of intrapersonal communication, eliciting intuitive insights that can be shared outwardly. The same could be true of channeling, if we acknowledged that we were channeling our higher self or intelligence rather than angels or ascended masters, or both expressed as Kripal's "Humans as Two." We may also listen inwardly when imaginatively in rapport with differing species, but in *Talking* each species comes across as the same species, with the same Oneness message, rather than sharing unique songs. Channeling, as an imaginative practice, may get the ego out of the way, allowing one to transcend the separate self-sense, but when Roads documents lengthy and literal conversations with differing species, he may be silencing them rather than allowing them to speak; again, backgrounding and incorporation. In another telling passage, Roads claims that a parched rose asks to be picked and placed in his home. But why doesn't the rose just ask Roads to water it? The response of the rose clearly benefits Roads more than the rose.

Like Boone, Roads claims that he is writing within the mystical tradition: "From an early age I have recognized a connection with Nature, a mystical thread moving beyond mind and heart, connecting me to some mysterious, unknown element. That I am not alone in this connection is proven by the mystical writings of the ages, yet for some reason mankind as a whole has failed to realize this inner truth."[50] Roads makes a salient argument here that we should traverse mystical traditions, and *Talking with Nature* does contain passages that reflect Wilber's descriptions of transrational experiences. Despite the literalness suggesting certainty rather than openness to Mystery, several passages reflect subtle state experiences of archetypal energies. He also claims to experience causal nonattachment while receiving this advice from the river: "Look not for form, but from within see that which energizes form. Hear not the sound, but the Silence which surrounds it. See not the form, but the space of which it is formed."[51] Roads swims in the river, and yet transcends it: "To my physical senses water engulfed me, and while swimming I was acutely aware of the river . . . yet, on some other plane of experience, I swam in space. Around me was vast emptiness, not sky blue or star filled, but sheer nothingness."[52]

Roads further provides passages that may be interpreted as expressions of nondual experience. After conversing with the river, he exclaims: "I could feel concepts and ideas quite new to me shifting and sifting as though we all floated interlocked in a deep, dark pool. My sense of Self expanded, and, with the expansion, a focus clarified . . . I was the I which I observed."[53] He later wisely alludes to both transcendent and immanent dimensions, arguing for integration: "A duality of experience emerged. There was wind—and no wind. There was sound—and Silence. Movement—and stillness . . . I knew my challenge was to synthesize the movements, denying neither experience. To move with both, accepting as 'IS.'"[54] And he even records a wu-wei experience while using a sledgehammer to demolish an unused milking yard after "dairying slid into economic decline."[55] Roads begins the task with too much effort, raising large blisters, but then he receives a revelatory thought: "If all life is connected, then the animal, mineral, and vegetable kingdoms are interconnected. Knowing the cooperation I have experienced with plants and animals, why should metal be so different?"[56] He proceeds to change his attitude, completing the job with effortless effort.

These experiences cause Roads to wonder why humankind has failed to realize a mystical awareness of Oneness, but despite finally acknowledging the need to synthesize transcendent and immanent movements, he fails to realize that poor interpretations of mysticism, expressed as pre-trans fallacies and pathologies, act as blocks to genuine transformation. After conversing with the Heron, Roads writes: "Nobody's going to believe this."[57] But, in actuality, plenty of New Agers will uncritically believe both his ability to channel and telepathically communicate, and the content of his communications, or the how and the what, as expressions of Oneness.

Boone and Roads do claim that projecting Oneness, and then actually feeling it, leads to spiritual humility and equality, which are needed when seeking dialogic relations with nature. Superiority is often a block, despite differences that make a difference, when beginning a conversation with humans or nonhumans and especially when listening. Yet overcoming this block does not lead to perfect understanding, but insights emerging from within catalyzed by the dance of both participants, from the spiraling flow of meaning. That, at least, is what "talking" with nature looks like; telepathy and channeling are of a different order, and more prone to self-deception. But, as Ramsey reminded, whatever the method, we remain responsible for the fittingness of what comes through. So much depends on what we bring. Projections should be consciously used as tools, and then we need to stop projecting and listen to start a dialogue. Projecting Oneness, as a spiritual practice, may begin the process of knowing more, but the danger is the assumption of perfect communication and mutual understanding.

Toward the end of the text, Roads asks the river, "What is it that responds," and the river replies: "Do not look for separation. The I that calls is not separate from the I which responds."[58] As mentioned, I have done, and still do, meditations of this type, recognizing the I that is all things, which usually makes me smile in response to how often we forget, getting lost in our isolated worlds. And then I forget and get lost again, but that is exactly why we need reminders and practices, including spending time near rivers, or swimming in rivers. Every college campus where I have taken classes or taught has been near a river, and they hold many lessons of history, economics, and trade, and the politics of abuse and protection. But, if we bring attentiveness and openness, they also speak, perhaps of flow and flux, inspiring philosophy, perhaps of unity and diversity, especially if we are immersed in its hopefully clean waters, inspiring poetry; but even if we have moments in which the personal "I" dissipates, we remain individuals in relation. And, I would not expect long-winded lessons.

Buber also perceives the mystical I, or Thou, in all relations, but he wisely brings distinctions between three spheres of relation: life with nature, which "vibrates in the dark and remains below language"; life with other humans in which the Thou is manifest in language and giving and receiving; and life with spiritual beings in which we feel addressed beyond language and respond via creating, thinking, and acting. Thus, Buber acknowledges the validity of pre-rational, rational, and transrational experience, and then how they are intertwined: "In every sphere, through everything that becomes present to us, we gaze toward the train of the eternal You; in each we perceive a breath of it; in every You we address the eternal You, in every sphere according to its manner."[59]

Buber gives an example of contemplating a tree. He can perceive it as a colorful picture or painting; feel it as a movement, growing and breathing along with earth and air; assign it to a species, scientifically classifying and constructing its life; abstract it as the expression of a law; and make it constant by dissolving it into a number. But despite the merits of these perceptions, the tree remains an It. Thankfully, we can also be drawn—via the integration of will and grace—into an I-Thou relation. Buber writes:

> This does not require me to forego any of the modes of contemplation. There is nothing that I must not see in order to see, and there is no knowledge that I must forget. Rather is everything, picture and movement, species and instance, law and number included and inseparably fused. . . . Whatever belongs to the tree is included: its form and its mechanics; its color and chemistry, its

> conversation with the elements and its conversation with the stars—all this in its entirety.[60]

Buber's analysis supports the dialogic interplay among differing modes of knowing and thus sympathy with intelligence expressed as an ecology of communication. Roads, on the other hand, too often neglects the fact that the mystic temperament must be tempered with other modes of knowing nature. I-Thou ontology respects reciprocity; others act on us as we act on others. For Bateson this acting while being acted upon is an aesthetic challenge, transforming epistemology and sense of self. Some adepts may well respond to this challenge by realizing the enlightened Self, or Isness, or the "I" that is all things, but, again, nondual moments do not negate others but allows us to perceive with uncluttered clarity. Roads is undoubtedly sincere, and probably having religious experiences of some type, but Wilber's model of development, including levels, lines, and states, demands a more complete accounting of what is actually taking place. In other words, what we bring may be dysfunctions. Perhaps Roads has tasted nondual awareness and interpreted it poorly, or perhaps he expresses more serious pathologies, like the egoic projection of oneness onto a materially diverse life-world, resulting in backgrounding and incorporation.

Roads defends his position by stating that he's found nothing better:

> If this is an illusion I am experiencing with Nature, if it is all imagination—then it's okay. I like it. Who can make me a better offer? Polluted food and air? Is that better? To maintain a belief in death, fear, greed? Are they better? A dogmatic religion with a judgmental God? Is that better? My experience is uplifting, expanding, loving, creative, intelligent. Who can offer me a better reality or illusion?[61]

This is an especially telling passage, reminding of angel-channeling Bob's turn to New Age thought and practice after suffering from living modernist techno-industrial capitalist narratives and Treadwell leaving society to "save" the grizzlies but in reality desperately attempting to save himself. But the choices of Roads, Bob, and Treadwell reflect false dichotomies—we can either live modernist lives filled with polluted food and air, death, fear, greed, and a judgmental God or communicate directly with Nature and create our own reality; or communicate with angels and create our own reality, or attempt to communicate with grizzlies and get eaten. A better offer is practicing an ecology of communication.

Despite his publisher's contention that thanks to Roads "a solution is in sight," there is little in *Talking with Nature* that directly engages ecocrisis. There are moments, like when he laments the loss of Red Cedar near his home: "In his folly, man plundered the valuable harvest of timber, leaving the valley barren of its former richness. Only a few Red Cedar survived."[62] He also laments the "cities with their choked masses" that "deny the space and time to find the Self which experiences further realities," while extolling his good fortune in having spent much time among "the crystal rivers which man has not yet poisoned."[63] But another telling passage occurs when he, or a Jade plant, argues that everything is perfectly predestined and ecocrisis is an illusion: "Man claims intelligence as a thought process, whereas in truth he measures only intellect. Intelligence makes no mistakes, there are none to be made. Even the apparent threat of mankind to the stability of your planet is no more than a movement within an uncomprehended dream."[64]

Roads may be having experiences beyond my comprehension, but there are plenty of reasons to think otherwise; in this passage, he once again takes the absolute plane as all of reality, neglecting to follow through with Brahman is the world, where mistakes, and unfitting responses, are made. This is troubling, but the most troubling aspect of Roads' communicative examples, and the New Age communi-verse generally, is the sense that telepathic connections will solve all our problems. Sheldrake is right that telepathy challenges the orthodoxy that mind may be reduced to the brain, and overcoming this bias is significant because it provides further evidence for overcoming the mind/nature dualism, but long-winded telepathic treatises or channeling, which Sheldrake doesn't touch, are another matter. The ecocrisis is indeed a crisis of self-understanding and perception, and a crisis of rationality and imagination, and still more fundamentally a crisis of communication, but telepathy and channeling, when married to the attempt to create our own reality, are more often unfitting rather than fitting responses to these crises.

I am both highly critical of and fond of Roads. After all, I communed with a verdant green piece of moss, feeling like I received an inner nonverbal response to my verbal query concerning whether the rest of nature was seeing what I was seeing. But it was one word, all is "Divine," and this experience on the trail was balanced with many more grounded ones. Still, I would be remiss if I did not express my fondness by highlighting more of the positives in *Talking with Nature*. Roads laments the fact that dualisms have become our habitual "accepted reality," expressing Thoreauvian transcendentalism by criticizing our separation from Nature as an intelligent, creative, and regulative force.[65] He even questions the literalness of his experiences, wondering whether his conversations with nature are real or imagined, although he does not delve into this question in a sustained or serious manner. And, in his best line, he displays respect for the existential density of life, as well as the integration of transcendent and immanent dimensions, when he remarks: Without the "falling to earth" there can be no "transcendent flight."[66]

CONVERSATIONS WITH TREES: GAGLIANO AND KAZA

Roads obviously goes beyond what non-New Agers would likely believe, but Gagliano also reports nonordinary states, asking readers of *Thus Spoke the Plant* to allow the "magical absurdity of your subjective experience to walk alongside the objective rationality of your logical mind." Gagliano's early research focused on damselfish amid the Great Barrier Reef, where she observed transparent babies turn into bright yellow youngsters "flickering in the water like wild confetti." Despite her objective scientific training, she perceived their journey from eggs to hatching to flourishing as a "true miracle," but protocol, for reasons she does not mention, required her to end the study by killing the same damselfish that had snuggled inside her hand as her "fingers gently curled around their scaly bodies and then opened again." But her intention of saying goodbye before doing the deed turned into an epiphany. She knew they knew what she was feeling, or what they had taught her: the empathy, kinship, and compassion of the open hand. The intimacy of her encounters with the damselfish had "broken down the taxonomic boundary," and from that day forward she committed to I-Thou science.[67]

Soon after this marker moment, Gagliano turned to plant studies, and, like Descartes, her journey began with a series of three dreams that take readers into hard-to-believe territory. These were not typical dreams, but intricate ones of entering a hut among the Shipibo people of the Amazon lowlands of Peru, encountering the piercing yellow-green eyes of a black jaguar, and then a young fire keeper, and then an older Shipibo man who instructed her to sing; no sounds initially came out but then she sang a bizarre tune she could not comprehend. After awakening from the third dream, she felt called to go to Peru, and months later she did, finding Don M, the Shipibo man from her dream and a plant shaman, and also Socoba, a 35-feet tall tropical tree also known as Bellasco-caspi. Both would become her teachers, with Don M called by the tree to instruct her in the ways of plant communication.[68]

Her lessons began by following the dieta, in which, on a regular basis and in solitude, she ingested a concoction made from Socoba bark, refrained from sex, and ate a strict diet of vegetables and rice. She drank the concoction for weeks, swinging in a hammock in a hut, or rather, the hut from her dreams, with the intent of becoming receptive to plant spirits. Gagliano was instructed to pay attention to her dreams, which are a vehicle of communication in her experiences. Her first "contact" with Socoba was via a recurring image of a "dark blood red background with thick black lines," which she drew in her notebook along with a revelatory message—todas las cosas estan juntas, or all things are connected—which led to a further revelation, "it is through blood that everything is connected," and then to a final one: "Socoba was a blood cleanser." This dream insight, emerging from within her when in dialogic relations with the tree, was confirmed in the scientific literature: Bellasco-caspi has positive effects on the vascular system and may be used in healing inflammation-related diseases.[69]

While Gagliano's experiences may seem absurd to some, they are magical to her while also revealing the absurdity of spending years in scientific trial-and-error experiments when one can learn, in "a few weeks or months," directly from the plants via a practice that has been confirmed over millennia by shamans, and TEK generally, in plant dialogues experienced via dreams and visions. Gagliano argues that such learning by listening to plant others instructed early humans on what was eatable or had healing properties, and what was poisonous, rather than having to figure things out from a trail of trial-and-error dead bodies. Also, while mnemonic practices were used to recall this vast amount of information, it could never be lost because the wisdom is embedded within the plants, and they are "always open for consultation" and respond to human needs.[70]

Gagliano went on to take Ayahuasca, known for its hallucinogenic qualities, under the guidance of Don M, but, interestingly, she does not say much about it, emphasizing instead the importance of the dieta, which connected her to the Socoba tree, and still connects her.[71] She writes that Socoba still teaches her six years after the experience, if she lets go of mental workouts and becomes receptive.[72] When she asks questions with their relationship in mind, or in heart, answers effortlessly emerge. This reminds of Boone's claim that Strongheart responded to his questions with telepathic answers, but Gagliano does not suggest telepathy, and as mentioned earlier, she is critical of ventriloquizing, arguing instead that what emerges comes from both Socobo and her, or rather, from an intersubjective uprising of insight, or speaking without words. Of course, she shares insights from their reciprocal relationship in words, which can be understood as a kind of translation. Translations can and do go awry, although Gagliano seems quite certain of what she receives, not questioning whether her subjective ego desires infiltrate the conversations. These conversations are not nearly as long-winded as Roads' "talking," but they do disclose a great deal of information, like how humans and plants "breathe each other in and out of existence, one made by the exhalation of the other," or the oxygen-carbon dioxide cycle. Thanks to plants, who rarely receive our gratitude despite making up over 99 percent of biomass on the planet, we are able to breathe with the biosphere in an inter-natural giving and receiving.[73]

There is much more to this tale, including a nondual moment in which she swung in a hammock, physically there but not there, not separate, and thus fully there with Socoba, which led to a vision of being on a luminous road with all the people she had ever stumbled upon each cheering "you got it."

She realized that each person, good or bad, had taught her a lesson, and thus all were embraced with love as catalysts for remembering "the radiance of the light that is in you." At this point, Socoba is more like a guru than a dialogic partner, and Gagliano does question what is happening to her, knowing it carries the "hallmark of egomania, or at least a good dose of delusional grandeur" and assures readers that she "did doubt my own sanity many times." But, ultimately, she interprets her experience as revealing a reality of unity-in-diversity that "we mostly agree to deny."[74]

Despite these experiences and questioning the worth of scientific investigation given that we can get information directly through dream dialogue, or from asking questions with a quiet mind and connecting heart, Gagliano goes back to science, devising clever experiments to explore plant sentience. She also incorporates deep listening into her methodology by emptying herself of indoctrinated belief systems and prescribed boundaries so that exchange of information may emerge. She is very clear that what blocks listening are claims for and attitudes of human supremacy. Like Boone and Roads, practicing humility and seeing nonhuman others as equals are necessary practices for dialogue (or telepathy and channeling).

Gagliano cites plant communication philosophers to support this claim, including Michael Marder and Matthew Hall. Marder, in "To Hear Plants Speak," argues that plants have their own language, which, at the least, speaks of phusis, or growth that forms a world for them but also for us. He slips in an admission, stating that the "meaningful articulations" of humans are more varied, which I took as more complex, but then he quickly dismisses this possibility, arguing that it is a mistake to make that assumption.[75] Hall, in *Plants as Persons*, turns to Darwin's early inquiries into plant sentience, and the recent work of Anthony Trewawas on plant plasticity or their ability to respond to changes in their surroundings. They move, and are aware, and also respond to previous experiences. Hall cites Trewawas' admission that these are not reflective responses, and yet decisions are made by the whole plant. Of course, this takes us into plant neurobiology and signaling, which takes Hall to the claim that plants and persons are not much different, but the long history of human supremacy has blocked us from this equalizing realization.[76]

Gagliano, Marder, and Hall are all in agreement on the many problems associated with supremacy, especially in regard to listening to the language of plants. They understandably want to lay the ladder of hierarchy down. But the developmental holarchy and differences that make a difference remain, including Ground, intrinsic, and instrumental value. Gagliano's science is based in the humility and equality of Ground value necessary for dialogue, but we must take care when interpreting insights that emerge from this orientation; otherwise, we are back to biocentric egalitarianism where a human child has the same value as a mosquito, or a plant for that matter. Like

Bateson, she privileges analogic knowledge, which communicates context and relationship, over analytical digital knowledge, with the latter serving the former. Her method, then, is inspired by the internatural realization that all species are nature, while also mirroring Emerson's inter-spiritual claim that nature speaks of Spirit and Thoreau's championing of the language that all things speak. Yet, rationality is still part of the equation, it still serves, and thus while Thoreau listened to the voice of nature, he also attended to nature's diverse voices. Gagliano does as well, but she, and Marder and Hall, must also embrace holarchy and differences in value.

As previously argued, we cannot create reality, but what we experience reflects what we bring to interactions. We need to bring awareness of the equalizing involution of Spirit, but also the evolutionary unfolding of differences that make a difference. We also need to bring an awareness of shadow-elements and pathology. What are the controls on projection and self-deception, especially when embracing objective science and the "subjectivity of transcendental experiences and intuition?"[77] I would venture to say awareness of the developmental holarchy and thus growing up, waking up, and showing up, but also cleaning up personal biases and shadows. Thus, there must be a return to rational interpretation after embracing feeling, intuition, and spiritual experience and a return to intrinsic and instrumental value after embracing Ground value. We must compare our objective scientific findings *and* our subjective and intersubjective practices with others doing similar experiments that combine all three; in other words, communal confirmation of results but also the methods used to arrive at results.

Goodall was an outlier, and a prophet of sorts, and, fortunately, she is still speaking out on behalf of the voice and voices of nature. Barbara McClintock was another outlier, winning the 1983 Nobel Prize for her investigations into maize cytogenetics by ensouling the corn under study to learn more than orthodox scientists. McClintock was a mystic-scientist, displaying "feeling for the organism," and I am a bit surprised that Gagliano does not mention her.[78] They not only have a shared epistemology but Gagliano references novel experiments where it was discovered that corn plants signal via sounds, or "loud and chirpy vegetal clicks," which were heard via sensitive scientific laser instruments.[79] Along with communicating via sound, Gagliano writes that experiments have disclosed that plants exchange information among themselves via touch, wavelengths of light, contact induced by gravity, changes in pressure gradients, the "vocabulary" of chemical molecules, underground root networks, and airborne cues.[80] Of course, human-plant dialogue may occur through dreams and visions and learning the songs of vegetal others, all inspired by the tradition of shamanism, or being in sensual rapport or nondual rapport, or just being in relationship and listening without and within.

These days, Gagliano may find plenty of other I-Thou scientists with whom to compare notes on methods and results, including Kimmerer, and I am sure she does at conferences. But, always, we must return to the question of sentience in relation to complexity and the three kinds of value. Gagliano's experiences and research further reveals that there are fewer differences that make a difference, as it is no longer mere anthropomorphism to conclude that plants have intelligence, feelings, language, and voice, albeit expressed differently. But those differences in expression still reflect differences in levels of complexity, despite the biosphere and noosphere being deeply intertwined, as Bateson argued, and the intelligence of evolution as a whole deserving our main focus. Human thought and language and tools are still more complex, still reflect more of the developmental holarchy, despite animals and plants and insects being able to do many things we cannot and despite humans desperately needing to respect nature as mentor, model, and measure. We must lay the ladder of dominating hierarchy down, but raise the integral ladder of holarchy for the sake of all species, as the complexity of human consciousness includes the possibility of alienation and the illusion of separateness, leading to the stupidity of fouling our own nest and acting like the clowns of evolution.

In many ways, Gagliano practices internatural, and inter-spiritual, communication more than interspecies communication, as she sees herself as part and parcel of nature, and states that she has been inspired by mystics. She has a deep identification with other forms of life, and this has led her into new terrains of rigor and imagination, and new discoveries, especially in regard to bioacoustics. However, unlike Plec's internatural communication, which focuses on animals and empathy, Gagliano focuses on plants and emptiness, as empathy is more of a projection; it is often a good projection of not just thought but feeling that leads to the good result of breaking down barriers, as Plec's anthology discloses, but as a projection it is prone to downsides like narcissism, or incorporation, of which Roads provides many potential examples. The practice of emptiness, on the other hand, leads to "knowing by deep listening" that emerges as "perfect surprise, pertinent to the moment of the encounter"; in other words, grace and gift.[81] Still, genuine empathy is also a gift, and may well lead to emptiness and the experiences of listening Gagliano describes.

What we need is the integration of a variety of practices, and thus the dialogic interplay among differing forms of communication. Gagliano goes a long way toward doing so, although she does not spend much ink questioning possible dysfunctions, and thus she does not spend much ink discerning the unfit. I may be expecting too much—Gagliano thoughtfully shares her experiences and her science, and readers may critically interpret—but Stephanie Kaza, a scholar and professor with a PhD in biology, provides a stellar example of both questing and questioning in *The Attentive Heart: Conversations with Trees*.

Published in 1993, one year after Roszak's *The Voice of the Earth* and before Abram's *The Spell of the Sensuous*, it a significant contribution to the history of ecopsychology. Kaza is also the author of *Green Buddhism* and the editor of two volumes on Buddhism and the environment, and *The Attentive Heart* makes it clear that Buddhism is not just a scholarly topic for her, but part of her spiritual practice and life. And a blurb on the back by Joanna Macy, one of the cocreators of the Council of All Beings, states that it reflects deep ecology writing at its best. *The Attentive Heart*, then, integrates insights from biology, ecopsychology, Buddhism, and deep ecology, as well as ecofeminism and other sources, covering a lot of eco-academic ground while providing a handbook of sorts on practicing an ecology of communication.

Kaza writes with strength and passion, but also humility, recognizing that attempting to converse with trees brings forth questions more than answers, which is in direct contrast to Roads' literal conversations bringing forth questionable certainties and the likelihood of self-deception. For Kaza, a widened awareness makes fitting responses more likely, and thus she integrates scientific inquiry with bodily empathy and the senses, critical reason, a heightened aesthetic sensibility, and spiritual perception to investigate the possibilities for dialogic relations with nature within the nondual context of Spirit unfolding.

The Attentive Heart was re-released in 2019 as *Conversations with Trees: An Intimate Ecology*. It is obvious that the publisher, Shambhala, noticed the proliferation and popularity of dialogue with nature titles, and that Kaza made an early contribution. I found her book long ago, immediately drawn to "conversations with trees," but I imagined that some environmentalists might knee-jerk judge it as a bit out there, a bit New Agey, even though a closer look discloses roots in science and Buddhism. Conversing with trees, rather than animals, also took it more toward *Secret Lives* and away from Goodall, who by that time had been honored and accepted. Regardless, all that has changed these days, making Kaza a bit of a trailblazer.

Critics of *Conversations* would likely cite purple passages of anthropomorphism. Kaza is quite clear about her affection for trees and admits to the possibility of psychological projection.[82] However, anthropomorphism expressed as sympathy is more easily defended within the context of the extended mind, although Boone and Roads have taught us that critical discernment is essential. For example, Kaza refers to branches that are "still full of delight," sympathetically projecting that the "corners of the mouth are smiling," but she does so within the context of a biology lesson on golden leaves being nourished by springtime sugars in the root system of a gingko tree as wintertime approaches.[83] Her descriptions, utilizing both scientific rigor and the sensuous imagination, reflect her experience of shared intelligence and aesthetic communication. We are wise to question her experience, as she does herself, but we will find a deeper wisdom by following her lead, bringing an attentive heart and range of responses to our encounters with the other-than-human world.

In a provocative introduction, Kaza writes that trees have always been a presence in her life, from childhood play to study as a naturalist, to spiritual practice inspired by Buber's articulation of I-Thou relationships. Kaza studied biology at UC Santa Cruz and has spent much of her life in the Pacific Northwest, listening to the call of specific places. Thus, her connection with trees is tied to this bioregion and has been intensified by the loss of California Oaks and the clearcutting of Northwest mountaintops. Her teaching and travels have also taken her to Thailand and Costa Rica, where she witnessed damaged forests. All these experiences confronted her with the tension between shortsighted economic policies and long-term ecological integrity: "I began to realize that the social ethics of industrialized culture did not include any thoughtful basis for mutually respectful relations with trees."[84]

Conversations with Trees is Kaza's response to the destruction of forests and ecocrisis generally. It chronicles her search for a method for conversing with trees and her search for meaning during difficult times. Kaza's writing practice with trees is based on "just sitting," or the Zen form of *Shikantaza*, spending time in silence, "close to trees, doing my best to be simply present with the tree as Other, aware of my thoughts, moods, and projections." She did not know what to expect, and not having expectations, or an agenda or preconceived story, was part of the practice as she "persisted in the experiment." Her writing was "an excuse to listen to the call from the trees," open to whatever unfolded, to whatever came through, "by being completely present in the specific place and moment."[85]

For Kaza, trees may be a catalyst to revelatory experiences and awakening, disclosing the truth of global interdependence. Such revelations penetrate to the core of who we are, eliciting awe, wonder, love, and most importantly, an ethical response. Her approach, however, is investigative rather than prescriptive, as "moral recipes for human-tree relationships tend to be oversimplified, cutting short the depth of transformation possible." The point is to not only investigate outward nature but our inner nature, becoming mindful of patterns that condition our thoughts and actions or both results and methods. This rigorous investigative approach leads her to be critical of our habit-bodies, including "habits of language and mind that block the flow of communication between person and tree." Kaza knows that her attempts to dialogue with trees may be awkward and that "anthropomorphic projections" as well as "stereotyping, objectifying, idealizing, and oversimplifying" are possible, but her Zen training leads to the conviction that "genuine contact at a core level," which would be a nondual level reflecting spiritual communication, is also possible.[86]

Despite her training and conviction, Kaza proceeds slowly and always with humility. The five sections of *Conversations with Trees* reflect steps to a widened awareness, or, following Bateson, steps to an ecology of mind, and then further steps to an ecology of Spirit. She also includes a disclaimer of sorts, stating that her experiences are her own—she is not speaking for anyone else. Still, while others may discover different encounters and different conversations, she hopes that her revelatory experiences are recognizable and useful.

In the first section, or her first step, Kaza acknowledges her "uncertainty and clumsiness in the conversation." She begins by practicing differing types of greetings—getting to know the land, observing with field guides, opening her senses—while exploring what it feels like to be addressed by the trees. In a chapter entitled "Called by Alders," she moves from a walking meditation to touching an Alder: "Listening through my hands, I meet this tree from my own experience of sunlight and stillness. . . . The tree itself is a manifestation of rhythm, of the way light works. . . . This is what my hands recognize—this movement of sun, earth, and water."[87] Kaza states that, at some fundamental level, she and the tree are "made of the same rhythms" and "share a common understanding" and that reaching out, offering her hand, and meeting via touch is "a simple way to begin a conversation."[88]

In the second section, Kaza confronts the complexity of her desire for dialogue. Trees tell many stories, including of "fire, agriculture, and commercial cultivation," as they are "victims and shapers of human activities." She engages the material realities of trees as an aspect of the spiritual, while recognizing the difficulties of dialogue. Or rather, she balances her desire for mythic-animistic, aesthetic, and spiritual communication with rational reflections. The third section continues her reflections, heightening the complexity and difficulties of dialogue at the same time that she more fully enters into relationship. She realizes that trees also tell "painful stories of fear, killing, unconsciousness, and objectification." Seeking a mythic-animistic, aesthetic, and spiritual rapport with trees entails encountering them in their wholeness, and, as much as possible, our wholeness, and that includes the rational revelation that there are no easy solutions to ecocrisis: "I am caught in dialogues of time and place that reflect a long history of habits that distance and kill the Other."[89]

It is only in the fourth section, after much attentiveness to the embodied qualities of trees, that Kaza's inquiry leads to a "stable and attentive mind." She experiences reciprocity, feeling compelled to act from a "context of mutual causality." While outer circumstances have not changed, her relationship to trees has changed through the cultivation of spiritual awareness. In the fifth section, she seeks ways to "restore spiritual as well as biological relationships with trees." In other words, she has moved from the material to the spiritual and then to their integration, more completely "recognizing the dynamic workings of mind" and "the awful truth of the damaged planet."[90] Kaza's sections mirror stages of growth, with successive chapters disclosing her struggles and successes in conversing with differing tree species embedded in specific places, all within the context of a world in need of ecological and psychological healing.

Kaza articulates this context in chapter 1, "Close to Water," by telling the story of a spring visit with California sycamores. She writes that to feel the lure of sycamores is to also "speak with water," as "sycamores mark the trail of watercourses through dry and wet years." Sycamores need a steady supply of groundwater, making them the "great trees of riverbanks, of wet places, of fertile valleys, of high water tables." She proceeds to supply a biology lesson on the relation between water, photosynthesis, trees, and the wider landscape, while remarking that sometimes the best way to seek dialogue is to lie on your back, with "nothing between you and the sky except the arms of the tree."[91]

Kaza notices the lack of leaves on a particular sycamore. She is typically drawn to sycamores for their "porous canopies of maple-shaped leaves"—the large leaves letting in light for photosynthesis while providing moisture-preserving shade. But this tree does not have large leaves, nor "flower strings of fuzzy golden ornaments." Instead, she perceives an "assortment of spherical seed pods and a few shriveled leaves" decorating the "tree's empty arms." She then surmises that a cold front must have caught the sycamore by surprise, leaving the leaves "limp and lifeless" and the tree as a whole to "try again under more favorable circumstances" Kaza goes on to ruminate on the importance of timing, along with water, to the sycamore's vibrancy. This leads to further reflection on both the vulnerability and resiliency of trees, and the "unpredictability of the ebb and flow of life force."[92]

Kaza states that she has always counted on the resiliency of trees, because they support "a certain confidence that life will go on, noticed or unnoticed." But she also admits to the limits of her scientific knowing and her ability to know generally. This humility, which in many ways is gifted by the otherness of trees rather than our connection, is then juxtaposed with the arrogant behavior that has long threatened sycamores, including 90 percent of sycamores lost in the last century due to land clearing, river channeling, and new housing, or human expansion and encroachment.

For Kaza, the remaining sycamores are reminders of a gracious, and therefore more humble, time when people sat on porches enjoying summer shade, perhaps realizing that sycamores nourish their lives, making them less likely to destroy them. But that is no longer the case: sycamores, as objects, are subjected to "cumulative hardships" of reduced water supply, agricultural pesticides, and air pollution. She laments: "I fear the subtle stresses will cut short my conversation with sycamores, reducing the possibility of long-term relationship."[93]

This early passage sets the stage for Kaza's inquiry. Ecocrisis is given specific features, shaking her resolve while propelling her forward in search of reciprocal converse with trees as one fitting response. She emphasizes not knowing, but also our ability to know more than our current monologic mindset permits. This desire for dialogic knowing—characterized by insights that only emerge through a willingness to practice, or do the experiments, exploring possibilities for converse with nonhumans—have a recollective purpose, providing wisdom from our mythic-animistic past. Kaza writes: "I want to know something about long-term relationships with trees that last over centuries. I want these sycamores to tell me stories from the living stream of their history over millions of years. . . . In a time when fragmentation dominates the landscape, I need to meet these trees in sycamore context, investigating the wholeness of their lives."[94]

Kaza's encounter with sycamores explores mythic-animistic communication and the practical, rather than theoretical, focus of ecopsychology. She is attentive to the specific place where sycamores flourish, or at least where they are meant to flourish, reading the story of the sycamore clan through recollective practices. However, while she states that the best way to know a sycamore may be to simply lie on your back underneath its glorious branches, she also explores knowing that comes from science and history. In other words, she is not only sitting with trees in silence but practicing transversal rationality, contemplating the mythic history of the sycamore through time and within place, leading to rational insights that carry over into aesthetic responsiveness to rhythms and patterns.

Kaza's desire for wholeness leads to a plethora of human-tree relational practices. When she feels called by alders, she takes deep breaths, "each foot touching the earth in slow, rhythmic strides, connecting small body to large body." This bodily empathy engages her sensuous imagination, causing her to wonder if the alder knows when it is touched. This animistic query leads her to imagine the other as alive and responsive. Kaza's intention is to invite initial contact, while acknowledging her not-knowing and limitations, and that trees have different minds and voices. She admits that she does not know the language of the alders, leaving her to "guess at the shape of a tree's mind and what it knows about the life on the edge of the pond" and ask more questions: "How does the water taste to an alder? How does the morning sun feel on its new leaves? How does the wind feel moving through it branches?"[95]

Such questions do not need direct responses in order to disclose insights. Simply asking the questions establishes connections that did not exist prior to the asking. Such questioning, in which we do not receive immediate responses, also undermines Roads' literal conversations in which he easily speaks the language of nonhumans, or nonhumans easily speak his language via his projections. There are no comparable passages in *Talking with Nature* exploring the complexity of dialogic relations with nature nor does Roads match Kaza's complexity and depth generally. Both *Conversations with Trees* and *Talking with Nature* reflect spiritual journeys, but Kaza is far more critically aware of the roots of ecocrisis and our psychological disconnection, and of the tendency for self-deception and easy answers.

The obvious difference between Kaza and Roads is that she more fully practices rational communication and this dialogically informs her further practice of mythic-animistic, aesthetic, and spiritual communication. Roads and Kaza both gain confidence in their communicative abilities as their narratives unfold, but Kaza always returns to the openness of not-knowing and questioning that is essential to dialogue, while Roads' sense of certain knowing balloons, which, ironically, likely leads to monologue within his head. Kaza's questions act as an invitation to the possibility of reciprocal I-Thou relations, and perhaps nondual direct contact, while she stays mindful of critical rationality. Her methods are also different than Gagliano's, although they both focus on emptiness and dialogue and include their scientific backgrounds and practices, which is a difference that makes a big difference. However, Gagliano, inspired by shamanistic traditions, especially the dieta, converses with Socoba via visions and dreams, while Kaza takes smaller steps, and more steps, which are demanding in the sense that one has to put in much time and effort, but also easier to follow.

Still, some would return to the accusation anthropomorphizing due to Kaza's willingness to even ask trees questions, or to imagine what trees feel and taste, or by claiming that trees have minds of some type, although, as Gagliano and other researchers have shown, plant neurobiology is quelling that debate. Of course, distinctions must be made between accusations from those who operate within a Cartesian monologic mindset and critical concern with the quality of our projections from those who are exploring dialogic relations with nature. Plus, there are numerous examples of panpsychism and experiences of the world soul that must be traversed before claims of anthropomorphism are given merit. Mind placed back in the natural world will likely lead to abuse, as Roads' example illustrates, but when rational communication is consulted, as Kaza illustrates, it also leads to a widened awareness and more fitting responses.

Kaza continually exhibits the attentiveness and openness of mythic-animism, while also respecting trees as both other and self by recognizing that the beauty of nature's rhythms and patterns are also her rhythms and patterns. Like Bateson, she takes both a scientific and aesthetic approach, especially when meeting, and greeting, trees in Yosemite National Park: "One part of my mind wants to identify each of these delightful beings scientifically by name; another only wants to gaze at their exquisite beauty." This delight overflows into exuberance: "*Hello, noble trees! You are magnificent!*"[96] and then transfixed description of red firs: "Abies magnifica—magnificent—magnificent firs. Deep chocolate-red coats of furrowed bark, stately branches in the somber darkness of the forest: even the fallen trees carry a grand presence."[97]

Differing tree species call forth differing responses from Kaza. The bark of red firs discloses the elegance of pattern, making her aware of timelessness, unity-in-diversity, and the "evolutionary lineage of tree intelligence."[98] She experiences a moment of deep connection that elicits a sense of responsibility to continue seeking insight through science and beauty. Her responses to redwoods are influenced by visiting them with children, to whom she gifts a biology lesson on the regeneration of seedlings that draw nourishment from decaying stumps, while noticing their biophilic responsiveness to the trees; the children naturally understand the "truth of conviviality and friendship."[99] The redwoods, in response, share their ancient and comforting presence. And maples speak to Kaza of eros and ecstasy, reminding her of the yearning to touch and be touched and the call to intimacy: "The desire to merge slowly and imperceptibly is the deeper impulse; adrenaline merely initiates the motion. I have been propelled into the company of maples, irresistibly pulled by the desire to meet. Now, in the actual presence of the tree, I soften with the tenderness of the dance between two beings."[100]

The stories that these trees convey to Kaza are more than personal projections; they are dynamic examples of mythic-animistic and aesthetic communication revealing eros, relationship, and reciprocity. By being present with trees with a listening attitude, we become privy to our long relational history and timeless connections to place, which then provide insights helpful to healing damaged habitat in the present. Restoration projects, for example, may be guided by practicing an ecology of communication, allowing us to re-story the land. For Kaza, dialogic relations with nature, while individually enlightening, also provide insights that support a collective response to ecocrisis. The trees and land speak via their presence, and we must be mindful of their presence within the extended mind in order to discern fitting and unfitting responses. However, rational reflection reveals that such mindfulness does not always lead to easy decisions, including in restoration projects, which Kaza discloses when exploring the instrumental realities of human-tree relationships.

The chapter "Fallen Tree" begins with the unnerving whirr of a chainsaw interrupting a peaceful day at home. Kaza surmises that it is probably the volunteer fire department removing a Monterey pine that had fallen across the road. It's a necessary task, but Monarch butterflies are just returning to

California and she laments the rude greeting. Monarchs choose the pines for overwintering, and she remarks that several hundred have moved in, and the previous year three to four thousand "clung to these trees in massive orange clumps." She wants returnees to feel welcome, but the "WhreeEEEeeeEEnn" of chainsaws are "not helping one bit." The piercing sound turns her attention to the fallen pine, which must be rotten at the base. Kaza expresses sadness at the death of the tree, as well as gladness that it died a natural death. However, its removal with the chain saw, instead of decomposing back into the soil, seems artificial and invasive.[101]

The situation is complex, especially since some biologists argue that the Monterrey pine is an invasive species, with its natural habitat 75 miles to the south. Kaza remarks that neighbors are concerned that there are too many Monterey pines in the area, then muses: "With this reasoning, should I rejoice at its death? On the other hand, cone fossils found in Marin County indicate the presence of earlier maritime pines in this section of the coast. One could argue that the current northward expansion of Monterey pines is a natural evolutionary trend toward reestablishment of a similar type of tree."[102] Kaza states that it is "difficult to sort out these responses without oversimplifying the situation," while also remarking that we must make choices.[103]

Kaza has also used a chain saw to cut firewood and is fully aware that her house is made of wood. In fact, being "surrounded by wood and the memory of trees" is where she finds respite from "the sadness and tenderness that speaks through me as I teach about how we are living with the environment, how we are dying with the environment." Still, chain saws, while a useful tool, are metaphoric for the abuse of trees: "It is so insistent, so aggressive, so capable of overriding the resistance of the tree." Like the tree, she begins to feel worn down by the relentless growls and snarls of the saws: "The penetrating vibration presses into my chest, biting away at my heart, carving my trunk into segments. WhrrEEEeenEEnn. The racing sound slices through the tree and neighborhood, dominating the quiet butterflies and the roaring ocean with its presence."[104]

Kaza's dialogic relations with trees does not recede when they are put to human use, and she wishes others would also maintain such relations: "I wonder who among the many people who deal with wood as a product have walked in the forests of these trees and listened to their voices."[105] For her, when connections are broken and memories vanish, the wood becomes like a corpse, or worse, "something that appears to never have been alive." Kaza laments forgetfulness of nature's aliveness here, and thus mythic-animistic communication, which, as always, is due to habits of objectification and calculation: "An object is something to carry around, to count, to purchase, to collect. It is something separate. The process of objectification begins with the first cut toward straightness. After the trees are felled, the conspiracy of object continues in the timber sales report, lumberyard accounts, and architectural plans."[106]

Kaza's reflections on the Monterey pine and chain saw are a prime example of her nuanced integration of modes of knowing and communication, as well as respect for complexity and difficulty. Such dialogic interplay, mixing and prioritizing based on particular situations, goes to the heart of an ecology of communication. In a poignant chapter on instrumental use and ecocrisis, "Overtures of Peace," she provides further evidence for this enlightened practice. Kaza revisits the majesty of redwoods, seeking refuge from the violence of human words carelessly thrown, and then refuge from the rain. She finds peace and shelter in the base of an enormous stump, but she soon realizes that she is resting within a tree logged for timber: "I cannot escape the presence of the mind that kills."[107] This causes her to ruminate on the mind of the redwood, juxtaposing the meditative stillness and silence of redwoods, or the "yogis of the forest," with the killing mind of humans.[108]

Like Boone and Strongheart, and Roads and all nature, and Gagliano and Socoba, Kaza perceives the guru-like qualities of nature in contrast to the follies of humankind, as it once took two weeks of noisy axe-work to fell a redwood, but the myths of technological progress and unlimited growth, and the machines that define them, have shattered the silence of the mythic redwoods. She provides a history lesson:

> With the invention of the motor and the automobile, it was a short step to bulldozers, trucks, and power saws. By 1929 the rate of cutting was up to 500 million board feet a year. The war was on, the war of the killing mind against the living. . . . The obituary report released by the National Geographic Society in 1964 revealed the shocking and enormous loss. Of the original two million acres of ancient redwood forest, only 300,000 remained—a mere fifteen percent of the stillness that once graced the coast of California.[109]

The killing mind, and machines of course, have marched on, with the percentage of old growth much lower today, as the mind of abstraction and extraction is not only divorced from larger mind, but body, earthbody, and sensuous imagination. In still another poignant chapter, "Bones in the Land," Kaza returns to Santa Cruz hoping to spend time with her two favorite oak trees. But the killing mind has gotten there first. During her college years, she often spent time with these older oaks, enjoying shade, backrests, napping, tree climbing, picnics, solitude, sleep-outs, journal writing, and companionship—all spiritual practices. Her desire to visit one of her old friends is ruined, however, when the landscape is absent one of its community members, having been cut by a chain saw. Kaza writes:

> The tree left the world of the living in the hands and minds of people who did not see it as a living being with a rich history of relationship. They did not see the warblers, kinglets, and woodpeckers that nested in its branches and cavities. They did not see the ancient patches of pale-green foliose lichen on the wizened trunk. They did not see the lifetime of storms and coastal winds that blew through these branches. They did not see the underground roots interwoven with the Douglas fir ten feet away. They did not see the human friendships that grew out of the life of this tree.[110]

Kaza's love for the oak and the landscape where it dwelled is marked by the joy of relationship, and her sadness emerged at the loss of this relationship, reminding of Leopold's "world of wounds." It would be easy to turn our backs on this wounding or to not want to feel it, but we must see it all if we are to see at all. We also cannot turn our backs after we have lain on our backs looking through the canopy of trees, learning about extended mind: "The landscape had taught me well, cultivating a mind of place, a mind *in* place, a place of mind."[111] But despite her learning, she admits to being influenced by the myths of unlimited growth and technological progress, the metaphoric power of the chain saw, and the habits of forgetfulness when she loses root connections that maintain her life and the life of the tree. She knows "something about moving too fast" and not noticing what is right in front of her. Like all of us, the faster she goes, the more she consumes unconsciously, eating "the ground below my feet" amid the "dizzying thrill of speed." The connection was still there, "speaking through" her, but she had stopped paying attention; when her attention returned, the "tree soul" in her "screamed with the shock of sudden loss."[112]

Kaza's criticisms of the killing mind and instrumentalism once again leave us with the central question: what is the fitting response? But her sensitivity to not-knowing prevents her from pontificating a final answer; instead, she ponders, points, and prods, turning the question into a Zen koan. More specifically, she contemplates the creative tension between instrumental use and mindfulness while stacking wood as if crafting a stone wall, placing logs by entering into communion, making it sturdy and long-lasting. She then attempts the same contemplative practice with the device of the chain saw: "Now I am the one cutting the wood; it is my hands on the chain saw. It is my fingers on the trigger making the whirring and biting sound. Can I stay conscious and aware of what I am doing? What conversation do I have with these trees as they fragment into firewood in quick, efficient slices?"[113] It would seem that dialogue is not possible in this situation, only one life-form monologically dominating another, but Kaza works to embody a humble attitude while working, akin to predator-prey reciprocity, becoming aware of and thankful for the gift of firewood: "I cut the limbs with a requesting heart. It is a cold day and I need the warmth from these trees. I am asking them to serve my life. I am asking them to enter my bones and blood and fuel my cells with fire, to make it possible for me to stay warm through the winter."[114]

Kaza, as always, continues to ask questions and seek insights, even when they are not readily available. These koan-like questions must be wrestled with, engaging the critical mind within the context of the extended mind: "The hungry chain saw sputters. I stop to refill the tank with a mix of gas and oil. Gas and oil from where? The Persian Gulf, Alaska, or the coast of Southern California?"[115] She asks more questions, following the trail of oil through war zones or ice floes before entering her saw, and then wonders about the cost in transportation and "energy to produce this liquid gold?" Questions multiply as she sinks further into the core of relationships that inevitably include instrumental ones. The koan of the fitting response "digs into conditioned thinking, unconscious patterns, habitual ways of seeing," working her like a teacher, opening the possibility for insight.[116]

Kaza experiments often with greetings, humbly investigating what differing approaches will call forth. But how does one greet a chain saw? When she approaches the gingko tree, she greets it like an old friend, since she has visited for many years. The gingko's companionship as it prepares for winter leads her to ruminate on change and impermanence and slow time, and she receives a revelation: "The deeper the stillness, the better the capacity to respond to change when it comes."[117] But is stillness always possible, or desirable, amid the many storms of ecocrisis? Do we sometimes need conscious rage? Is there such a thing as wu-wei rage against the literal and metaphorical machine? After much mental wrestling, Kaza seems to find such stillness with the chain saw, while seeking and then experiencing a deeper relationship with the wood she is cutting.

As previously described, I know such stillness and relationship with wood when using the technology of the axe, but I have found that the aggressiveness of the chain saw, which Kaza has compellingly described, elicits power in the user—it feels good, so good that it provides insight into why loggers like to log. My dialogue with wood, therefore, is different based on the bias of the technology utilized, leading me to wonder if less efficient technology may provide a deeper stillness and relationship. Robert Pirsig wrote of Zen stillness when maintaining his motorcycle, greeting it as an old friend while seeing its basic elements as earth elements, but axes definitely elicit less noise, or differing sounds, including my grunts on the downward strike, and perhaps a differing experience of reciprocity. But Kaza's realization has caused me to question my biases while questioning hers. Koans are designed to elicit questioning and paradox that awaken the stillness of the enlightened mind as we realize the futility of thought. The fitting response is impossible to figure out, even though we learn much from trying. But after the trying, or rather, after another Zen koan—Stop trying, stop trying not to try, stop stopping—we may settle into the deeper stillness of a fitting responsiveness, bringing wider awareness to particular situations, like cutting wood for winter.

For Kaza, the chain saw inspired koan furthers her rational respect for complexity and nuanced positions concerning the attentive heart, "the heart that feels the presence of others and the call to respond, the heart that lives in relationship with other beings." The instrumental realities of life are not brushed aside, but engaged, hopefully with mindfulness, but always with rationality informed by dialogic relations with nature. Despite finding some peace with cutting wood, she admits that practicing an attentive heart is "painstaking and demanding" and prone to self-deceptive projections and habitual disuse: "In each moment of observing a leaf, a squawk, a firm touch, there is a temptation to make it something more than it is—an object of fascination, a delirium of nature bonding, a symphony of deliberate orchestration."[118] But a deeper danger is not trying, and perceiving nature as less than it is, "missing the context and history of the tiny event striking the senses." Bare attention is an "impossibly thin razor's edge," and we are bound to fall off: "Fall and return, err and correct."[119]

An attentive heart takes effort, and then release from effort, allowing us to become more mindful in specific moments and open to the grace of spiritual communication. But mindfulness, if it is to include all modes of communication, must be grounded in, well, the ground, the body and earthbody, and the deep history of the ecological unconscious, which includes listening to significant dreams. Gagliano's dreams provided specific messages, but we may have more general dreams of contact and connection with nature. Greenway states that hikers often have more vivid and meaningful dreams after a day of hiking has quieted their mind and they sleep beneath the stars and within the wilderness. Kaza writes of such experience: "I woke up this morning under the graceful, arching branches of bay laurels and Douglas firs. All night the trees have been conversing under the full moon, weaving me into their stories, capturing my dreams with their leaning limbs and generous trunks."[120]

Kaza knows the peace of dreaming in wilderness, as well as the ecological unconscious's desire for wildness. She seeks nondual heights, but does not seek to transcend without including the ordered chaos of trees and their root systems: "The roots go where they find passage; they make their own way through the orderly routes of least resistance—along rocks and hard places out to the softer spots."[121] Kaza also wishes to find passage, to "go down into the dark and know the labyrinth of support and nourishment," creating "pattern and design and history like turns of the root." She does not fear but embraces the dirt, going down in the "burrows of beetles, through the tunnels of worms into the maze of root hairs." Going down, descending, she finds more of who we are: "The mind of chaos is the mind of wilderness—pure expression of uncontrolled and untameable life force."[122]

Conversations with Trees challenges readers to begin their own practice of conversing within the more-than-human world. Such practice is informed by the project of ecopsychology, including experience of a wild and chaotic life force, as well as the relational selves of deep ecology and ecofeminism; however, she also goes beyond them by acknowledging experience of a transcendent dimension that is fully immanent. She writes:

> There is a tension in my attraction to the infinite. Given a grand view, one becomes a visionary. The sky calls me out to see beyond the ordinary limits of life in human form. I stand in the wind, on the edge of the rock, stripped by the elements to what I share in common with all I see. The shock of this expansiveness awakens the truth of interpenetrating realities. I am the sky, I am the rock, I am the pine. The wind is in me, in my breath; the stone is in me, in my body; the tree is in me, in my mind. The truth of this is real; I recognize the depth of it.[123]

This passage reflects nondual spiritual communication, with Kaza's experience of unity speaking to her. But her ecopsychologically informed connection to the ecological unconscious, wildness, and the earth's diversity, as well as the practices of mythic-animistic, rational, and aesthetic communication, once again leads her to celebrate the immanent:

> At the same time I exist in the limit of who I am, distinct from all other beings and forms of matter. The wash of the infinite clarifies the sharp edge around my form, naming me the recognizer. In the same moment I experience formlessness and form—complementary tensions, each constantly changing into the other. I flow into the tree, inhabiting a shared conversation, and at the same time flow back into myself, watching the edge of the self change.[124]

Kaza's embrace of emptiness and form provides the context for all her conversations, but in the end, she has "no grand conclusions." Instead, she desires more conversations, and thus the opportunity to listen and learn from the stories of nonhuman others, allowing her, and all of us, to better fit within the eco-spiritual narratives that sustain us, rather than habitually conforming to the myths of unlimited growth, linear progress, and technology as savior. Kaza is aware of material and spiritual processes, aware of Plumwood's deconstruction of master-slave dynamics and interpersonal dysfunction, aware of bias and personal dysfunction, aware of the difficulties of crossing boundaries and making contact, and aware of the need to critically interpret once we do. There is no dieta, no ayahuasca, no specific converse via dreams and visions—which have merit, depending on the practitioner, as Gagliano has wisely shown—but plenty of greetings and meetings that reflect the dialogic interplay of an ecology of communication.

In this process, she provides plenty that Peters would applaud in regard to perceiving communicative blocks and admitting fallibility, while also gifting a grounded responsiveness that we may also practice. If there are imaginative flights, like enlightened redwoods, her grounding brings them home, making them less like anthropomorphism and more like feeling-with inspirations of Spirit in the world. We may take such moments as moments, integrating them with the transversal play of so many others as we move to meet the next.

When writing this chapter, my beloved dog, Rufus, who arrived to our family from a shelter in Tennessee some twelve years earlier, passed away. We made the difficult decision of putting him down, at home and outside, after his back legs continued to fail him. We tried to prepare, but there is no preparation, only the trying that gives way to responding to the mystery of the other, the mystery of death, and the mystery of that moment. Rufus, in his 100 pounds of soft black fur with white patches and big paw glory, was not like an enlightened guru, but he was saint-like to us, responding to sadness by nuzzling with his big body, nose, and inviting eyes, wanting only to give and receive love.

After many tears, I have already begun talking to him in my head, and sometimes aloud, as our bonds change from mostly physical to something more. While the "something more" must be questioned and interpreted like everything else, seeing Rufus as saint is not mere projection, or bias, but one description of how we met, or how he often met us, and what matters most is a continual interspecies, internatural, and inter-spiritual reaching out, learning more by being open to mystery.

NOTES

1. *Grizzly Man*, directed by Werner Herzog (U.S.: Discover Docs/Lion's Gate, 2005).
2. Harries-Jones, *A Recursive Vision.*
3. Christopher Riley, "The Dolphin Who Loved Me: The NASA Funded Project That Went Wrong," *The Guardian*, June 8, 2014, https://www.theguardian.com/environment/2014/jun/08/the-dolphin-who-loved-me.
4. Ibid.
5. Ibid.

6. Peter Tompkins and Christopher Bird, *The Secret Life of Plants* (New York: Harper & Row, 1973), 135–144.

7. SF Environment, "Celebrating Black Environmentalists During Black History Month," https://sfenvironment.org/article/celebrating-black-environmentalists-during-black-history-month#:~:text=%20Celebrating%20Black%20Environmentalists%20during%20Black%20History%20Month,his%20career%20working%20to%20solve%20social...%20More%20.

8. Tompkins and Bird, *The Secret Life of Plants*, 3–19.

9. Derrick Jensen, *A Language Older Than Words* (White River Junction, VT: Chelsea Green Publishing, 2004), 292–304.

10. Monica Gagliano, *Thus Spoke the Plant: A Remarkable Journey of Groundbreaking Scientific Discoveries and Personal Encounters with Plants* (Berkeley, CA: North Atlantic Books, 2018), 1–2.

11. Ibid., 6–7.

12. Emily Plec, "Perspectives on Human-Animal Communication: An Introduction," In *Perspectives on Human-Animal Communication: Internatural Communication*, ed. Emily Plec (New York: Routledge, 2013), 4–5.

13. Gagliano, *Thus Spoke the Plant*, 34–35.

14. Plec, *Perspectives on Human-Animal Communication*, 6–7.

15. Klimo, *Channeling*, 178.

16. J. Allen Boone, *Kinship with All Life* (New York: HarperCollins, 1954), 26–27.

17. Ibid., 35.

18. Ibid., 44.

19. Ibid., 59.

20. Ibid., 68.

21. Ibid., 71.

22. Ibid., 74–76.

23. Ibid., 83.

24. Ibid., 82.

25. Ibid., 131.

26. Ibid., 144.

27. Ibid.

28. See Bill McKibben, *The Comforting Whirlwind: God, Job, and the Scale of Creation* (Cambridge, MA: Cowley Publications, 2005).

29. Jane Goodall and Marc Bekoff, *The Ten Trusts: What We Must Do to Care for the Animals We Love* (San Francisco: Harper San Francisco, 2003), 171.

30. Boone, *Kinship with All Life*, 41.

31. There are numerous translations of the *Tao te Ching*, including online at egreenway.com, http://www.egreenway.com/taoism/ttclz23.htm. The Thoreau quotation comes from The Thoreau Society Facebook group, which sends daily quotations. See The Thoreau Society at thethoreausociety.org.

32. Michael Roads, *Talking with Nature/Journey into Nature: A Michael Roads Reader* (California: H J Kramer, 2003), v–vi.

33. Ibid., 9–10.
34. Ibid., 10–11.
35. Ibid., 15.
36. Ibid., 16.
37. See the "Thinking Like a Mountain" chapter in Leopold's seminal *Sand County Almanac*.
38. Roads, *Talking with Nature*, 16.
39. Ibid., 17.
40. Derrick Jensen, *A Language Older Than Words*, 18–24.
41. Roads, *Talking with Nature*, 18.
42. Ibid., 68.
43. Ibid., 66–67.
44. Ibid., 67–68.
45. Ibid., 68.
46. Ibid., 19.
47. Ibid., 89.
48. Ibid., 88–89.
49. Wilber, *Grace and Grit*, 46–47.
50. Roads, *Talking with Nature*, 131.
51. Ibid., 114.
52. Ibid., 113.
53. Ibid., 112.
54. Ibid., 116.
55. Ibid., 19.
56. Ibid., 18–19.
57. Ibid., 11.
58. Ibid., 111.
59. Buber, *I and Thou*, 57.
60. Ibid., 58.
61. Roads, *Talking with Nature*, 87.
62. Ibid., 121.
63. Ibid., 131.
64. Ibid., 78.
65. Ibid., 75.
66. Ibid., 133.
67. Gagliano, *Thus Spoke the Plant*, 3–4.
68. Ibid., 10–11.
69. Ibid., 12.
70. Ibid., 14.
71. Ibid., 20.
72. Ibid., 14.
73. Ibid., 15.
74. Ibid., 18-20.

75. Michael Marder, "To Hear Plants Speak," in *The Language of Plants: Science, Philosophy, Literature*, ed. Monica Gagliano, John C. Ryan, and Patricia Vieira (Minneapolis, MN: University of Minnesota Press, 2017), 122–123.

76. Hall, *Plants as Persons*, 145–146.

77. Gagliano, *Thus Spoke the Plant*, 2.

78. See Evelyn Fox Keller, *A Feeling for the Organism* (New York: W. H. Freeman, 1983).

79. Gagliano, *Thus Spoke the Plant*, 35.

80. Ibid., 27–28.

81. Ibid., 16–17.

82. Stephanie Kaza, *The Attentive Heart: Conversations with Trees* (Boston: Shambhala, 1993), 9.

83. Ibid., 89.

84. Ibid., 4.

85. Ibid., 5.

86. Ibid., 9–10.

87. Ibid., 29.

88. Ibid.

89. Ibid., 12–13.

90. Ibid., 13.

91. Ibid., 18–19.

92. Ibid., 20–21.

93. Ibid., 21–22.

94. Ibid., 22.

95. Ibid., 26.

96. Ibid., 34, her italics.

97. Ibid., 35

98. Ibid.

99. Ibid., 40.

100. Ibid., 51.

101. Ibid., 104.

102. Ibid., 105.

103. Ibid., 106.

104. Ibid., 107.

105. Ibid., 113.

106. Ibid.

107. Ibid., 130.

108. Ibid., 131.

109. Ibid., 134.

110. Ibid., 124.

111. Ibid., 125.

112. Ibid., 122.

113. Ibid., 170.

114. Ibid., 171.

115. Ibid.

116. Ibid.
117. Ibid., 89.
118. Ibid., 160.
119. Ibid.
120. Ibid., 157.
121. Ibid., 84.
122. Ibid.
123. Ibid., 235–236.
124. Ibid., 236.

Chapter 7

The Call to Responsibility

Thoreau and the Voice(s) of Nature

WHY THOREAU?

Thoreau as a person and thinker and spiritual explorer has been a slowly unspooling thread weaving this project together, especially his call for sympathy with intelligence and listening to the language that all things speak. However, not everyone is fond of the guy. In *Dark Thoreau*, a deeply critical review of Thoreau's life and writings, Richard Bridgman runs with the theme of his aloofness, stating that "the uncertain, lonely young man enlarged his own isolation into a trope of independence." He then goes on to cite passages in which Thoreau is less than generous in his appraisal of fellow human beings, including his dismissal of the old as having no advice to give, their lives being a "miserable failure."[1] Of course, Thoreau is also famous for saying the mass of men lead lives of quiet desperation.

A more recent scathing interpretation of Thoreau is "Pond Scum," a 2015 New Yorker piece by Kathryn Schulz, which, if one has read Thoreau well, can be easily countered, and many scholars did just that. However, Schulz makes some interesting points that make one think, like criticizing his ascetic practices in the "Economy" section of *Walden* and thus an adherence to an ideal purity that leaves out the desperate masses, but this point leaves out the fact that Thoreau was experimenting—the Walden experiment—exploring what one actually needs to survive, but also how to live mindfully and well.[2] Did he go too far? Perhaps. Make mistakes? Sure. Thoreau was a human being, not a myth, and certainly not a wilderness hermit; he stayed at his tiny, self-built cabin on *Walden Pond* for two years and two months, a couple miles from town, which he visited, occasionally having a meal with friends and getting his laundry done by mom, and he had visitors at his cabin. He never claimed otherwise. But he also read, wrote, walked, so much walking in the woods, hoed his bean fields, and spent plenty of time in solitude doing his ascetic practices, which resonates deeply as response to today's over-consumptive world where little distinction is made between wants and vital needs.

Another Schulz criticism is directed at Thoreau's contradictory nature and arguments, defending civilization in one moment, arguing against it in the next, extolling vegetarianism and then describing an impulse to devour a woodchuck whole, or dismissing the old while learning much from the classics. But to disregard differing sides of arguments would be to disregard the truth he was feeling at the moment, and thus differing sides of himself as a complex human being. Thoreau was a flawed man who attempted to express a diversity of experience: that's what Schulz doesn't get, or want to get. Thoreau is not contradictory because he is hypocritical; he's contradictory because humans and life are contradictory, and exploring contradictions leads to higher truths.

I first perused *Walden* when living in Maine. My roommate, and the "owner" of free spirited Mac the Newfoundland, watched me turn red-faced from chopping wood and made the perfect remark: "Chop your wood and it warms you twice." I immediately asked where that phrase came from, and of course it was Thoreau. I later read *Walden* twice in graduate school, once fast, and the other time slowly, leading me to realize that the latter was much preferred, much more rich. But not all my fellow graduate students were as rapt as me. One, a high-school English teacher, made the "too dismissive of the masses" argument, claiming he was an out-of-touch elitist who did not honor people that need to earn a living. As someone who had grown up in a blue-collar family, and had worked for years in a factory, print shop, book shop, and as a landscaper, I wasn't seeing it. For starters, my classmate missed Thoreau's humor, but also his heart, or how he filtered his head through his heart, and how that led to honest feelings often poetically expressed.

I made my points in his defense, fearful that the English teacher would squash the hearts of rapt students like me, who found in Thoreau much needed wisdom and solace in a world gone mad, including the madness of meaningless jobs. But when I began to teach Thoreau to my college students, it once again became clear that he does not speak to all, especially those who put in no effort, or read him too quickly, or misread him, or struggle to read him because of his language and references to an unfamiliar time, despite timeless passages that jump off the page, or refuse to read him because he challenges their existence—cognitive dissonance, as necessary as it is, is not pleasant. But there are always the rapt students who always spark good discussion on the many timeless passages. What I most want, however, is for them to get outside and explore after being inspired. Of course, some of my assignments lead them to do just that.

But, unless one reads some biographies, especially Laura Dassow Walls *Henry David Thoreau: A Life*, you don't get the whole man.[3] Or, as Robert Sullivan, another biographer, puts it: "the Thoreau you don't know."[4] In *Young Man Thoreau*, Richard Lebeaux analyses Thoreau's early years via Eriksonian psychology, arguing that he went through an early identity crisis of intense vulnerability that resulted in a radical change in perspective, followed by an unwavering search for meaning and growth.[5] If we are to attempt to find the core of the man, that is a good place to start: the conscious desire to move up the developmental holarchy.

Thoreau's writings are taught and read worldwide due to the fruits of this search, which unfold in differing iterations: Thoreau the nature lover and defender who listens to the language that all things speak, or the voice of Spirit within nature; Thoreau the in-sympathy scientist, writer, and activist who listens and responds to diverse voices, both nonhuman and human; and, Thoreau the futurist, who warns of increasing quiet desperation as industrialization takes over all facets of our lives. In all these iterations, he speaks to modern masses, more than ever in today's climate crisis and environmental injustice times. But there is something more that makes him eternally relevant: his practice of an ecology of communication.

OVERCOMING NATURE?

In the "Higher Laws" section of *Walden*, Thoreau makes the following claim: "Nature is hard to overcome, but she must be overcome."[6] This is a startling, and seemingly contradictory, statement, since the majority of his writings record in considerable detail his sensuous sympathy with nature. In this moment, however, Thoreau is expressing his sympathy with transcendentalist principles that privilege Spirit over matter, or Emerson's argument that nature always speaks of Spirit. But, to consider Emerson's view of nature and what he thinks we must overcome, we need to unpack the fuller quotation: "Through all its kingdoms, to the suburbs and outskirts of things, it is faithful to the cause whence it had its origin. It always speaks of Spirit. It suggests the absolute. It is a perpetual effect. It is a great shadow pointing always to the sun behind us."[7]

Emerson is indebted to the philosophical legacy of Plato and the claim that material existence is a reflection of ideas or archetypal forms, writing that mysticism "finds in Plato all its texts."[8] But is nature merely a shadow? Is its sole purpose simply to point toward some superior reality? As we saw, criticisms of such arguments are as old as the arguments themselves, with Aristotle's more this-worldly focused philosophy being the earliest and most influential. But, again, Shepard, Griffin, deep ecology, ecopsychology, ecofeminism, and nature writers generally have also criticized transcendent orientations that direct attention away from our heavenly home on earth. However, if we agree with Wilber that a transcendent dimension is essential to philosophical inquiry and the developmental holarchy, and by extension an ecology of communication, our attention turns to how we may overcome the dualism of dualisms, harmonizing a transcendent dimension with immanent nature in all its sensuous glory.

In his essay "Nature," Emerson is not wholly successful at this, despite its influential insights on commodity, beauty, and language, and perceiving the miraculous in the common; in many ways, he remains anthropocentric, although from a spiritual perspective, finding value in nature for humans only because it reflects Spirit.[9] Thoreau, on the other hand, is more grounded in pastoral farmlands, swamps, and wilderness, and thus provides a provocative exemplar for probing this age-old dualism. How, for example, are we to harmonize his demand that nature must be overcome with his more grounded claims, such as the clear and direct statement in "Walking" that we are part and parcel of Nature?[10] Despite this seeming contradiction, Thoreau's writings, and life, disclose many examples of such harmonizing, as he revels in the glories of transcendence, and thus listening to the voice of nature via spiritual communication, as well as the music of nature's myriad voices via mythic-animistic and aesthetic communication, although he is not immune from potential shadow-sides of "overcoming."

Thoreau's commitment to spiritual development is made explicit in numerous passages of *Walden*. In the opening section, "Economy," he argues that true philosophers are "progenitors of a nobler race of men."[11] And in a later passage, which Bridgeman might claim reflects more darkness than light, he compares the majority of humankind to a torpid snake who lay unmoved in the water for more than a quarter hour: "It appeared to me that for a like reason men remain in their present low and primitive condition; but if they should feel the influence of the spring of springs arousing them, they would of necessity rise to a higher and more ethereal life."[12] For Thoreau, this "spring of springs" is our developmental potential.

Wilber's work helps us make sense of Thoreau's criticisms of torpid masses, given the analysis that 60 percent or more, today, remain at the Mythic level, as well as the claim that nature must be overcome, if nature is taken as the biosphere, which is then transcended and included in the noosphere and theosphere. Also, as Wilber's criticisms of deep ecology disclosed, the higher transrational experience of the world soul transcends, includes, and transforms the lower, pre-rational experience of biospheric immersion, providing a more expansive spiritual perspective from which to understand it. Thoreau, while living before the full articulation of the developmental holarchy, had a clear sense of it, valuing bodily sensual, contemplative, and spiritual experience, and thus all modes of communication. He writes: "I found in myself, and still find, an instinct toward a higher, or, as it is named, spiritual life, as do most men, and another toward a primitive rank and savage one, and I reverence them both."[13]

And yet, Thoreau continues to be contradictory. In "Higher Laws" he makes another surprising claim, "He is blessed who is assured that the animal is dying out in him day by day,"[14] but in "Walking," we find the dictum, "In Wildness is the preservation of the World."[15] So, which Thoreau are we to believe? Do we celebrate our spiritual nature or our wild, animal nature? The answer, of course, is to reverence both, especially since to disregard either statement would be to disregard the importance of integrating differing levels of development and the responses called forth at differing moments. Thoreau may be maddening to some readers, but there is method; he engenders readers to practice vision-logic, and thus the logos of rational communication, by supplying paradox and seemingly contradictory truths, prodding readers to search for higher integrations. His contradictory passages reveal not only a complex man seeking diverse experiences but the need to embrace complexity when attempting to express a diversity of experiences and multiple modes of knowing.

Thoreau's commitment to and experience of higher developmental levels is robustly declared in the aptly named section of *Walden*, "Where I Lived, and What I Lived For," where he compares the morning to spiritual awakening: "Then there is least somnolence in us. . . . To him whose elastic and vigorous thought keep pace with the sun, the day is perpetual morning."[16] He continues by stating that the masses are suitable for labor, with few desiring an intellectual life, or rather "only one in a million," and still fewer, or "only one in a hundred millions," seeking a poetic or divine life.[17] As we've seen, such elitism, which is more accurately interpreted as exaggeration to make a point, turns some readers away, even though he labored plenty and respected hard work. Too many readers also miss the humor in Thoreau, like in "Walking" when he questions whether shopkeepers sitting on their bums all day know what their legs are for, impressed that they have not "committed suicide long ago."[18] But why torpid humans seem to refuse to wake up to a larger life was a source of frustration, since to "be awake is to be alive."[19] He understood the danger of societal structures influencing our habit-bodies, wondering whether we ride trains or if they ride upon us, but, despite intuitions, he did not have the stage-step precision of the developmental holarchy to more fully make sense of it all.

Still, Thoreau shares his devotion to the perpetual dawn of spiritual awareness, arguing that developmental potential lies in our ability to elevate our lives through conscious endeavor: "It is something to be able to paint a particular picture, or to carve a statue, and so to make a few objects beautiful;

but it is far more glorious to carve and paint the very atmosphere and medium through which we look, which we morally can do."[20] For Thoreau, affecting the quality of the day is "the highest of arts," and he even softens his pessimism concerning the mass of men, allowing for the possibility of spiritual growth, although only if such growth is made conscious. However, the high art of affecting the quality of the day should not be equated with the desire to create our own reality, although our ability to elevate "the medium through which we look" discloses why New Agers are entranced by this claim. We can, and must if we are rationally and spiritually awake, embrace our subjective power to elevate perception and our lives, but we cannot, nor should we desire to, subjectively create reality without being affected by the inter-objective influence of circumstance and the intersubjective reciprocity of others.[21]

The call to awakening is expressed often in *Walden*, especially in regard to the present moment, as the true and sublime are "now and here," and God culminates in current times and places that "will never be more divine in the lapse of ages."[22] As we saw, Thoreau experiences this eternal now most revealingly when hoeing his bean fields. While he honors hard work, and worked plenty as a surveyor and at his family's pencil business where he displayed engineering skill, jobs too often divert time from spiritual pursuits like walking and contemplation; still, he expresses affection for working his bean field that he discusses as a dialogic relationship: "What shall I learn of beans or beans of me? I cherish them, I hoe them, early and late I have an eye for them; and this is my day's work."[23] Thoreau soon discovers that hoeing and contemplation are compatible activities, and this becomes a stepping stone to his loss of a separate "I" doing the work, a thoroughly now moment disclosing an "immeasurable crop."[24]

This experience, of course, takes us back to wu-wei spiritual communication, in which there is physical distance, and thus relationship, but no psychological division between Thoreau and the beans, or Thoreau and his work, and in this reverie he is the best "bean-hoer" on the block. There is no past, no future, only the eternal present of right action, and just so his plea to "simplify, simplify, simplify" our wants and needs is transfigured into its ultimate meaning: we are called to simple, clear nondual awareness. Or, following Dogen's *Mountains and Rivers Sutra*, at first mountains are simply mountains viewed through our social conditioning, but when we wake up to the interdependency of clouds, rain, trees, animals like wolves and deer—or thinking like a mountain—they become more complex, and then, if will and grace coalesce, we fall into the simplicity on the far side of complexity, whether with mountains, rivers, bean fields, or any and all diverse forms emerging from emptiness; or, Brahman is the world with mountains becoming mountains again, rivers becoming rivers, and there is nothing we must not see in order to see.[25]

Dogen, or any Zen master, will claim that there is nothing extra-ordinary about this nondual awareness. Rather, it is our ordinary awareness when our perceptions are not filled with the habitual clutter of unproductive desires. Desire, as Ramsey has compellingly argued, is not our enemy; rather, when understood within the context of paneroticism, it is the precondition of much communicative praxis. But there are moments of transcendence, or the overcoming of desire, in which our responsiveness reaches new heights in the form of spiritual communication.

Thoreau's glimpse of the highest level of spiritual state experience provides context for reading his contradictory assertions. Ultimately, these contradictions, including the seemingly insurmountable psychological distance between subject and object, dissolve amid a more direct apprehension of reality, and then reappear, albeit differently, within the expanse of a widened awareness. But how can this awareness be attained, especially when adepts claim that we must transcend the desire for spiritual attainment? This is the classic koan-like paradox of the spiritual path. Adepts working within spiritual traditions provide students with some type of practice, which, when done with diligence, is supposed to lead to the radical realization that we are always already awake and enlightened and that our desire for awakening has actually kept us from our goal. After all, how can we desire something that we already have? Or rather, what we always already are.

Yet, we are still given a practice, and there are still stage-steps, and, as Peters asserts, we still spend most of our communicative lives in a variety of messy existential circumstances that take effort, not tinkling our hoes against rocks creating music that results in the communication of nondual reverie. Thoreau, aware of the doubleness of divine and mundane moments, continues his practices of sauntering amid nature and contemplation that includes slowing down and living deliberately, making us more deliberate in our choices and pursuits.[26] He also holds up the life of the mind and reading: "To read well, that is, to read true books in a true spirit, is a noble exercise, and one that will task the reader more than any exercise which the customs of the day esteem." For Thoreau, reading requires training similar to skilled athletes, and thus the same deliberation and reserve exercised by authors when they write their books.[27] And yet, despite his reverence for reading deliberately, this practice does not go far enough. Books and reflection are a stage that leads to deeper revelations: "What is a course of history or philosophy, or poetry, no matter how well selected, or the best society, or the most admirable routine of life, compared with the discipline of looking always at what is to be seen? Will you be a reader, a student merely, or a seer? Read your fate, see what is before you, and walk on into futurity."[28]

These passages celebrate rationality and vision-logic, while also previewing higher levels of awareness, like reading the intelligence of transcendent

Spirit and immanent nature, as well as our inner nature or genius. For Thoreau, different types of knowing and communication, when integrated, disclose the higher law of unity-in-diversity. However, practicing sympathy leads us to respect our not-knowing: "Nature is a personality so vast and universal that we have never seen one of her features."[29] Thus, we are beset with another contradiction: we must be seers and yet Nature's personality is beyond being seen. But, again, contradictions and paradoxes are not obstacles as much as catalysts for contemplation, until, perhaps, paradoxical desires unravel and we experience the seer and the seen as not-two, and the question of paradox or contradiction no longer arises. But until such awareness becomes self-evident, Thoreau chooses a wise path: not complete knowing or not-knowing, but sympathy with intelligence.

There are many passages in Thoreau's writings that illustrate this sympathy, but the opening paragraph of "Solitude" is one of most poetic:

> This is a delicious evening, when the whole body is one sense, and imbibes delight through every pore. I go and come with a strange liberty in Nature, a part of herself. As I walk along the stony shore of the pond in my shirt sleeves, though it is cool as well as cloudy and windy, and I see nothing special to attract me, all the elements are unusually congenial to me. The bullfrogs trump to usher in the night, and the note of the whippoorwill is borne on the rippling wind from over the water. Sympathy with the fluttering alder and poplar leaves almost takes away my breath; yet, like the lake, my serenity is rippled but not ruffled.[30]

This is a beautiful articulation of both the sensuous and the spiritual, and thus the beauty of bullfrogs and whippoorwill, rippling wind and water. It is a "delicious evening, when the whole body is of one sense" echoes synesthesia, or the overlapping of senses, which echoes Abram's preverbal bodily converse and sensuous imagination, and mythic-animistic communication generally. But Thoreau also calls forth the divine in sympathy with the body, the senses, the trees: all are integrated, revealing a "strange liberty." This integration, this liberty within Nature, "a part of herself," blossoms at the Integral level of the world soul, the home of graced aesthetic communication, where spiritual unity amid material diversity is initially apprehended.

Thoreau does not sound like someone who wishes to overcome nature. But if we read "overcoming nature" within the context of his glimpses into nondual awareness *and* his claim that he "loves the wild not less than the good," it is not really nature that he wishes to overcome, but an understanding of nature that does not also acknowledge Spirit.[31] However, this desire to overcome without complete overcoming gets complicated when he articulates an experience of nonattachment. "By a conscious effort of the mind we can stand aloof from actions and their consequences; and all things, good and bad, go by us like a torrent. We are not wholly involved in Nature. I may be either the driftwood in the stream or Indra in the sky looking down on it. I may be affected by a theatrical exhibition; on the other hand, I may not be affected by an actual event which appears to concern me much more."[32] This experience of not being affected by the things of this world is a clear articulation of causal-witnessing consciousness, and Thoreau continues witnessing: "I only know myself as a human entity; the scene, so to speak, of thoughts and affections; and am sensible of a certain doubleness by which I can stand as remote from myself as from another. . . . This doubleness may easily make us poor neighbors and friends sometimes."[33]

Previously, we discussed Thoreau articulation of doubleness as his recognition of the spiritual heights in relation to more frequent mundaneness, although once one glimpses the nondual, or other forms of spiritual communication, the experience never leaves us. But this passage suggests negatives, while further unraveling Thoreau's contradictory nature. This kind of doubleness seems to lead to both good and ill, or a healthy nonattachment and a problematic detachment. Again, causal-witnessing state experiences may guide us to discover a healthy nonattachment that lessens the influence of the isolated ego, as well as our addictions and habits, allowing a deeper integration of transcendent and immanent dimensions. Thoreau does seem to be accomplishing some of these goals, especially when he later describes his ego identity as an "illusory spectator" and a "kind of fiction."[34] However, he makes a key admission when he states that this experience easily degrades into a detachment from other human beings that may cause him to be a poor friend and neighbor.

In one of Wilber's earliest books, *No Boundary*, he offers a meditation that I practiced for a number of years, which goes something like this: I have a body but I am not my body, I have emotions but I am not my emotions, I have thoughts but I am not my thoughts; I am the witness of my body, emotions, and thoughts.[35] For me, this meditation was an effective means for not getting lost in emotions, which is a hallmark of emotional intelligence, or my thoughts, which is a hallmark of mindfulness, but over time I felt distant, not just from emotions and thoughts that can make one anxious, but from flesh and blood relationships. I became hyper-self-conscious, always aware of being aware, which is a stage-step to greater and wider awareness, but can easily degrade into a damaging detachment. I changed the meditation to "I have a body but I am also more than my body, I have emotions but I am also more than my emotions, I have thoughts but I am also more than my thoughts," in an attempt to transcend but also include. But these meditation experiments disclosed interesting results.

While Thoreau intuited but did not have the developmental holarchy to better interpret his neighbors, he also did not have it to better analysis himself. It is worth remembering that each level, and each state, has associated dysfunctions, which may block further growth or cause us to regress, and there are numerous lines that run through the various levels, and some may be repressed. After all, New Ageism disclosed that the desire for spiritual communication is littered with dysfunctions, especially when "overcoming" occurs without integration. And, as we saw with Boone and Roads, interspecies communication becomes problematic when divorced from the critique, articulation, and disclosure of transversal rationality and rational communication, but also the grounding of aesthetic and especially mythic-animistic communication.

Thoreau's admission of detachment and poor social relations is borne out, sometimes in a rather severe fashion, in his more caustic passages, as Bridgman notes, as well as in his life. Emerson, for example, stated in his eulogy for Thoreau: "There was somewhat military in his nature not to be subdued, always manly and able, but rarely tender, as if he did not feel himself except in opposition."[36] Emerson also once remarked that Thoreau had no friends.[37] It is true that Thoreau describes meetings with farmers, hunters, and passersby in *Walden* but they are mainly portraits of acquaintances. Yet, true to his contradictory nature, and the complexity of exploring a life, biographers of Thoreau clearly show that he had plenty of friends, including Edward Hoar, with whom he traveled in Maine wilderness and with whom he accidently set fire to the Concord woods by trying to cook within a rotted out tree stump, a very non-nondual moment that further rattles the myth he never asked for. He also had friends who had the good fortune of accompanying him on his walks, and was particularly liked by children, including Emerson's, despite his love of solitude, or rather, likely because of it—he certainly was no phony, sharing anecdotes from his walks and observations, playing around, and going on huckleberry picking parties.[38] And Emerson's son Edward remarked that when he left his Walden cabin and returned to town for a meal, he went for "friendship, not food."[39]

Not surprisingly, Emerson and Thoreau had a complex mentor-student relationship, and then friendship—yes, the same Emerson who once remarked he had no friends was a friend—which included deep affection as well as mutual annoyance at differing times.[40] Communication, even among those committed to awakening, is often hard. But despite his somewhat negative eulogy remark, Emerson also stated that if Thoreau "slighted and defied the opinions of others, it was only that he was more intent to reconcile his own practice with his belief."[41] In other words, he was always true to himself and higher laws. And the eulogy as a whole was a rousing tribute that included these prescient words: "The country knows not yet, or in the least part, how great a son it has lost."[42]

Still, Bridgeman runs with the theme of Thoreau's dark detachment, arguing that his use of imagery is frequently marked by cruelty, citing a passage from the "Spring" section of *Walden* as a prime example: "I love to see that Nature is so rife with life that myriads can be afforded to be sacrificed and suffered to prey on one another; that tender organizations can be so serenely squashed out of existence like pulp—tadpoles which herons gobble up, and tortoises and toads run over in the road; and that sometimes it has rained flesh and blood!"[43] Thoreau certainly uses stark prose, and, once again, he does not hold back what he is experiencing in the moment, not caring how he might be judged; but does this passage reflect a problematic detachment or a healthy nonattachment? It could be argued that witnessing predator-prey realities reflects a healthy fascination rather cruelty, and such passages are also balanced by his numerous poetic and empathetic ones. The beauty of the created world, for example, is frequently referenced in his journals, and in the Conclusion to *Walden* he extolls us to see light within the darkness: "However mean your life is, meet it and live it; do not shun it and call it hard names. It is not so bad as you are. It looks poorest when you are richest. The fault-finder will find faults even in paradise. Love your life, poor as it is. You may perhaps have some pleasant, thrilling, glorious hours, even in a poorhouse."[44]

It is easy to trash Thoreau, as well as praise him. I am most interested in making sense of his contradictory nature in relation to an ecology of communication and a fitting responsiveness. Thoreau often sought isolation away from his fellow townspeople and was capable of emotional detachment when observing nature, but there is little doubt that he also experienced the depths of solitude, discovering the further reaches of his spiritual nature via experiences of aesthetic and spiritual communication, as well as grounding via mythic-animistic communication, feeling his body in relation to sensuous surrounding. He was also a townie, involved in the social doings of his beloved Concord, as well as a virulent abolitionist. However, it is also likely that some lines of development, including interpersonal, were regressive at times. Thoreau often uses blunt prose when speaking of people and society and poetic prose when speaking of and for nature. It would seem that his identification with diverse nature is stronger, but for understandable reasons. He saw the destructive potential of industrialization more clearly than fellow, not-so-diverse townsfolk, who busied themselves with making a living within the machine rather than living a life of continuous awakening. And yet, it is these very complexities of his life in Concord, and his personality, that led him to the Walden experiment, to write, walk, and contemplate, and to discover a solace and spirit in the natural world that few have equaled, confronting essential facts and making sure that before he died, he had lived well.

There have been some psychological studies of Thoreau, including Carl Bode's Freudian analysis in the epilogue of an early edition of *The Portable Thoreau*, where he argues that his unconscious was marked by an Oedipus complex that "aborted his emotional life but richly informed his writing."[45] It is true that Thoreau failed miserably in love and had a propensity for older women, including Emerson's second wife, Lydian. But Freudian psychology is often reductionistic, accounting for behavior via discussion of the libido and the sublimation of the sexual drive. I don't discount the possibility that sexual dysfunction and Oedipus fantasies were influential in Thoreau's psyche. However, a strict Freudian analysis does not fully explain Thoreau. Instead, it is in danger of explaining him away.

An ecology of communication corrects this tendency, providing a more expansive context for psychological, and ecopsychological, analysis that discloses the importance of mythic-animistic, rational, aesthetic, and spiritual communication to moving up the developmental holarchy. Thoreau's life is complex and resoundingly unique, but his self-development and struggles have universal appeal. He fails at continuously living his spiritual awareness; but rather than this making him a failure, it makes him human. That we cannot overcome, but we can overcome perceiving nature as mere shadow, seeing it instead as a shining light saturated with Spirit. Thus, the ultimate goal for the transcendentalist should not be to understand nature as pointing toward the superior reality of Spirit but to grow into a heightened reality here and now on this sweet earth.

While Thoreau provides another cautionary example of the potential shadow-side of spiritual communication—detachment and repression—his legacy is ultimately defined by the awakening potential of such communication, such as in this passage from "The Ponds" in which he poetically describes via an analogy with fishing the integration of transcendent and immanent, spiritual and material, subtle and dense:

> It was very queer, especially in dark nights, when your thoughts has wandered to vast and cosmogonal themes in other spheres, to feel this faint jerk, which came to interrupt your dreams and link you to Nature again. It seemed as if I might cast my line upward in the air, as well as downward into this element, which was scarcely more dense. Thus I caught two fishes as it were with one hook.[46]

SYMPATHY AND SCIENCE

Thoreau knew that nature speaks, or is silenced, in relation to what we bring: by our ability to bracket dominant narratives and habits of perception, by openness to multiple modes of meeting nonhuman others, exploring expressive systems that include a range of verbal and nonverbal competence, and, ultimately, by listening to the language that all things speak.[47] Thus, while his vision encompasses the transcendent, his focus is consistently on the immanent: "My profession is to be always on the alert to find God in nature, to know his lurking places, to attend all the oratorios, the operas, in nature."[48] References to nature's music occur often in his writings, whether the "ambrosial" creaking of crickets, the "cool bars of melody" of the wood thrush, or the mystical tinkling of his hoe against a rock.[49] For Thoreau, sympathy with the voice of nature and diverse voices within nature are a balm to human awareness, awakening our potential for transcendentalist "self culture" and the call to responsibility. And this sympathy extended to the practice of science, to which he turned in later life.

In *No Man's Garden*, Daniel B. Botkin uses the example of Thoreau measuring the depths of Walden Pond to show his early and adept scientific practice. Using a compass, sounding line, and weight, Thoreau made over one hundred measurements which he translated into a map. Botkin writes that his curiosity concerning his quantitative measurements led him to formulate hypotheses about other ponds that required further measurements and hypotheses.[50] Thoreau followed the scientific method, and Botkin argues that facts brought him closer to nature, as well as to praise the ways of science: "Science is always brave; for to know is to know good; doubt and danger quail before her eye."[51] However, Thoreau's scientific practice also included collecting species for Louis Agassiz, a leading scientist at Harvard, and the "bravery" of science was not always apparent, as he records his killing of a box turtle in his journal: "I cannot excuse myself for this murder, and see that such actions are inconsistent with the poetic expression. . . . I pray I may walk more innocently and serenely through nature."[52] In another section of his journal, he puts the matter bluntly: "The man of science, who is not seeking for expression but for a fact to be expressed merely, studies nature as dead language."[53]

Thoreau's scientific practice evolved to be more poetic-aesthetic than rigidly objective. He acknowledged the unavoidable subjective dimension of science, arguing that the biography of the scientist often reveals more than what is studied. This was both a rational criticism and a deeper truth, or a more complete insight into reality and living nature. As a result, he was fond of Goethe, both for his literature and his investigations into plant metamorphosis. Goethe practiced a subjective involvement with the plants under study, and it led him to recognize the law of the leaf, or the directing and animating life force underlying matter. Goethe was interested in apprehending the growing process and pattern and intelligence that manifested itself throughout the plant. Thus he perceived that the pattern of the leaf was present in the plant as a whole, and by extension, in nature as a whole, which, of course, sounds a lot like Bateson's conception of mind or the pattern that connects.[54]

In *Natural Life: Thoreau's Worldly Transcendentalism*, David M. Robinson writes that Thoreau's scientific practice must be interpreted within the context of Goethe's law of the leaf, as well as transcendentalism and the poetic-aesthetic imagination, to which he still adhered, even though his focus was increasingly on cataloguing nature's diversity. He admired the "old naturalists" who "sympathized with the creatures they described."[55] According to Robinson, Thoreau's naturalist tendencies led him to recognize "effluence," a concept he derived from attending to the fragrance of fruits and blossoms. In other words, the sensuous and aesthetic allure of flowers led him to claim that effluence linked the observer with the observed, making the observer receptive to the expressions or utterances of the species under study. Thoreau wrote in his journal, "Only that intellect makes any progress toward conceiving of the essence which at the same time perceives the effluence."[56]

In *Seeing New Worlds, Henry David Thoreau and Nineteenth-Century Natural Science*, Laura Dassow Walls writes that Thoreau's early scientific practice was also influenced by Alexander von Humboldt's rational holism and the concern, influenced by Emerson, of how to be a whole individual. As he grew older, he strove to keep differing fields of knowledge integrated, transforming facts into wisdom.[57] In doing so, he participated in the debates of his time on the possibilities yet limits of science, including deductive versus inductive methods, the laws of nature versus higher laws, and the unattainability of complete objectivity given that we are part and parcel of the nature we are studying, resulting in the inevitable subjective recording of experience, whether from poet, philosopher, or scientist. Ultimately, he advocated for a "true sauntering of the eye," in which we see what we are prepared to see, including "love for earth and all its creatures" rather than "human arrogance," and relational and thus "moral knowledge" instead of only data.[58]

Still, Thoreau's practice was rigorous—his dated records of wildflower blossomings have been used by current scientists to track climate change, and thus collecting raw data clearly matters—yet he was also guided by sympathy, or effluence guiding him closer to essence.[59] His spiritual and scientific practices often merged, marked by the rational yet empathetic orientation of the naturalist within the context of spiritual experiences. Scientists need not have spiritual experiences to listen to the voices of nature and receive telling information on the primacy of relationship—close observation, receptive senses, and contemplation may overcome the rigid dualism between mind and nature—but nondual glimpses more fully place us in sympathy with the voice of Spirit within nature, allowing us to know more. Thoreau wrote, "Perhaps the facts most astounding and most real are never communicated by man to man," a claim that Kaza and Gagliano would most assuredly support.[60]

Thoreau practiced a subjective intimacy with the pond, woods, and bean field, and also animals, who were often like friends, although he would not be so foolish to try to befriend a grizzly, as his Mt. Katahdin "contact" experience led him to respect nature's power, plus he was more savvy than that. But if the masses were desperate, it was because they did not have ears to hear nature's concert, or had forgotten how to listen, or worse, been socialized out of it, dismissing such listening as romantic yearning rather than what it truly is: a profound openness to mystery, to wildness, and the ground of a life well lived. Raw data matters, but does not matter enough, or mean enough, or inspire enough, unless it is understood within the wider awareness of multiple modes of knowing and an ecology of communication.

For Thoreau, exploring the "one hook" way of things was the work of both the poet and the scientist. Emerson, in his eulogy, stated that his observatory powers, marked by silence and stillness, becoming a "log among the logs," suggested additional senses and sympathy, such that a "snake coiled around his leg; the fishes swam into his hand . . . ."[61] In Walls' recent biography, she writes that word spread of his knowledge and abilities, and he became in demand among locals, especially farmers, whose land he seemed to know better than they, due to being adept at listening to inner genius and genius of the place.[62]

Walls' biography provides telling details of Thoreau's scientific ventures, along with expressions of sympathy. In earlier years, he sent lots of species to Agassiz, including some he had never seen, filling him with questions before feeling dismay over killing for science. Later, in 1860, after being one of the first in America to read Darwin, only five weeks or so after publication, he found a theory to match his studies. Ironically, Agassiz' Christianity led him to reject Darwin's evidence for natural selection in favor of God creating each species individually and placing them on the planet, with white men like himself at the top of the hierarchy, while Thoreau saw that everything evolved from the same source, everything was entangled, mirroring the Native view of "all my relations."[63]

Thoreau, as it turns out, knew a few things about agriculture beyond hoeing his bean field. After reading Darwin, he gave a talk on forest succession for an agricultural fair, stating "I have great faith in a seed," which included the seeds of justice. The speech was well received; his audience praised him as a model citizen contributing to the community, but also as an educator. "The Succession of Forest Trees" was placed in the *New-York Weekly Tribune*, and became his mostly widely read piece in his lifetime. Farmers knew they were destroying their woods and needed practical advice. Applying Darwin, he explored the origin of all things in relation to the dispersion of seeds, including via animals and wind instead of God, yet divine intelligence was implicated, expressed as nature's unfolding patterns, including cycles of life and death; when the stuff of seeds died, they nourished the soil, becoming receptive to more seeds, more growth, slow and steady, over a long timescale.[64]

Thoreau once considered killing a snake for Agassiz, but came to an I-Thou realization that it was "not the means of acquiring true knowledge."

Science, he realized, made the mistake of treating communicating subjects merely as voiceless independent "Its," an error he no longer made, his more poetic science perceiving interdependency and intimacy.[65] Animals were killed all around Concord, for science and from hunting, while Thoreau communed with a flying squirrel and reached into a knothole to pet a screech owl. Teased by some about his refusal to kill, he responded: "Do you think I should shoot you if I wanted to study you?"[66] Thoreau also told a Harvard librarian that science made another mistake: not listening to and learning from Indians, who had fifty names for cedar, could locate a snake by its call and call animals to them, knew more habits of fish than Agassiz, and, generally speaking, listened to and learned from nature. For Thoreau, Indians were "damned, because his enemies were his historians."[67]

Thoreau's sympathy seemed to increase as he increasingly turned to science, with his journals taking on added importance as he catalogued the details of divinity on daily walks. Animal minds became a model, similar to Shepard's mythic-animistic articulation of the birth of the human mind and Benyus' aesthetically inspired biomimicry: how to walk like a fox, how to have one's mind and senses wholly open. On the muskrat, he again displayed his contradictory nature, not overcoming but equalizing, or simply expressing Ground value, stating "he is a different kind of man, that is all," and when someone shot the summer ducks he cared for, he lamented that they considered it more important to "taste the flavor of them dead than I should enjoy the beauty of them alive."[68]

Thoreau became a member of the American Association for the Advancement of Science and a president of Harvard appointed him to the Harvard Committee for Examination in Natural History, making him part of scientific institutions, and leading him to write in his journal, "I am a mystic—a transcendentalist—& a natural philosopher to boot."[69] However, the commons are what nourished and educated him, even while they were being lost as the masses became starry-eyed over new technology.[70] The economics of the railroad demanded logs for train fuel and railroad ties, at the same time sparks from trains set fire to woods.[71] By 1854, when *Walden* was published, most woods had been clear-cut, leading Thoreau to ask: "how do you expect the birds to sing when their groves are cut down?"[72] Yet, contradictory Thoreau—true to and honest about differing experiences—could be starry-eyed as well: he was fascinated by the making of locomotives, sometimes rode trains, praising them for making him feel like a citizen of the world, while also questioning whether technology's improved means led to unimproved ends.[73]

The integration of science with other modes of knowing can be traced back beyond Thoreau, but he was prescient because he appealed to conscience when approaching questions of his time that remain questions of our time, like the techno-science-inspired myths of bigger, better, and more. In *Climate: A New Story*, Charles Eisenstein argues for, well, a new story that overcomes what is worth overcoming: our illusory sense of separation. He also argues that while data is exceedingly important, it is not nearly enough, especially since it reflects the dominance of a quantitative bias.[74] And thus environmentalists are faced with a conundrum when invoking science as sole response to climate crisis: we buy into this bias as fitting, yet this same bias has "presided over and defended our ecocidal system."[75] Again, we bang our heads against the wall when we use the same consciousness that created the problem to search for solutions. While hypothesis, experiment, and communal confirmation works for both scientific and meditation practices, quantitative methods used for the measurable do not disclose significant insights about what is often immeasurable, like sacredness and love, or the grace that comes from being willing to risk losing ourselves so that we may find new gifts. Just so, as important as Thoreau's recording of wildflower blossoming are to climate scientists, of more importance are lessons learned from practicing sympathy.

Although Thoreau praised science, stating "the man of most science is the man most alive, whose life is the greatest event," as we've seen, he also criticized it plenty.[76] He knew that scientific practice, as wonderful as it is, is not enough, especially when nature is studied as dead language. And he knew that we need to invite dialogic wisdom, listening to the voice and voices of nature, and ultimately, a divinely inspired love. Thoreau, while flawed and contradictory, as humans are, taught children, not just botany, but wonder. Such is the beginnings of sympathy expressed as agape, eros, and biophilia.

Emerson wrote: "Beware when the Great God lets lose a thinker on the planet."[77] Thoreau was a graced thinker because he thought within the contexts of Spirit, cosmos, earth, and human cultures and communities, and thus within the contexts of spiritual, aesthetic, mythic-animistic, and rational communication; he was a scientist among poets, a poet among scientists, living the sciences and humanities embodied within the whole individual, one of the few, if not the only, of his time who could turn two into one.

SPEAKING FOR NATURE: THOREAU, NATURE WRITING, AND ANTHROPOMORPHISM

Thoreau's transcendentalist vision led him to seek dialogic wisdom and fitting responses, but the first line of "Walking," while seeming to be a straightforward celebration of freedom, wildness, and identification with nature, raises critical questions. He writes: "I wish to speak a word for Nature, for absolute freedom and wildness, as contrasted with a freedom and culture merely civil—to regard man as an inhabitant, or part and parcel of Nature, rather than a member of society."[78] The simple phrase, "I wish to speak a word for nature," is also a combustible one, especially for Nancy Craig Simmons; in her essay, "Speaking for Nature: Thoreau and the Problem of Nature Writing," she argues that claiming to speak for nature imposes ourselves on nature, controlling and reducing it via language for human consumption.[79] In other words, she shares Plumwood's concern with backgrounding and incorporation.

This is hardly the way we typically think of Thoreau, but Simmons raises difficult questions for not only his writings, in which he consistently turns to nature as a symbol for human potential—the Concord River would not flow for millions of years if it was not happy; the clouds concentrate his thoughts, focusing attention on heaven on earth; a fallen, centuries old tree deserves the ringing of the village bell and the town's mourning; etc.—but any nature writer practicing an ecology of communication.[80] Can Thoreau, or anyone, speak *for* nature? Gagliano asserted that her book was written as a co-authored dialogue with plants but did not claim to speak for them. Should Thoreau even try, or is the attempt an act of hubris reflecting monologic domination? Or New Age channeling? Or Roads-like telepathy? Or perhaps it is simply destined for failure?

These questions lead to still more: Can we perceive nature directly, on its own terms, via the virtues of humility, care, and reverence, and the practices of empathy and emptiness, or are we always describing, translating, recording, reporting, and interpreting?[81] Said differently, is language ever transparent, revealing the world of non-words with the world of words, or is it always laden with cultural meanings that keep us separate and apart rather than part and parcel of the natural world? Or rather, despite being part and parcel of nature, when we speak and write, we still speak and write as humans, not nature—right?[82]

To Thoreau, being in sympathy with a natural world saturated with Spirit and meaning allows him to speak a word for nature. He does not claim that nature is literally speaking through him; rather, he is speaking on nature's behalf as a fellow inhabitant, rather than only a member of society, and this provides insight into what is wild and free. But Simmons is not having it, finding such sympathy to be a manifestation of instrumental use or another monologue. For her, encountering nature spiritually still leads us to use nature for our benefit. In fact, she states that Thoreau habitually uses nature in his writing for his own poetic purposes.[83] Thus, he does not let nature speak; rather, he makes use of it for the sake of symbol, language, and the craft of nature writing, creating a dualism between mind and nature.

Thoreau is an exemplar for practicing an ecology of communication, and thus Simmons' criticisms deserve attention, especially since her questioning mirrors likely critical concerns directed toward this project: New Age solipsism masquerading as spiritual communication and anthropomorphism resulting from mythic-animistic and aesthetic communication. I have already explored and countered many shadows and unfit responses, but continuing to mine, without undermining, Thoreau's rich work supplies further guidance into what constitutes a fitting responsiveness.

As we saw via Roads, a likely shadow-side of claims for nondual experience and spiritual communication are long-winded literal claims of channeling nature's voices. Thus, if nature writing can be transparent, and we can speak a word for nature, wouldn't this group Thoreau with Roads, and nature writers with New Age writers who claim they can commune with crystals and channel dolphins and dead yogis? This assertion must be resisted, as should the knee-jerk counterclaim that language is a human construct placing an unbridgeable gulf between nature and humanity. After all, nondual experiences have been documented in all cultures, language also emerged from the earth, or as Emerson declared, "Words are signs of natural facts," and our bodies and psyches participate in ongoing nonverbal communication with the nonhuman world.[84]

Thoreau, as well as Emerson, write of such nonverbal communication. Emerson, for example, had his grounded moments: "The greatest delight which the fields and woods minister is the suggestion of an occult relation between man and the vegetable. I am not alone and unacknowledged. They nod to me, and I to them." He continues by stating such nods, along with the waving of boughs, takes him by surprise, and yet is common: "Its effect is like that of a higher thought or a better emotion coming over me, when I deemed I was thinking justly or doing right."[85] Delight, then, resides in our dialogic rapport with nature and openness to what comes through, like higher thoughts and better emotions. Emerson's "occult relation" with vegetal others is consistent with Gagliano's experiences, and many other examples of dialogic relations explored in this project, but it is not the same as speaking for nature. Instead, it reflects listening and then sharing, in a human voice, what we have learned, or what is more just, more right, and therefore more fitting.

Thoreau, of course, does plenty of listening and learning, and Simmons does praise some of his writing, particularly his journal description of tracking a fox for several miles. A light snow has betrayed the path of the fox, and Thoreau follows and records its movements. The writing is pure observation and detail, but then ends with the following: "How dangerous to the foxes and all wild animals is a light snow . . . betraying their course to hunters . . . I followed on this trail so long that my thought grew foxy."[86] Simmons praises this entry because to think like a fox and let it speak through writing, to the degree that one can, demands close observation and recording, not poetic flourishes or inspired spiritual heights that suggest anthropomorphism.

Simmons further argues that Thoreau himself was concerned with the difficulty of speaking for nature. In *Walden*, he writes: "I fear chiefly lest my expression may not be extra-vagant enough, may not wander far enough beyond the narrow limits of my daily experience, so to be adequate to the truth of which I have been convinced . . . I desire to speak somewhere without bounds; like a man in a waking moment, to men in their waking moments."[87] Simmons states that this passage reflects Thoreau's desire to speak beyond the boundaries and limits of culture, although he is mostly concerned with the narrowness of his bias. For Simmons, he comes closest to ridding himself of bias, and "enabling nature to speak," when recording events like the path of the fox.[88] To her, field note-like nature writing tends to have a greater ability to let nature speak because it supplies a running record of our observations, not writing crafted for human use. She also argues that Thoreau backs her up, as he stated that a simple recording of his perceptions rather than artful language was best for his journals.[89] But I don't think she fully gets contradictory Thoreau, and this interpretation assumes that there is nothing to be gained from reflecting on our experiences and carefully crafting our language. A running record also does not remove cultural habits of thinking and seeing; rather, it unthinkingly reproduces an objective, scientific bias.

Thoreau is clearly referring to spiritual awakening by wishing to speak extra-vagantly. He desires to share nondual experiences of spiritual communication via his writing to others who are also awake and aware. But Simmons dismisses Thoreau's spiritual writings as reflecting the bias of transcendentalism, as they reflect Emerson's claim that nature is a symbol of Spirit, and a symbol can only be used as a writer's tool.[90] But this response, while not without critical insight concerning the dangers of too much overcoming—nature is *only* a shadowy symbol of Spirit—fails to account for the evocative power of Thoreau's spiritual writings and the possibility that both Spirit and nature are given voice, to some degree, via poetic prose, and that while symbols may divide us from nature, they don't always do so. In a journal entry, he provides a further distinction: "Nature will not speak through but along with him. His voice will not proceed from her midst, but, breathing on her, will make her the expression of his thought."[91]

Thoreau understood the struggle of speaking a word for nature, yet the experience of and commitment to the language that all things speak is expressed often in his writings: we are called to the simple, clear awareness of awakened consciousness and to listen to nature's chorus. The passages cited thus far scratch the surface of Thoreau's deep inquiry into the relations between Spirit and nature, or the voice of unity and diverse voices, yet these passages speak with power and immediacy because of Thoreau's craft as a writer. He attempts, as mystics always have, to communicate "extra-vagantly," articulating experiences beyond words and thoughts with words and thoughts—an attempt that inevitably falls short, but that may inspire others to spiritual practices, to listen to the voice and voices of nature. As with his I-Thou scientific practice, he knew that there is no complete objectivity, and thus facts became frames for his writings, in which he attempted to disclose the sacred and our place within nature.

Simmons, on the other hand, criticizes the "transcendentalizing" of nature, concerned that the particular (voices) will be lost in the totalizing of experience (Voice). But she does not acknowledge the possibility of nondual experience, and that the universal may be found in the particular, or that the particular may lead to the universal, and that integrating them leads to deep reverence rather than incorporation. Still, Simmons' core questions remain: Can we speak for nature? What constitutes good nature writing? And is it possible to not use nature when writing? It is more important to ask these questions than respond with rigid answers. Still, I would respond by arguing that we are part and parcel of nature and cultural animals, and thus writing is never fully extra-vagant, or beyond the bounds of language. But this merely means that nature writing should interpret via multiple forms of language, including scientific field notes, story, critical rationality, the grammar of animism and aesthetics, contemplative/poetic imagination, and spiritual, and thus multiple forms of experience and perception. Sympathy with intelligence demands nothing less. If such sympathy constitutes use, then there are clearly different types of use, with the listening and writing practiced via embodied virtues reflecting a more responsive use. Still, we may experience nondual moments where we are empty of use, empty of thought, empty of division, which we may attempt to communicate via multiple forms. We merely need to begin a practice of deep listening, comparing findings with other competent listeners.

Thoreau attended to nature like few others, developing skills of observation, embodied virtues, and spiritual reflection. If anyone can speak a word for nature, blurring the boundaries between science and poetic imagination, it ought to be him. Language is always of the world, and thus there is always a cultural bias, but Thoreau's bias is integrative, informed by experience beyond words and thoughts while remaining open to not-knowing and free from totalization, and thus it reflects a better bias. And anthropomorphism, when abused rather than reflecting genuine empathy that reveals relational insights, should be questioned via critical rationality, or rational communication informed by mythic-animistic, aesthetic, and spiritual communication. Sympathy with intelligence does not lead to anthropomorphism, but to the awake and widened awareness of an ecology of communication.

Thoreau's desire to speak a word for nature speaks to us today because nature needs defending, and our narratives of growth, progress, and technology need to be redefined within the context of attentive listening to the voice and voices of nature. Such practice returns us to a familiar critical question, however: Does toxic waste and polluted rivers also speak of Spirit? Are the voices of nature still present? Yes and no. Spirit is within every material manifestation, but humans obviously have the destructive ability to ignore the ethical constraints called for, screamed for, by this realization—the Ganga River, which is considered sacred, is polluted, extirpating or chasing away the voices of diverse species, not to mention what this pollution does to human voices, with nearly half a billion depending on the river for water. Yet, recognizing Spirit leads us to more clearly see what was, and the potential of what could be. Thus, listening to the voice and voices of nature has a practical result: we are more likely to refrain from doing damage in the first place, but if we have done damage, we may more wisely restore habitat by attending, as best we can, to the voices that were once there. Still, as Awiaka, a Cherokee poet, writer, and activist, articulated, it is "difficult to listen through concrete."[92]

Most certainly there are constraints on voice—social and economic structures, habit bodies, one-dimensionality—and yet Ramsey argues that becoming aware of these constraints discloses fissures for response and liberates our capacity to act. There is much that can be done to free nature's voices, as well as our own, allowing them to be heard. Such liberation provides an expanded array of responses, whether confronting the insanity of polluted rivers, discussing the insanity of climate crisis, or working to restore sanity by participating in protests or other forms of activism. Yet, as always, no single perspective is final, and some contradictions resist situational integration rather than stimulate higher order thinking and synthesis; but we may find the resolve to keep the conversation going, including with nonhuman others. Rather than being voiceless, then, we must recognize that living means being addressed, and then expand our capacity for listening and agency, which includes critical assessment of our eco-existential predicament.

In his journal, Thoreau laments the extirpation of species: "Is it not a maimed and imperfect nature that I am conversant with? . . . I listen to a concert in which so many parts are wanting."[93] Sadly, this wanting has exploded into the sixth "great" period of species extirpation, the other periods occurring long ago due to asteroid hits and volcanic explosions, while today's decimation is due to exploding human expansion and overconsumption. Thoreau, in response to the maimed nature of his time, offers the wood thrush's song, both in its actuality and as symbol, which tells the story of the "immortal wealth and vigor" of the forest; when we hear it, the "gates of heaven are not shut" such that new worlds may be created and institutions amended.[94]

Listening in sympathy, then, is a radical act that has the power to create new narratives, new institutions, and new worlds. Yes, the beginnings of a new age. In the manner of Thoreau, many need to speak a word for nature, or many words, and we must speak out and support fellow marginalized humans, and speak for future species of all kinds, and speak for ourselves; but what we say, and how we act, will be more fitting when we practice an ecology of communication.

SPEAKING OUT: THOREAU AND ACTIVISM

Thoreau is undoubtedly a mythic figure, but the biggest myth is that he hermited himself away from society and people, unattached, uninvolved, uncaring, spending his days sauntering in deep wilderness. He did spend plenty of time practicing deep listening, befriending nature, being awed by the beauty of nature, and overcoming nature in the good sense, but the higher laws he discovered, like the language that all things speak unfolding into sympathy with intelligence, led to ethical principles that informed not only his criticisms of industrialization and the masses of men, but also his activism, especially against slavery.

As we saw, the transcendentalists who met at Emerson's home were philosophical idealists, but this led to ideals they worked to make real in the world. They were not passive, but active souls who not only criticized institutions, but were ready and willing to kick ass, in the deliberative sense of merging logos with action. Fuller's journalism even took her abroad to Italy, where she continued to fight the powers-that-be as an early feminist and social reformer, and the whole lot were virulent abolitionists, among much other activism and social experiment, including the short-lived but instructive for our time Brooks Farm commune, which, along with advocating for living in community, balancing work, leisure, and learning based on the principle of equality, ran a school for all ages, from pre-school to adult education.[95]

Emerson's essays and lecturing were often directed at reforming the church and religious education, making them responsive to the active soul, while Thoreau's activism was a bit different. He wrote and gave lectures, but his radicalism was mainly expressed in what he refused to do. In "Life without Principle" Thoreau once again assailed bustling businessmen, suggesting that most jobs are like throwing a rock over a wall and then throwing it back again. We would be insulted if this were our actual job, he claimed, but most are "no more worthily employed," or worse, they find employment cutting down the forest, "making it bald before its time."[96] Thoreau was not against jobs, or logging for that matter, but he was against destructiveness, and he thought an awakened community would be as well, if only the bustling would stop long enough for citizens to reflect on what truly matters. Meaning is what we crave and we could have it if we lived a life of simplicity, without the clutter of extraneous needs. In Thoreau's hands—hands that built cabins, hoed bean fields, engineered pencils, and surveyed land—the call for simplicity challenged the established order.

Thoreau's activist heart is fully expressed in "Civil Disobedience," although most initially read it only as a treatise against big government. But this seminal essay criticizes life without principle, as well as life without conscience, which ends up being no life at all. For Thoreau, to be free we must first and foremost be subjects of individual conscience rather than governments or any majority for that matter. We must respect the collective good and the higher good within ourselves, listening without and within, separating the diabolic from the divine, and doing what we must in many moments of daily decision-making rather than what laws or social convention tell us to do.[97]

The arguments in "Civil Disobedience" have their roots in Thoreau's refusal to pay a poll tax for six years, a protest against the government's support of slavery and war with Mexico. His defiance led him to spend a night in jail, thanks to going to town to get his shoe fixed during his Walden stay and running into the jailer—another example of his non-hermit-like life. It would have been more than one night but his aunt bailed him out, to his chagrin, but looking through his jail house window at oblivious passersby, who he ventured gave little thought to justice or the jailed, provided him with a different perspective. The result was his influential exhortation: "Under a government which imprisons any unjustly, the true place for a just man is also prison."[98] To Thoreau, the only way to freely serve the State is with conscience. Politicians, legislators, lawyers, ministers, and office-holders tend to serve with their heads only, making them poor in ethical decisions. They patronize virtue instead of living virtuously, sold to institutions that make them rich, beholden to free trade instead of true freedom. Thoreau greatly respected the courage of the soldier, who served with their bodies, but argued that they too often follow the orders of the State, marching counter to head, heart, and ethical sense.[99]

True heroes and patriots, then, serve their conscience—their inner genius—expressed outwardly in the pursuit of justice. Such citizens are friends of the State, although treated as enemies. Thoreau understood that the machinery of government can't be perfect, that there is a degree of injustice in its workings, but if this makes us an agent of injustice against another, conscience demands that we make ourselves a "counter-friction to stop the machine."[100] Thoreau wrote "Civil Disobedience" in 1849, but we must ask ourselves during climate activism and Black Lives Matter movement times: Does any of this sound familiar? Politicians and others in power privileging profit over principle? The inner demand to counter the machine of the powers-that-be? The call to follow conscience as a radical act?

Thoreau's activism became more radicalized as he countered the machinery of slavery. In 1854, inspired by Frederick Douglass and the abolitionist orator Wendell Phillips, he gave a fiery address, "Slavery in Massachusetts," before an estimated crowd of 2,000, one of the largest anti-slavery gatherings, in Boston in July 4th heat, appealing to higher law and calling for secession from the union.[101] He had lived for months with a feeling of "vast infinite loss," unsure what it was, and then realized: "At last it occurred to me that what I had lost was a country." Running with this theme, he stated, "The remembrance of my country spoils my walk. My thoughts are murder to the State, and involuntarily go plotting against her," a revolutionary remark, before ending on a hopeful note: the water lily is rooted in the "slime and muck of the earth," and thus a lotus may yet bloom, despite the "sloth and vice of men, the decay of humanity."[102]

Walden was finally going to the printers in 1854, yet higher laws were being trampled upon and discarded. How could he write of freedom in nature, of beauty and justice as part of the framework of the cosmos, when so many were enslaved? Haunted by this question on the eve of his biggest triumph, with the annexation of Texas and the further march of Manifest Destiny stirring the pot by adding more slave states in the South and West, Thoreau stewed on the unethical, illogic of slavery, stoking a rage not yet fully formulated, while he and his family continued to help runaway slaves escape as part of Underground Railroad.[103]

When Anthony Burns was captured and tried in Boston in accordance with the Fugitive Slave Act of 1850, Thoreau wrote that it was actually Massachusetts on trial: "every moment that she hesitates to set this man free—she is convicted." Thoreau had become a radical abolitionist, his address run in newspapers, with one using the headline "Words that Burn." Emerson was also speaking out against slavery, stating "we must get rid of slavery, or we must get rid of freedom," and yet Thoreau's burning words signified a break from Emerson in the public's eye.[104] The defender of nature and simple living had joined Frederick Douglass and others in declaring that the Declaration of Independence was a lie.[105]

And then, John Brown. Brown and his followers murdered five proslavery Kansans, dragging them out of their homes, in retaliation for the overthrow of Lawrence by proslavery militia who killed up to 200 men and boys, pillaging and burning down the town. Brown then traveled east, including to Concord, dining at the Thoreau home, to raise funds for the abolitionist movement. Thoreau was suspicious, but eventually won over by his passion, even though Brown had turned to violence in response to violence.[106]

After Brown's imprisonment due to his failed raid on the federal arsenal at Harper's Ferry, Thoreau simmered, and then, two days after hearing the news of Brown's capture, ready to explode, passed the word among fellow townsfolk that he would speak. Friends and some family begged for silence. He replied to one friend, Sanborn the schoolmaster, who had been involved in Brown's plot but now feared for his life: "I did not send for you for advice but to announce that I am to speak." Brown was on trial for treason, newspapers denounced him, and Thoreau was the first to stand up in public in his defense. The church was filled to capacity, and Thoreau read his "Plea for Captain John Brown" before a curious but doubtful audience, with "no oratory, as if it burned him." Young Edward Emerson, Ralph Waldo's son, said he sounded like he was speaking about his own brother, and summarized the reaction: "Many of those who came to scoff remained to pray."[107]

Word spread, thanks to Emerson. When Douglass, scheduled to speak in Boston, was implicated in the plot by a letter found on Brown, a warrant was put out for his arrest and he was forced to flee to Canada; Thoreau stepped in, before a crowd of 2500: "The reason why Frederick Douglass is not here is the reason why I am," he began, the audience rapt and applauding for an hour-and-a-half. An abridged "Plea" was printed in newspapers across America; he hoped to get it published as a pamphlet, but no publisher would print it. Yet Thoreau felt lost, again, while waiting for Brown's death by hanging; nature was not its usual comfort with the soul of a nation at stake: "So great a wrong . . . overshadowed all beauty in the world."[108]

Thoreau planned a memorial service on the day of Brown's execution; not all in Concord were happy about it, not all were in favor of Brown. Thoreau later helped Francis Jackson Merriam, a Harper's Ferry plotter and financial supporter, and the most wanted man in America, to escape to Canada, although his identity was unknown to him at the time. And the Thoreaus helped take care of Brown's children, such aid an act of defiance to the government. Brown was a man of action against physical institutions, but those speaking out, like Douglass, like Phillips, like the Concord Female Anti-Slavery League, and like Thoreau, disclosed a larger lesson: the call to responsibility turns ideas into action, but also actions into ideas that can drive a wedge into institutions.[109] In the growing response to slavery, Thoreau wrote: "The *living* North . . . is suddenly all Transcendental," giving him hope in the power of words and ideals and speaking the truth.[110] The government was on the wrong side, maintaining slavery and killing "the liberators of the slave."[111]

Brown's turn to violence alienated many, but contradictory Thoreau, non-violent by disposition, learning, and experience in nature, was having none of it, calling the government a brute force, and worse, "a demoniacal force!" while questioning what citizens were willing to do.[112] Jefferson wrote that "Rebellion to tyrants is obedience to God," and Brown justified his actions as God's will, or rather, a response to the fact that slavery could not possibly be God's will, and thus action was needed, equating caution with cowardice.

Thoreau agreed, stating, "The man this country was about to hang appeared the greatest and best in it."[113]

In "The Last Days of John Brown," Thoreau argued that Brown was a transcendentalist above all, revering higher law over law, embodying principle and conscience. But, in a kind of compliment to himself, and nod to differing levels of development, he stated only the noble perceive the noble: "How can a man behold the light who has no answering inner light?" And if anti-Brown citizens were Christians, they were neither citizen nor Christian, having no genius for listening to inner light or being a free man: "They seem to have known nothing about living or dying for a principle."[114] And that, for Thoreau, was the lesson of John Brown: The public now had an example of living and dying for principle, if their inner light was not completely dimmed.

In a provocative move, Thoreau then turned to liberal education, stating that for the Romans only free men were worthy of such education, while slaves learned trades and professions. Working from this premise, he argued that only the liberally educated are truly free and thus able to revolt: "In a slaveholding country like this, there can be no such thing as a *liberal* education tolerated by the State." In other words, a liberal arts education, which moved the learner up the developmental holarchy, is dangerous to those in power, as those contented under tyrannies "have received only a *servile* education."[115] Brown, as martyr, revealed the State's character: it could choose violence or admit its failures and culpability. It chose violence, or murder, yet for Thoreau government was not only a brute force, but also a human one capable of growth—the water lily emerges from the muck—and thus appeal to protect the rights of dissenters was possible.[116]

Slavery infected everything and caused Thoreau to continue to question the worth of *Walden* and love of nature in the face of the baseness of man. But this sets up a false dichotomy, which Thoreau in his actions overcomes: higher laws, just like extended mind or the pattern that connects, transcends, and includes nature and culture. A revelation provided guidance: Being one with Spirit and part and parcel of nature leads to an ethic of least harm, whether toward moose, snake, or man. "We must ask ourselves weekly," he wrote, "Is our life innocent enough? Do we live inhumanely—toward man or beast—in thought or act?"[117] Thoreau even advocated thinking like a bream, and writing from such thinking, as the bream is "another image of God . . . a provoking mystery."[118] Freedom, for Thoreau, included the rights of nature and humans, with the freedom to be, to habitat, and to become having universal resonance.

Such resonance provided him with hope, as it should today, but his revelation leads to more queries: what to do when an ethic of least harm is not universally held, or not held by those in power? Thoreau lived questions we still ask: Are there limits to freedom? Is violence ever an appropriate response to violence? What are our responsibilities? Or, more specifically, why care about nature when black men and women are being killed by police? We are also steeped in culture wars, with Mythic, Rational, and Pluralistic level citizens hacking rather than hashing it out, including vitriol over the value of coronavirus masks, and the taking down of confederate flags and statues: confederate generals go down in the south, to Pluralistic glee, but then, in Mythic anger, so too does a statue of Frederick Douglass in Rochester, New York, the site of his most famous speech, "What to the Slave is the Fourth of July?"

For Thoreau in the mid-1800s, the confederate flag waved and the Civil War loomed, and he wondered how he could do more: "I do not wish to kill or to be killed, but I can foresee circumstances in which both of these things would be by me unavoidable."[119] The commons were also increasingly being lost—he bemoaned no trespassing signs, forcing him to pay for berries he used to pick for free—and the financial panic of 1857 led to mass unemployment, leading him to further question an economic system that distains simple living, and then the Transatlantic telegraph arrived in 1858, a great invention but for what purpose?[120] What would be the quality of our communications? Would it be an improved means to an unimproved end? This, of course, is another question that has reverberated throughout the decades to the digital age, along with: How can we be women and men of conscience?

Reading Thoreau, and about his life and times, once again reveals the wisdom of the developmental holarchy: we communicate and act from our stage center of gravity, and as our consciousness is raised we are better able to raise the consciousness, and conscience, of others. When we have the freedom to inquire, to dare to know, we are more likely to take next steps, and, eventually, embody a degree of spiritual maturity, recognizing our common home and humanity, including with adversaries, and thus our own alienation and desire to lash out in violence, as well as historical circumstances that cause conflict.

Martin Luther King, who read Thoreau and was emblematic of such maturity, stated he just wanted to do God's will. But King had moments of doubt, illustrated by his "kitchen table experience" when he appealed to God, with deep humility, to give him strength. He was considering leaving the struggle for liberation due to threats on his life, but God's presence descended, and his inner voice spoke giving him peace with his calling. King was able, despite the threats and hatred he confronted daily, to forgo the desire to humiliate others in retaliation, understanding with great love that the true enemy was the bondage of ignorance. As we saw, Archbishop Oscar Romero also turned to civil disobedience fueled by conscience, and also described his allegiance to God's will, or seeing with God's eyes, leading him to expand the focus on salvation from the individual soul to the liberation of a community of souls, freeing them from economic, political, cultural, and religious oppression.

Again, God's will, at higher levels of development, manifests as unity uniquely expressed within the individual will; or, if all life is one, or not-two, we must serve the all. Surrendering to something larger puts us in touch with this higher law, which is a creative force, silencing fears that block listening, growth, and action. Mahatma Gandhi, also influenced by Thoreau, actively listened via satyagraha, translated as truth, love, or soul force. Satyagraha, the core of nonviolent practice, is a tremendous force for the simple reason that following conscience inspires others to do the same, breaking down ego-barriers and the illusion of separateness—which, of course, is a root cause violence—and potentially creating a wave of grassroots transformation. Or as Gandhi put it, in a moment of great optimism and hope: "Things undreamt of are daily being seen, the impossible is ever becoming possible."[121]

I am not much interested in advocating for hope, but I am interested in response and responsibility. When teaching college students who seem despondent from too much bad news, I remind them of the record of nonviolent, and sometimes violent, resistance bringing down oppressive dictatorships and changing governments, and enlightenment rationality in service of freedom, equality, rights, and the common good—higher laws, all—bringing progressive change generally, with citizens speaking out because their conscience demands it. We discuss the 1848 Declaration of Sentiments, principally penned by Elizabeth Cady Stanton, which declared that "all men and women are created equal"; the 100 signees—68 women and 32 men, including Frederick Douglass, were called to use every instrument within their power to do the great work of fighting for equality, despite misconception and ridicule. And we discuss Douglass' 1852 "What to the Slave is the 4th of July?" a stirring example of logos, inspired by principle, taking on unfulfilled ideals of a nation and thus the grand irony of celebrating independence but not dismantling the institution of slavery.

We discuss the spirituality of Thoreau and Gandhi and Romero and King, but also King's "Letter from a Birmingham Jail," which, along with proving Thoreau's point that jail is the true place for the just in an unjust society, is another stirring example of logos and passionate rhetoric inspired by principle, as King takes on the wrong-headed arguments of fellow clergy wary of civil rights and fearful of protest and disorder. But when human law does not mirror higher law, conscience may demand that we break the law. How, and when, are never-ending questions of situational fittingness, as is the nonviolence versus violence question, but we can traverse the past for lessons. We also discuss Malcolm X's virulent rhetoric, "change by any means necessary," and claim that nonviolence is fine, if it works; Thoreau, despite influencing Gandhi and King, was sometimes more rhetorically aligned with Malcolm X, especially in his defense of John Brown, but also in his defense of Anthony Burns; after all, his thoughts were murder to the State.

Thoreau may be a dead white male, but as the Black Lives Matter movement grows and climate crisis protests grow, he still has plenty to offer, especially since their intersection is also growing as those most negatively affected by rising temperatures will be those who are already most negatively affected by toxic pollution and undrinkable water and increased cancer rates: people of color and the poor.

Contradictory and far from perfect, Thoreau defended planet and people, especially the marginalized, despite criticizing the mass of men. He was not always the most progressive with women, unless they stood for principle, like his mother and sisters, and Emerson's aunt, Mary Moody Emerson, who was a feisty conversationalist. He quarreled a bit with Margaret Fuller when he wasn't published in the Dial, but learned from her, and he was the one who traveled to Cape Cod to potentially recover her body when she died in a shipwreck. He also supported the Female Anti-Slavery League, which included his mother and sisters, having them over when he lived on Walden Pond.[122] He also learned from Joe Polis and defended Native Americans against prejudice, unfortunately dying before he could turn his "Indian Notebooks" into a published work. And, along with his abolitionist activism, he wrote with sympathy in *Walden* of his black neighbors, who were forced to live away from town, and the Irish, the latest immigrants to suffer prejudice, whose labor laying down the railroad was most certainly ridden upon and forgotten by most riders.[123]

In all these relations, he was way ahead of his time, and would be a force in our time, and the reason why is because of his appeal to the universal, to higher law, which was expressed as an ecology of communication. Mythic-animistic, aesthetic, and spiritual communication do not only focus on the natural world; rather, they inform the dialogic interplay of nature and culture, or wherever there is relationship: they inform principle, they inform conscience, and they inform rational communication, the mode of communication where our daily decisions are formulated and given voice. But Thoreau the activist is driven by Thoreau the inner listener who responds based on what emerges within. In other words, Thoreau was a model of Emerson's call for self-reliance: "No law can be sacred to me but that of my own nature. Good and bad are but names very readily transferable to that or this; the only right is what is after my constitution, the only wrong what is against it."[124] This is rousing rhetoric, which Thoreau mirrors in *Walden*: "The greater part of what my neighbors call good I believe in my soul to be bad, and if I repent of anything, it is very likely to be my good behavior. What demon possessed me that I behaved so well."[125] Both mentor and student declare that we must follow inner promptings, or the soul within the context of the world soul, no matter the outward mores of society. This is sound advice, the only advice for producing true genius and creativity, as well as discovering the depths of Spirit and the call to responsibility.

An ecology of communication includes such inner listening, but always balanced with listening to human and nonhuman voices and the voice of Spirit in nature, in sympathy with the language that all things speak. Roszak directed us within, making us mindful of wildness and the ecological unconscious. Thoreau explored such wildness, but his responsiveness also included the inner listening of spiritual traditions, where God and genius may be found, where we may divorce the diabolical from the divine, finding a path forward despite the many pressures of the world: "If one listens to the faintest but constant suggestions of his genius, which are certainly true, he sees not to what extremes, or even insanity, it may lead him; and yet that way, as he grows more resolute and faithful, his road lies."[126]

But freedom is interpreted differently at differing levels of the developmental holarchy, and so is responsibility. One person's divine right is another's demon, whether fighting over masks during a pandemic or the removal of statues. As the culture wars rage on and protests grow, with some Pluralistic level victories, the greatest fight, and what would be the greatest victory, is moving more folks up the developmental holarchy, where one finds greater freedom, stepping out of the cave of shadows and one-dimensionality, and greater responsibility to others by aiding their stage steps. That takes many forms, but listening is key.

Listening inwardly and outwardly leads us to fully participate within the dialogic system that sustains us. Listening inwardly and outwardly leads us to realize that we are nothing without the gifts of Spirit and the earth. And listening inwardly and outwardly, via mythic-animistic, rational, aesthetic, and spiritual communication, inevitably leads to a life of activism, expressed in a myriad of ways, in what we refuse to do, and in what we feel we must do, when confronting eco-social crises of our time.

Or, to quote Frederick Douglass: "Power concedes nothing without a demand. It never did and it never will."[127]

After Thoreau got out of jail, he joined a huckleberry party. He was definitely not ascetic about picking berries, and he liked to play; his call to simplify, simplify, simplify was a call to more joy, not less, and his pursuit of freedom in nature mirrored pursuit of freedom in culture. And that is how we listen to the voices of others and find our voice. The lesson of Thoreau: We are called to freedom and joy in all relationships, and called to responsibility in all relationships, because when such freedom and joy are denied, we must respond.

NOTES

1. Richard Bridgman, *Dark Thoreau* (Lincoln, NE: University of Nebraska Press, 1982), 80.
2. Kathryn Schulz, "Pond Scum: Henry David Thoreau's Moral Myopia," *The New Yorker*, 91, 32, October 19, 2015, https://www.newyorker.com/magazine/2015/10/19/pond-scum.
3. Walls, *Henry David Thoreau.*
4. Robert Sullivan, *The Thoreau You Don't Know: What the Prophet of Environmentalism Really Meant* (New York: HarperCollins, 2009).
5. Richard Lebeaux, *Young Man Thoreau* (Amherst, MA: University of Massachusetts Press, 1977), 9, 13.
6. Thoreau, *Walden*, in *The Portable Thoreau*, 462.
7. Emerson, "Nature," 53.
8. Ralph Waldo Emerson, "Plato; or, the Philosopher," in *The Portable Emerson*, edited by Carl Bode (New York: Penguin Books, 1981), 296.
9. Emerson, "Nature."
10. Thoreau, "Walking."
11. Thoreau, *Walden*, in *The Portable Thoreau*, 270.
12. Ibid., 296.
13. Ibid., 457.
14. Ibid., 466.
15. Thoreau, "Walking," in *Nature, Walking*, 95.
16. Thoreau, *Walden*, in *The Portable Thoreau*, 342.
17. Ibid., 343.
18. Thoreau, "Walking," in *Nature, Walking*, 74–75.
19. Thoreau, *Walden*, in *The Portable Thoreau*, 343.
20. Ibid.
21. Ibid.
22. Ibid., 349.
23. Ibid., 405.
24. Ibid., 408.
25. Kaza, *The Attentive Heart*, 11–12.
26. Thoreau, *Walden*, in *The Portable Thoreau*, 351.
27. Ibid., 353.
28. Ibid., 363.
29. Ibid., 625.
30. Ibid., 380.
31. Ibid., 457.
32. Ibid., 386.
33. Ibid.
34. Ibid.
35. Ken Wilber, *No Boundary: Eastern and Western Approaches to Spiritual Growth* (Boston, MA: Shambhala, 1979).

36. Emerson, "Thoreau" in *The Portable Emerson*, 576.
37. Quoted in Bridgman, *Dark Thoreau*, 10.
38. See Walls, *Henry David Thoreau.*
39. Ibid., 193.
40. See Smith, *My Friend, My Friend*. See also Jeffrey S. Cramer, *Solid Seasons: The Friendship of Henry David Thoreau and Ralph Waldo Emerson* (Berkeley, CA: Counterpoint, 2019).
41. Emerson, "Thoreau" in *The Portable Emerson*, 574.
42. Ibid., 593.
43. Quoted in Bridgman, *Dark Thoreau*, xi.
44. Thoreau, *Walden*, in *The Portable Thoreau*, 566.
45. Carl Bode, "Introduction by the Editor," in *The Portable Thoreau*, edited by Carl Bode (New York: Penguin Books, 1975), 690.
46. Thoreau, *Walden*, in *The Portable Thoreau*, 424.
47. Ibid., 363.
48. Henry David Thoreau, *The Heart of Thoreau's Journals*, edited by Odell Shepard (New York:Dover Pub., 1961), 58.
49. Ibid., 91–92.
50. Daniel B. Botkin, *No Man's Garden: Thoreau and the New Vision for Civilization and Nature* (Washington, DC: Island Press, 2001), 65–67.
51. Ibid., 75.
52. Thoreau, *The Heart of Thoreau's Journals*, edited by Odell Shepard, 135.
53. Ibid., 112.
54. Richardson, Jr., *Henry Thoreau*, 29–31.
55. David M. Robinson, *Natural Life: Thoreau's Worldly Transcendentalism* (Ithaca, NY: Cornell University Press, 2004), 182.
56. Quoted in Robinson, *Natural Life*, 181.
57. Laura Dassow Walls, *Seeing New Worlds, Henry David Thoreau and Nineteenth-Century Natural Science* (Madison, WI: The University of Wisconsin Press, 1995), 7.
58. Quoted in *Material Faith: Thoreau On Science*, edited by Laura Dassow Walls (New York: Houghton-Mifflin, 1999), ix–xii.
59. See Science Daily, "Using Thoreau, Scientists Measure the Impact of Climate Change on Wildflowers," March 14, 2019, https://www.sciencedaily.com/releases/2019/03/190314192645.htm.
60. Thoreau, *Walden*, in *The Portable Thoreau*, 463.
61. Emerson, "Thoreau" in *The Portable Emerson*, 585.
62. Walls, *Henry David Thoreau: A Life*, 442.
63. Ibid., 458–460.
64. Ibid., 470–472.
65. Ibid., 345.
66. Ibid., 374.
67. Ibid., 420.
68. Ibid., 433.

69. Ibid., 307.
70. Ibid., 439.
71. Ibid., 236.
72. Ibid., 237.
73. Ibid., 288.
74. Charles Eisenstein, *Climate: A New Story* (Berkeley, CA: North Atlantic Books, 2018), 34.
75. Ibid., 250.
76. Quoted in *Material Faith: Thoreau On Science*, edited by Laura Dassow Walls (New York: Houghton-Mifflin, 1999), xii.
77. Emerson, "Circles," in *The Portable Emerson*, 232.
78. Thoreau, "Walking," in *Nature, Walking*, 71.
79. Nancy Craig Simmons, "Speaking for Nature: Thoreau and the 'Problem' of Nature Writing," in *Thoreau's Sense of Place: Essays in American Environmental Writing*, edited by Richard J. Schneider (Iowa City: University of Iowa Press, 2000), 223.
80. Members of The Thoreau Society use a public Facebook group, "Thoreau's Journal," to send out daily entries.
81. Simmons, "Speaking for Nature: Thoreau and the 'Problem' of Nature Writing."
82. Ibid., 225.
83. Ibid., 224.
84. Emerson, "Nature," in *Nature, Walking*, 22.
85. Ibid., 9.
86. Simmons, "Speaking for Nature: Thoreau and the 'Problem' of Nature Writing," 231.
87. Quoted in Simmons, 230.
88. Simmons, "Speaking for Nature: Thoreau and the 'Problem' of Nature Writing," 229.
89. Ibid., 225.
90. Ibid., 224.
91. Quoted in Jeffrey S. Cramer's "Introduction" to Thoreau's *The Maine Woods*, xviii.
92. Quoted in Carbaugh, "Just Listen: 'Listening' and Landscape Among the Blackfeet," 118.
93. Henry David Thoreau, *The Heart of Thoreau's Journals*, edited by Odell Shepard, 157.
94. Ibid., 92–93.
95. Lawrence Buell, "Introduction," *The American Transcendentalists: Essential Writings* (New York: Random House, 2006), xiv–xviii.
96. Thoreau, "Life Without Principle," 633.
97. Henry David Thoreau, "Civil Disobedience," in *The Portable Thoreau*, edited by Carl Bode (New York: Penguin Books, 1975), 111.
98. Ibid., 122.
99. Ibid., 112.

100. Ibid., 120.

101. Walls, *Henry David Thoreau: A Life,* 313.

102. Henry David Thoreau, "Slavery in Massachusetts," American Transcendentalism Web, https://archive.vcu.edu/english/engweb/transcendentalism/authors/thoreau/slaverymass.html.

103. Walls, *Henry David Thoreau: A Life*, 314, 317–318.

104. Ibid., 445.

105. Ibid., 345–349.

106. Ibid., 446–449.

107. Ibid., 451–452.

108. Ibid., 452–454.

109. Ibid., 454–456.

110. Henry David Thoreau, "The Last Days of John Brown," in *The Portable Thoreau*, edited by Jeffrey S. Cramer (New York: Penguin Books, 2012), 549.

111. Henry David Thoreau, "Slavery in Massachusetts," American Transcendentalism Web, https://archive.vcu.edu/english/engweb/transcendentalism/authors/thoreau/slaverymass.html.

112. Walls, *Henry David Thoreau: A Life*, 449.

113. Thoreau, "The Last Days of John Brown," in *The Portable Thoreau*, edited by Jeffrey Cramer, 547.

114. Ibid., 550.

115. Ibid., 552.

116. Henry David Thoreau, "Slavery in Massachusetts," American Transcendentalism Web, https://archive.vcu.edu/english/engweb/transcendentalism/authors/thoreau/slaverymass.html.

117. Walls, *Henry David Thoreau: A Life*, 348.

118. Ibid., 435.

119. Ibid., 451.

120. Ibid., 432–433.

121. See M. K. Gandhi, *Non-Violent Resistance (Satyagraha)* (New York: Dover Press, 2001).

122. Walls, *Henry David Thoreau: A Life*, 320–321.

123. See Elise Lemire, *Black Walden: Slavery and Its Aftermath in Concord, Massachusetts* (Philadelphia, PA: University of Pennsylvania Press, 2009).

124. Emerson, "Self-Reliance," in *The Portable Emerson*, edited by Carl Bode, 142.

125. Thoreau, *Walden*, in *The Portable Thoreau*, 266.

126. Ibid., 462–463.

127. See the opening of *Disruption*, Climatechange12.com, *Disruption*, September 7, 2014, YouTube video, 52:27, https://www.youtube.com/watch?v=uWPj6CxtsGo.

Epilogue

A Fitting Responsiveness: Communicating Our Way into the Future

When teaching my *Freedom of Speech* course, I sometimes ask students: who do you trust more, government or the press? While a provocative question, is also a bit of a false dichotomy logical fallacy. More broadly, the question is who or what can we trust at all, and on what grounds. So many voices are speaking out: who to listen to?

Among the true, good, and beautiful, one might think that "the true" would be easiest to teach—just focus on facts—but our age of ecocrisis is exacerbated by the age of spin. Post-truth used to mean postmodernists wisely deconstructing absolute Truth; now it seems to mean the lie of "alternative facts" or that everything is up for grabs because everything can be spun to suit one's preferred, and profitable, version of reality.

But what of my bias? I teach *Disruption*, a documentary on the 2014 People's Climate March in New York City, which "deniers" would consider climate propaganda expressing a one-sided "alarmist" point-of-view; but the talking heads in the film have impressive credentials, including Naomi Oreskes, a historian of science, who argues that climate research is quite old and well-established, and thus it's not a two-sided issue, with a 97 percent consensus among current climatologists that warming is happening and significantly anthropogenic.[1] In other words, some claims are supported by the weight of evidence and credentials matter; I have a bias toward such weight and things that matter.

I often highlight my bias by telling students why I am teaching their class: my degrees, sure, but the main reason is I am older than they and have been researching and thinking about our course topic for decades. I am there to share what I have learned and to keep learning along with them. I also start some classes with the oft-quoted section from the preface of Walt Whitman's *Leaves of Grass*, beginning "This is what you shall do; Love the earth and sun and the animals," and then continuing with a lively list, but I focus student attention on "re-examine all you have been told at school or church or in any book, dismiss whatever insults your own soul, and your very flesh shall be a great poem."[2] My point is this: Respect my experience and credentials, and listen to the evidence, but, ultimately, find out for yourself what is true. The easy appeal to authority is another logical fallacy, after all, at least when the "expert" does not back up their arguments.

Whitman, by the way, along with Joe Polis and John Brown, and of course Emerson, probably had the most influence on Thoreau; he carried *Leaves* around for weeks, saying it had done him more good than any other reading for a long time, and met him in New York with two others, making the meeting short, but he remarked that Whitman was a great fellow who spoke more truth than any other American.[3] No doubt his praise was due to Whitman's passion for both sensuous and spiritual freedom, and, I imagine, for lines like these: "I think I could turn and live with animals, they are so placid and self-contain'd, I stand and look at them long and long."[4]

I hope to have a Whitman-like influence on students, but my spiritual practice while teaching is to let go of expectations, and thus my advice to them has long been to not label yourself, and to realize that you are larger than you think, containing multitudes beyond the limits of liberal or conservative or any political persuasion; be free by practicing logos, taking it issue by issue, and then see what happens. Create your own list, I say, and transform yourself into a great poem, having "the richest fluency not only in its words but in the silent lines of its lips and face and between the lashes of your eyes and in every motion and joint of your body."[5] But then I let go and soldier on.

Early in my teaching career, an Agriculture major gave a speech on the benefits of pesticides, castigating Rachel Carson. His facts seemed accurate—increased food production, lower price, in a hungry world—but, of course, I don't let an argument go without the class exploring counter-arguments. During the Q&A, I asked about the environmental and social costs of pesticides, and if the distribution of food, rather than the amount, was a root problem. I also stated, without rancor and with a smile, that if Carson could hear his spin on pesticides she would be spinning in her grave; what, I wondered, would she say in response? My bias says that Carson is an in-sympathy, logos-wielding eco-hero, but I was also anxious to see his sources to see if I was missing something. Later, I looked over one-sided citations despite requiring a diversity of sources.

Researching diverse sources is an internet-easy fix, but evaluating the quality of sources has become increasingly digital-difficult. Thoreau's question of whether new technologies are improved means to unapproved ends may be the most prescient thing he ever said, as the web has become a means for tumbling down, rising up, or getting siloed into differing levels of the developmental holarchy. Seemingly sound attacks on Carson are readily available on websites, for example, and thus I am not sure if I should blame my student for what he did not know, even though we spend quite a bit of time on media literacy.

In *Merchants of Doubt*, Oreskes and Erik M. Conway, a NASA historian, tell the fuller tale of Carson's legacy and resistance to it, providing a case study of post-truth politics—good for students, teachers, and all citizens—but also on applying logos and rational communication to the wrong-headed debate on climate crisis, and thus what we collectively face as fallible humans dependent on a biosphere.

Carson discovered that DDT and other pesticides were accumulating in the food chain killing a host of species: birds, ironically including the Bald Eagle, as well as fish, squirrels, pets, and beneficial insects who kept other insects in check. Her findings held up after being reviewed by respected science organizations, like the President's Science Advisory Committee, and the weight of evidence was sufficient to influence policy. In 1972, ten years after publication of *Silent Spring*, action was taken: a ban on DDT in the United States, with exceptions granted for public health emergencies. The sale of DDT to the World Health Organization for use in countries struggling with insect-born epidemics like malaria was also permitted.[6]

Ten years is a long time for science to influence policy, but *Silent Spring* was a science and democracy success story: Carson combed through scientific studies, alerting citizens to an environmental hazard via compelling writing, and thus free inquiry led to action based on what is true. As we've seen, she was called an extremist, and yet the response by the chemical industry was the extreme hysterical-emotional woman catcall, with Carson charged with sounding a false alarm not supported by science and ignoring many benefits of pesticides, like increased food production at lower prices.[7]

Years later, the controversy was reignited by free market advocates at The Competitive Enterprise Institute, The American Enterprise Institute, and The Heartland Institute, among others. Carson, not DDT, was the killer; she was not only hysterical and emotional, but accused of the murder of millions in countries like Africa who, responding to evidence, stopped widespread spraying to control malaria. Some compared Carson to Stalin and Hitler, and mainstream newspapers printed opinion pieces condemning *Silent Spring*'s influence on the public mind.[8]

A few facts, however, were missing.

DDT had seemed to be a miracle chemical, with Paul Muller, who resynthesized it in 1940, making it cheap and easy to spray, winning the Nobel Prize for Physiology or Medicine in 1948 after it was used in World War II. Wartime technology was then used in agriculture, with DDT sprayed from planes turned into crop dusters. Indiscriminant use was soon widespread—DDT was everywhere, and bioaccumulation had begun. Species like eagles did not need direct exposure; they were infected and affected by eating small rodents who had eaten exposed insects.[9]

Bioaccumulation is established science, as are the basics of natural selection: insects that survived DDT mated, making their offspring increasingly resistant to spraying. Bugs breed a lot, and so it only took seven to ten years for DDT to largely lose its effectiveness.[10] Thus, outdoor spraying only worked as a deterrent against disease in the short term, and even then it didn't work alone. Malaria was eradicated in so-called first-world countries due to spraying *along with* education, good nutrition, health care, and the removal of breeding grounds. And that fact leads to more irony given the attacks on Carson: colonization, along with free market economics are often responsible for the lack of good education, nutrition, and health care in poorer so-called third-world countries.[11]

It also became clear that precise indoor spraying was more effective than indiscriminant outdoor spraying and had the benefit of not invading the ecosystem. But again, spraying was losing its effectiveness, including before publication of *Silent Spring* and thus long before the 1972 ban, which, once again, did not ban the use of DDT in emergencies or the sale to other countries, who could make up their own minds.[12]

Carson argued that DDTs use in agriculture was greatly speeding up bioaccumulation and growing insect resistance and also suggested that DDT and other biocides might cause cancer and other ill-effects, as they last long in the environment, although no conclusive evidence and no human, as far as was known, had died or suffered adverse health from DDT. Now there is evidence. One review published in the *Lancet*, a leading peer-reviewed medical journal, stated that high exposure to DDT damages reproductive health, including preterm birth, low birth rate, and possible birth defects. DDT in breast milk could also lead to infant and child mortality. Another study found that women exposed to DDT in childhood were five times more likely to have breast cancer. Unless, of course, these are not facts, just more "hysterics" from emotional women.[13]

After Carson, there was "hysterical housewife" Lois Gibbs, along with diverse others residing in the toxic town of Love Canal in 1978, including low income black women who rented federal housing, speaking out about chemical waste;[14] and Erin Brockovich, a legal clerk with little formal education taking on Pacific Gas & Electric's chemical contamination of drinking water in the early 1990s; and Marjorie Richard, the first African-American to win the Goldman Environmental Prize, who fought Shell Oil in "cancer alley" in Norco, Louisiana, from the 1980s to the early 2000s by using a range of methods, from starting a community group and conducting research to installing a web camera on her home to broadcast nonstop the spewing of chemical byproducts and approaching Shell officials at a conference and inviting them to inhale from a bag of Norco air;[15] and Sandra Steingraber, a cancer survivor who, following Carson, became a biologist and found links between pesticides and her cancer, and cancer in her family and community, documented in her 1997 book, *Living Downstream.*[16]

Steingraber has since been arrested twice, in 2013 and 2014, for acts of civil disobedience to protect water in the Finger Lakes region of New York. She wrote letters from jail, including one in which she stated she was held in 24-hour lock up for five days, not having access to her children, but that their tears, and the tears of her grandchildren, will be greater in the future "if we mothers do nothing and allow the oil, coal, and gas companies to take us off a climate cliff."[17] We all live downstream, which is why we need to be active with the activists, honoring Carson's legacy while being open to new information via the never-ending process of critique marked by distanciation and participation, the articulation of interpretations, and discernment of the unfit. Some medical experts, for example, argue that DDT could help control malaria in some parts of the world today; if so, evidence, the lessons of history, and the fact of interdependency—revealed through ecological science (the true), I-Thou ontology (the good), and aesthetic and spiritual experience (the beautiful)—disclose that indiscriminate use, which was what Carson warned against, would be a big mistake.[18]

Carson's main claim was undoubtedly true—DDT and other biocides unbalanced ecosystems—while free market institutes, who also muddied the science documenting the effects of smoking, the ozone hole, and acid rain, are merchants of doubt, guilty of revisionist history calculated to favor big industry. Oreskes and Conway explore the dystopic, Orwellian feel of all this, and another irony: right-wing defenders of corporate freedom turned to anti-freedom Cold War Soviet-style oppression, spinning facts for political and financial gain.[19]

Oreskes and Conway also catalogue the misdeeds of "respected" scientists who worked for free market institutes to poke holes in the science, not in the peer review sense, but in the politically motivated distort the findings sense, for personal financial gain, no doubt, but mostly because they were true believers in freedom as free market, despite the fact that such unfreedom unbalances ecosystems via biocides, species extirpation, the destruction of rain forests for palm oil, and climate disruption, and on and on. Still more irony: When the science of interdependency challenged the rights of corporations to pollute for profit, former Cold warriors were willing to use their scientific credentials to go to war on science despite lacking evidential proof, proving the dangers of the appeal to authority.[20]

But here the tale gets a bit murky. The validity of *Merchants of Doubt* was challenged by Nicholas Nierenberg, the son of William A. Nierenberg, a scientist criticized in the book, in a peer-reviewed paper and on a website. The elder Nierenberg, now deceased, chaired a 1983 National Academy of Sciences report entitled "Changing Climate," and Oreskes and Conway contend that Nierenberg downplayed the findings of fellow scientists in the summary and conclusion for political reasons. The younger Nierenberg, along with his sister and brother-in-law, counter that the report reflected the input of all committee members and showed concern for anthropogenic CO_2 but reasonably stopped short of recommending policy changes due to uncertainty. Nicholas further argued that false claims about his father constitute an ad hominem attack, and given this fallacy, the entirety of *Merchants of Doubt* should be called into question.[21]

Oreskes has stated she somewhat regrets the tone of an article on Nierenberg, in which she compared him to Voltaire's deluded Dr. Pangloss in *Candide*, who in the face of dire circumstance sees the best of all possible worlds but stands by every fact. Other critics have found minor errors in her writings, and Dr. Fred Singer, another scientist criticized in the book, claimed he was libeled but has not sued due to the expense and Oreskes, making so much money from her book and lectures, being able to outspend him.[22]

Who to trust?

Oreskes laughs at the outspending claim—she began her career as a geologist working in mining, where she could return and make serious money—and when asked, Singer was unable to name specific errors that constitute libel. He has also changed his tune over the years, first stating that global warming was not happening, then stating that of course it's happening but not anthropogenic, and then admitting that it's anthropogenic yet arguing that rejecting fossil fuels is too damaging to the economy. And while John S. Perry, who helped oversee the "Changing Climate" report, stated that the Nierenberg chapter was unfair, he also enthusiastically praised the book.[23] At issue, it seems, is Oreskes' bias versus Nierenberg's bias in regard to interpreting facts, which can be difficult to measure, but calling the whole book into question is a hasty generalization logical fallacy, and focusing on minor flaws, working them to work the media, instead of respecting scientific consensus, is a red herring.

The rational communicative art of unpacking contentious debates with logical fallacies, or bullshit detectors, which is what I call them with students, is much needed because much is at stake. It took ten years for *Silent Spring* to be a catalyst for action; it is now over thirty years since the publication of McKibben's *The End of Nature*—also providing scientific findings to a broad audience, also a best seller—but there has been little significant action in response to warming despite thirty-four climate conferences and several major international agreements. The steady march of the myths of bigger, better, and more are an obvious reason, but Carson's case provides further reasons why: the spin machine has had plenty of practice, is more skilled at and better funded for creating doubt, and now has the internet at its disposal.

When I googled James Hansen's historic 1989 climate testimony before Congress, the first link was not the transcript but a blog by a global warming denialist claiming that Hansen spun the facts to get funding for NASA and line his pockets. More bullshit, this time a false equivalency logical fallacy: We live in a moneyed economy where people get paid for doing both good work and being merchants of doubt, yet there is obviously a world of difference between working for the common good and working for your own good, or corporate good, or the good of free market ideology. And while the easy appeal to authority is a logical fallacy, Hansen has impressive credentials and data on his side, along with consensus—Oreskes argues that climate advocates do not appeal to the authority of an individual scientist, as deniers do, but a community of scientists and evidence subjected to scrutiny.[24]

In fact, it was Oreskes who initially counted published scientific papers, finding 100 percent consensus that anthropogenic warming was occurring while the media presented the issue as if there was significant disagreement. Later, several more counting studies were done with 97 percent of climatologists supporting anthropogenic warming. Deniers point to a cherry-picked, ignore-the-complete-data study, claiming there is no consensus; another red herring, and also a straw man logical fallacy.[25]

Such unpacking is a lot for students, or anyone, to wade through, but there is an obvious link between a systematically destroyed biosphere and systematically distorted communication, in which the narratives-that-be have power over our lives and conversations, which, these days, occur more in posts and tweets than in person. It is worth returning to Ramsey here, who, while arguing that habits are part and parcel of living, states that they are often ethically unreliable, as they are intertwined with consumer capitalism, racism, and sexism, which divide us from people and planet, and may speak for us, making our responses unfitting. Also, along with Orwellian illogic and obfuscation, Neil Postman argued long ago that Huxleyian distraction is often worse. In *Brave New World*, citizens are not manipulated by "Big Brother" but sedated by the pacifying drug "soma" and other amusements, willfully giving up freedom to habits of overconsumption and endless entertainment. For Postman, the problem, as he presciently saw it in 1985, is not the banning of books or information, but that no one would want to read a book, happily lost in a sea of irrelevant information in the age of television, or, fast forwarding, a visually arresting, always available social media age of the click-baiting untrue. Orwell or Huxley, pick your poison; both dystopian visions are available, if you hurry to buy, buy, buy.[26]

My students resist these interpretations, maybe because they hit too close to home but maybe because they better know the merits of social media. I resist it too, not only because I do not want to be reduced to habit, but because the learning and activist potential of social media is obvious, whether images and information about Extinction Rebellion protests, Greta Thunberg scolding world leaders who do nothing to stop climate crisis, or the eight-minute and forty-six second video of George Floyd's murder and the subsequent focus, not just on promoting protest but anti-racist education. Yet we are living in the age of Cambridge Analytic, a billion dollar "In Trump We Trust" re-election disinformation campaign, and the Facebook folly of not monitoring falsehoods in political ads—a tricky business, to be sure, but relying on the marketplace of ideas when lies lead to skewed elections, and online hate leads to murder in mosques, synagogues, and churches, is a fool's dream.

I do not enter the social media fray often, but honor those who do with thoughtfulness, and provide assignments in which students practice social media activism in the form of sharing accurate under-reported stories, correcting inaccurate ones, and disseminating some good news for a change, in both senses of the phrase. I figure that is where they so often live, move, and have their being, and so I encourage quality freedom of speech, which is the whole point of having it, as well as disconnecting from the fray and reconnecting with non-mediated sources of information, like, you know, the living and breathing biosphere. I did attempt to thoughtfully enter into Facebook conversation once, defending the Green New Deal after a "friend" trashed it, and then one of my friend's friends responded with a meme of Soviet-style socialism and called me stupid. My first impulse was to respond with insult, and then a long explication on the invalid analogy between democratic ecosocialism and Soviet-style socialism, which was actually State-run capitalism, but then I viewed his home page out of curiosity and he looked like someone's grandpa, and so I decided to let it go.

In *I'm Right and Your're an Idiot*, James Hoggan, who co-founded the DeSmog Blog website to counter spin with quality information on climate crisis, explores the toxic state of public discourse by interviewing a host of credentialed authorities on psychology, spirituality, and communication. Not surprisingly, he argues that there are links between polluting the public square and physical pollution and that social media too often makes it worse. The book provides plenty of counter-strategies to clean up the mess by committing to listening and dialogue and the flow of meaning, but confirmation bias is hard to overcome, and cognitive dissonance, which is uncomfortable but necessary to take stage-steps up the developmental holarchy, is often resisted, especially when one can find respite in a silo, or when you never leave it, or when we are interpolated by master-slave power relations and rhetoric.[27]

The ecocrisis, then, is a crisis of communication because mythic-animistic, aesthetic, and spiritual communication are so often missing from our lives, or not even considered to be forms of communication—if they were, we would be in much less of a mess—but also because of the systemic distortion of rational communication. Andrew Postman, writing after the 2016 U.S. election, argued that his dead father predicted Trump, or rather, we feared Orwell's nightmare while Huxley's in-broad-daylight vision walked right through the "amusing ourselves to death" door, with public discourse steadily degrading from 1980s soundbites to truths daily drowned in a sea of alternative facts, or Trumpian contradictions on nearly every issue—he says whatever is expedient in the moment and can never be wrong—which, of course, are quite different from Thoreau's contradictions.[28] For Thoreau, there is method, with each side of the contradiction being a partial truth; for Trump, there is the madness of all his utterances likely being falsehoods or partially false. In sympathy with Thoreau, we cascade up the developmental holarchy; in Trump we trust, we tumble down. After all, Trump supporters with their center of gravity at Mythic who deny the 2020 election results also likely deny climate crisis and the severity of coronavirus.

So, we can't fully trust double-edged social media, or any media, yet the communicative potential is enormous, as we've moved from the global village to conversing within the global mind, although discerning the unfit has become a full-time job. But there is still more to unpack: Can we always trust science?

Science is mostly solution, but part problem when divorced from ethics. After all, pesticides are a scientific wonder, having merit under specific circumstances, but their overuse leads to a treadmill of abuse. Or what of cloning, or Star Wars turned to Space Force, or GMOs? GMOs are considered safe by current science, but I am not buying it, figuratively and literally, not because I distrust data and current science, but because it feels too much like a grand experiment with all the data not in, and because it continues monologic agriculture when there are better ways, like dialogic permaculture and regenerative agriculture, which reduce rather than contribute to CO_2.

Science and technology are also often bound up with corporate and military interests, not necessarily the public good. And then there's the dark history of science and technology in Nazism: social Darwinism, eugenics, holocaust. Or, consider when technology wounds and kills in so-called democracies: start with any oil spill, chemical disaster, and on and on. Or what of the technology as savior dream of geo-engineering in response to climate crisis? All these examples come from the Rational level scientific mind, with every problem having a calculative solution, but as Bateson thoroughly disclosed, I-It epistemology ultimately destroys our home and quality of life.

Resistance to scientism, then, has good roots, while resistance to good science is quite dangerous. Martin Heidegger, for example, argued that objective science conceals as it reveals, with particular questions leading to particular answers, not whole truth. And love resists reductionism, yet exists and is central to existence, and not just any existence, to robust, sustainable, sacred living within a living earth. Beauty also resists reductionism, although science tries, concealing as it reveals. For Heidegger, objective science and calculative thinking commands, or challenges-forth, while meditative thinking, characterized by "releasement toward things" and "openness to the mystery" brings forth the world. Much withdraws and hides from being challenged-forth that may be revealed via the deep listening of meditative thinking.[29]

Unfortunately, good science and scientism may get confused in the public mind, with resistance to scientism, to calculative thinking becoming thinking as such, conflated with resistance to listening to scientific data. Perhaps rejecting the consensus on warming, then, is not only due to spin by big oil and free market think tanks, although most of it is, but also a knee-jerk and often unconscious fear: by accepting consensus, we accept scientific narrative as the only path to truth, and we all have lives that resist calculation. Gus Speth, the former Dean of Forestry and Environmental Studies at Yale, stated that he used to think that our top problems were global warming, environmental degradation, and eco-system collapse and that "scientists could fix those problems with enough science." But, he later realized he was wrong: "The real problem is not those three items, but greed, selfishness and apathy. And for that we need a spiritual and cultural transformation."[30]

Of course, I would add that communicative transformation is key to spiritual and cultural transformation and that includes the communication of science.

We are all familiar with changing scientific paradigms—what was once considered true ended up not being true—and studies that tell us one thing, like about diet, and then the opposite soon after; of course there was further study, but not understanding how science works, and the limits of science, leads to mistrust. According to Bruno Latour, a philosopher of science, there is a difference between science-in-the-making, like studies of diet, and ready-made science, with the latter reflecting consensus and established fact, including gravity, which you don't want to test by jumping off a building, or the physics and chemistry of the greenhouse effect, which, as McKibben often declares, are "bad at haggling." Yet Latour, as postmodernist, further argues that we are ultimately dealing with interpretation, not Truth. Science is a social enterprise, not just objective, and facts, or ready-made science, still needs to be interpreted, and for that we must include other modes of knowing what we can know.[31]

Science, as a socially situated set of practices, does not lead to Truth, but science has a method of inquiry that approximates reality, getting us closer to truths. Science, however, must always deal with this fact, revealed by science: we are nature studying nature. Gaia theory, in particular, details the depth of interdependency, of earth as self-organizing living system, and we must embrace the perceptual and ethical implications, and another irony: "objective" I-It Gaia science supports an inspirited mythic-animistic and intersubjective-aesthetic communicative experience of our place on the planet.

Another consideration: I-It calculative science provides data, but I-Thou meditative science is quite different, and often reveals more, whether illustrated by Goodall versus Harlow, or Kimmerer, Gagliano, McClintock, or Kaza, and so many more. Thoreau wrote that we should not "underrate the value of a fact," as "it will one day flower in a truth."[32] But for blossoming to happen we must not only be a scientist, but a scientist-poet like him, midwifing truths through sympathy and love, which he equated with Native wisdom, or TEK, learned from living within intersubjective nature and listening to the diverse voices of all our kin. For the transcendentalists, the higher law of unity-in-diversity, which fully emerges at the Integral level, blossoms through study of particular facts when married with the pursuit of the good and beautiful. For Vine Deloria, a Native American scholar, Natives were as systematic as modern scientists when learning about the natural world, although with a subjective bias in contrast to science's objective bias, and in their meeting blossoms a moral response.[33]

The gulf between scientific investigation and spirituality may be overcome, and thus good data crunching science at the Rational level, while worthy of trust, is not enough. The gulf between politics and spirituality may also be overcome as compassionate action, and thus facts, as necessary as they are to good policy, are not enough. And new technologies, even when they serve sustainability, are not nearly enough, because once calculation becomes our dominant epistemology, our value system is off, and we accept what we should never have accepted. We should not need climate science to wake us up; we have already received a host of messages prior to and along with climate data that indicate that an I-It attitude, and economy, is at the root of our despair. We should not need a climate emergency to respond with action; we have already been despoiling the planet without it, it's just the latest and greatest.

My first inclination is to fully support new tech like solar power, wind power, and electric cars as mostly fitting responses, despite the destructive mining needed to make them. On the other hand, the race to mine lithium for batteries on a worldwide scale will undoubtedly lead to much abuse, not to mention Elon Musk's abusive tweet that "we coup whoever we want" in support of the U.S. fueled overthrow of Evo Morales in Bolivia to get access to more of the metal. Thus, what we most need are new narratives of degrowth, transversal progress marked by participation *and* distanciation, and technology as tool. Alternative energy sources, if they maintain the status quo, get us nowhere; if they operate within new narratives that set us on the right course, they are a piece of the puzzle.

Buber wrote that I-It relationships are inevitable—we are not perfect and the earth supplies resources—but what would it be like to live in an I-Thou dominant world, a world where eros, agape, philia were not platitudes but lived reality? Where communication was not only a human affair, but the means for living in sympathy with intelligence, and thus in sympathy with all species? So much to consider, so much to fight for, if we step out of the cave and perceive the dystopian spin machine.

Thoreau's exhortation at the end of *Walden*, in which he implores readers to "go confidently in the direction of your dreams," was once co-opted in a Merrill Lynch commercial imploring consumers to buy into financial systems and a materialist dream. And his phrase in "Civil Disobedience" that "government is best which governs least" is often championed by conservatives and libertarians. More out-of-context logical fallacy spin: less government when government upholds slavery is one thing; it does not follow that Thoreau would argue for least government in the age of climate crisis when the so-called free market threatens all that he loved. If Thoreau was alive during Carson's time, or today, one can imagine what spinmeisters would say about his continual call for simple living—they would probably call him pond scum—or speaking out against tyrannical government or any tyranny that degrades true liberty. And one can imagine him responding by saying that government is best that protects the commons, the ecosystem upon which we all depend, as there is no freedom without a sustainable home. Or, "What is the use of a house if you haven't got a tolerable planet to put it on?"[34]

Still more irony: In Oreskes and Conway's *The Collapse of Western Civilization: A View From the Future*, a science fiction essay based on current facts, anti-regulation right wingers who delay a governmental response to climate crisis pave the way to the inevitable: a huge expansion of government as the crisis is not addressed and continues to worsen.[35] But "no governmental oversight" free market think-tanks and internet trolls are immune to irony, continually attacking good science and logic, as well as Oreskes, who does not enjoy being a lightning rod but states "the whole purpose of a lightning rod is to keep people safe."[36]

Carson was attacked by the right on the fiftieth anniversary of *Silent Spring* in a book written by three academics, non-scientists all, yet with impressive credentials and undoubtedly high cognitive lines, and published by the Cato Institute, still another free market institute. The authors argue that *Spring* is written within the tradition of technophobia, and thus was an overreaction that led to the toxicity of the precautionary principle, which, to me, is as rational as we can get. They also state that Carson falls back onto the too easy greedy businessman trope, and ask, why would an industrialist poison the food that they also eat?[37] Hmm, why would they? And why would we eat

GMO food or agribusiness meat, for that matter? Or why would Exxon spin the data on climate from their own scientists for decades when their children, and their children's children, will have to live with the consequences of a heated world? For starters, greed is not merely a trope, but a systemic driver of an often immoral capitalism—stockholders still trump stakeholders—but mostly, what I think these authors miss are the links among pesticides and the myths of unlimited growth, linear progress, and technology as savior, which are killing us and must be resisted.

For these authors, the rewards of pesticides, within the context of a calculative bias toward bigger, better, and more, is greater than the risks. I get this argument; if I did eat standard supermarket faire or GMO foods, like most habitually do, and most have no choice but to do because of structural availability, lack of education, and price, I don't necessarily think it would lessen the length of my life, but I prefer to distance myself from these modern myths rather than participate in them, and prefer precaution over risk. True to the master-slave dynamic, these authors also don't consider those who work in and take the most risks in pesticide-riddled fields. To see the links between narrative, habit body behavior, and eco-social destruction, or that everything is hitched together, is not a slippery slope logical fallacy, but cause and effect logic. We perceive the need to resist and reverse the steps that benefit industrialists in the short term but put all of us at increased risk in the long term of climate crisis, especially the disenfranchised and most vulnerable: the poor and people of color.

More recently, Chad Montrie, continuing his earlier work on environmentalism and the people's history, rethinks the influence of Carson from the left, arguing that the often repeated claim that *Silent Spring* started the U.S. environmental movement is a myth, as the roots are much more complex and include labor leaders, among others. He does not castigate Carson and does acknowledge the influence of *Spring*; however, he undermines the long-held narrative of Carson as eco-hero, or the only or most prominent hero, and her book as the sole source of the activism that came later.[38]

And so it goes. Who do you trust?

What I do know is that my bullshit detector is not only informed by logical fallacies and pursuit of what is scientifically true, but never-ending pursuit of the good and beautiful. To Thoreau, we are slaves of institutions when we surrender our rights of reason and conscience, and the depths of conscience emerge from deep listening that transcends and includes science. But even if we compare our reasoned findings, and our intuitions, with other competent listeners, practicing an ecology of communication does not mean that we will all agree; however, it does mean that we will explore together at higher levels because we are responsive to higher laws. Such higher laws take us to Schrag's fertile in-between rather than rigid Truth or tumbleweed relativism;

they unfold rather than cement, breathe rather than calcify, and inform rather than culminate in certainty.

Carson, while not espousing transcendentalist spirituality, does marry her science to the pursuit of the good and beautiful via experiences of wonder, and thus aesthetic communication. After all, DDT may silence the spring voices of song birds. If looking for guides, then, it wouldn't hurt to ask: What would Carson do? Or Whitman, or Thoreau. Or anyone who has lived or is living a more encompassing logos, open to listening to inner genius and genius of place, and to Spirit unfolding as matter, life, and mind. And then take the good risk, the necessary risk, of doing the same.

The practice of rational communication, or attempts at it, is where we spend most of our communicative time, but it is not nearly enough, especially when it continues to be informed by systemic distortion and anthropocentric bias. And that, of course, is why we desperately need to become critically aware of the intersections of social structures, interpellation, and habit-bodies, opening fissures and embracing our capacity to act, especially in support of the intersectionality of climate activism, environmental justice, and indigenous rights, or, more generally, the health of the landbase upon which diverse species depend. And that is why, along with the transversal logos of rational communication, we desperately need to practice mythic-animistic, aesthetic, and spiritual communication, and then embrace their dialogic interplay expressed as an ecology of communication and a fitting responsiveness.

The climate crisis, and communication crisis, stakes are high—BP spent $500 million on PR to "contain" the 2010 Deepwater Horizon Gulf of Mexico oil spill—but students, the next generation of activists, may not be ready to turn the boat over.[39] Or maybe they will. Frederick Douglass is right that power concedes nothing without a demand—what would Douglass do? Revolution is needed from the young, who are mostly wrestling with identity at the Pluralistic level, but, of course we need revolution as personal and social transformation from all ages, as we creep and claw, and sometimes artfully leap, our way up the developmental holarchy.

A key question, then, is: What will you do? Or, a better question: How will you respond and what gets in the way? John Lewis, who marched with Martin Luther King and later became a congressperson known for conscience and continuing the fight for civil rights, was arrested forty-five times in acts of civil disobedience, or what he called "good trouble."[40] I have never been arrested, and I am not planning to—we all find our unique paths when we practice inner and outer listening—but who knows what the future holds, what steps we will all take toward an ecology of mind and Spirit.

I began this project by contrasting Peters' focus on communicative gaps between self and other with Buber's I-Thou reciprocity. Peters would appreciate the existential realities of my AT hiking experience, opening myself to others but finding that boundaries, questions, and culture are not transcended. For him, our goal should be creating community, as best we can, and we are most receptive when we acknowledge the "impossibility of communication," out of which "the blessing comes."[41]

Yet my experiences of I-Thou reciprocity with moss and moose, and sauntering amid the sensuous allure of nature, disclosed what Peters himself concludes: that communication remains the ultimate border-crossing concept. To me, we are most receptive when exploring permeable boundaries and finding our place within human and ecological communities. My experiences, and the similar experiences of so many others, disclose the causes of and possible solutions for ecocrisis. We must perceive the damage done by monologic I-It attitudes structured by rigid dualisms, which may be discerned via rational communication but are more fully overcome and felt via experiences of mythic-animistic, aesthetic, and spiritual communication. And out of such experiences flow care, love, and a deeper call to responsibility.

Both Peters and Buber demand that we become responsible, with Peters extolling the existential hard work of creating community via rational communication and Buber focused on being spiritually addressed by others. And yet, this is too rigid. Peters is open to I-Thou ontology and other modes of communication while remaining critical concerning the dream of losing self in other, and Buber practices rational communication, respecting otherness while acknowledging that we cannot continuously live a I-Thou mode of existence. Respecting the dialogic interplay between their views, rather than positing them as another dualism, is needed to more fully face the communicative crises of our times: ecocrisis, but also systemic racism and gender inequality, a pandemic, financial crisis, a crisis of leadership, and a crisis of media. Some things you don't want to integrate, but we do need an integrative response.

This is admittedly a theoretical work but with much practical promise. I have suggested a few specific possible responses to ecocrisis, like deep listening in agriculture, which is already a permaculture practice; in I-Thou science, of which I have given several telling examples; and in restoration, allowing us to better re-story the land. But there is nothing to which an ecology of communication cannot be applied. In a dialogic universe, earth, and socio-cultural life-world, problems and solutions are inevitably communicative in nature. Or, communication as culprit and cure.

In *Promise Ahead: A Vision of Hope and Action for Humanity's Future*, the sustainability and voluntary simplicity proponent Duane Elgin argues that "adversity trends" like climate crisis, population growth, poverty, species extirpation, and the depletion of fresh water are all fundamentally communication challenges. Climate crisis, for example, demands that we dialogue globally to reduce carbon dioxide and other emissions, while addressing population focuses on illiteracy and educational opportunities for women. Poverty challenges us to live equitably in a world where the global communication of advertising promotes unsustainable consumerism, as well as despair at the immense gulf between rich and poor; this immense gulf and first-world over-consumption, more than overpopulation, drives ecocrisis. And responding to the extinction of species by supporting biodiversity and ensuring that there is enough fresh water, as well as food and energy, will take much planning and cooperative effort. While these examples may seem to only be rational communication challenges, rational communication is most effective when practiced within the context of the widened awareness of an ecology of communication and the call to responsibility.[42]

Ecological design is also a communication challenge, as well as a communication opportunity. Design principles must be propagated communicatively, but once they are put into place in buildings, transportation, landscaping, and infrastructure generally, they will speak to us. The natural world communicates but so does our created environments, and working with this understanding is especially important in cities where the majority of humans already live. If the majority of world citizens are not exposed to places that elicit mythic-animistic, aesthetic, and spiritual communication, making us feel connected or alive or peaceful, we must question if we are living lives of quiet desperation.

Rooftop gardens, alternative energy and energy-efficient buildings, sustainable and sane forms of transportation, and green space that elicits mindfulness but also mindful converse, need to become the norm, engendering dialogic experiences like mine when hiking. Many love urban environments; I too am drawn to the excitement and beauty of cultural life in cities. A concert, play, or piece of art enlivens our senses, as does the collective energy of unique and interesting people. But when urban locales fail to liberate our bodily empathy and sensuous imagination, and thus mythic-animistic communication, or the beauty of nature's diversity, and thus aesthetic communication, we begin a downward spiral. When we don't make communicative connections on a daily basis, we tend to get lost in abstractions and ignore fundamental ecological principles: when we pollute the lifeblood of the earthbody—the air, water, and soil—we poison the wellspring that nourishes our bodies.

All transformations to sustainable living assumes new narratives, but they are a tough sell when to monologue is to make money and most enact whatever myths they are given, especially when supported by social structures. Nonhuman species maximize their growth, but are curtailed by other species maximizing their growth, forcing each species to find their niche, or survival of the fit-ins. Humans, because we are more complex, have the foolish freedom to attempt to live beyond natural limits. And once we create and live myths of bigger, better, and more, our bodies, which Abram has compellingly explored as pliable, sensate, and communicative, learn lessons of the dominant culture all too well; such lessons lead to armoring—we are like turtles hiding in their shell—expressed as bodily rigidity, inhibition, and a failure to open ourselves to new communicative possibilities. It is truly amazing how often humans are unreceptive to bodily empathy, the sensuous imagination, and mythic-animistic listening. We are the only creatures capable of dismissing what lies before our senses. When pollution burns our eyes and chokes our lungs, it is providing information about healthy and unhealthy relationship in the Batesonian sense, and knowledge of the good and not good in the ancient Greek sense. And yet armoring deflects these messages.

That's the bad news, the good news is we are free, or may learn to be free, to practice an ecology of communication, and such practice sensitizes us to eco-social problems we have only dimly perceived. And we only need a small percentage of grassroots, take-it-to-the streets, awake souls to fight for change for change to happen, like changing infrastructure that opens up pathways to genuine growth. Practicing an ecology of communication also enables us to listen to warnings, although Roszak is critical of the overuse of shock and shame by scientists and activists wishing to alert the general public to eco-destruction. Such awakening strategies can exploit fear and guilt, closing down responsiveness and receptivity generally. The result is often increased resistance and deepened apathy; perhaps armoring is partially due to these kinds of messages concerning ecocrisis. On the other hand, the likelihood of a 2° or more hotter world has caused Jem Bendell and the Deep Adaptation Facebook group to embrace bad news rather than propagate false hope, arguing that such embrace does not lead to inaction, but to doing what is necessary: degrowth, creating community, and preparing for the strife that is to come.[43] Bendell's deep adaptation views have found support from the Initiative for Leadership and Sustainability (IFLAS), who have put out a plea, thus far signed by over 250 scientists and academics, to have difficult conversations about possible collapse.[44]

For Elgin, how we respond to all these communicative challenges will tell the tale: we will either call forth an evolutionary bounce to a new consciousness and new ways of living—the promise that New Agers have long pined for—or we will slam into an evolutionary wall.

Near the end of his life, Thoreau surveyed twenty-two miles of the Concord River, trying to figure out the effect of a dam, used to power a Lowell textile mill, on flooding, which threatened fertilized farm land. Farmers had called on Thoreau, the only one from town who could take on such a grand task. He studied the river as a human-natural system, while also reading it using math talents, data coming alive. Turned out to be more complex than a dam; farmers were at fault for the loss of wetlands and denuded hillsides. And the more the plains flooded, the more they cut down hillsides, making it worse, changing the Concord landscape. Thoreau did not figure it all out, but saw the damage, leading him to forcefully argue for public parks and land, which would become part of his legacy.[45]

But, of course, I-It attitudes in the mid-1800s were not confined to war on the natural world. In 1861, with Thoreau's health worsening, the Civil War would begin when Confederate troops fired on Fort Sumter in South Carolina, with Union soldiers surrendering less than thirty-six hours later. In 1862, the year of his death, starved Dakota Sioux living on a Minnesota reservation didn't receive promised aid from the government, and killed five white settlers. Conflict followed with many dead, the governor vowing to exterminate or drive Natives out of the state. Treaties, which were already not being honored, were nullified, and thirty-eight were killed in a mass hanging, the largest in U.S. history. Little Crow, a Dakota Sioux chief, fled west and was killed by a white settler while picking berries with his son; his head and arms displayed at the Minnesota Historical Society until 1971.[46]

The human family has so much yet to learn, so much growth awaits.

In 2020, we are 64.20 ppm CO_2 over the "safe" threshold of 350. By 2025, with CO_2 rising on average 2.5 ppm per year, we are approaching planetary conditions not seen in 3.3 million years, when sea levels were twenty meters higher and long before the first humans appeared in the Pleistocene 1.6 million years ago. And that does not include other greenhouse gases like methane. We have also reached the point where the human material outweighs the Earth's entire living biomass, according to a study in *Nature*.[47] Imagine if the majority were mature enough to perceive such data with clarity and emotion. Teenaged Greta Thunberg gets it, Sunshine Movement youth get it, and anyone practicing a larger logos gets it. But what the numbers truly indicate is a world governed by dead rather than alive, inspirited matter, separation rather than interbeing, the loss of the fire of love as a larger force, and a lack of love for specific places and others, including other-than-humans. The numbers also indicate a world governed too much by numbers, even as we need to learn from them.

And climate disruption is clearly both an ecological *and* social problem. Thoreau's walks were ruined by slavery for good reasons, and not only because of disturbed thoughts but also due to long links between racism, economic inequality, and eco-destruction. A hotter world is already bringing drought, a food and water issue, and rising waters from melting ice will lead to refugees fleeing coastal areas for already overcrowded cities, which will likely lead to more discontent and anger, more crime, more terrorism. Those living in the mostly poor and black neighborhoods in New Orleans during Hurricane Katrina know the ecological and social are not separate; some residents were forced to rooftops, others to the Super Dome, the first time most had been inside due to not being able to afford high ticket prices.[48]

In Hop Hopkins article, "Racism Is Killing the Planet," he reviews the long intertwining history of racism and ecological destruction in the United States, beginning, of course, with the genocide of Native Americans and land theft, with both justified by the narratives of Manifest Destiny and white superiority, and then "progressing" to slavery, once again justified by white superiority. This story of dehumanization continues today, with the same mindset bound up with techno-capitalist industrial structures or neo-liberalism generally, in which the pollution and poisons generated by the system are disproportionately disposed on people of color and the poor who are considered disposable. The result is the Navajo Nation poisoned by uranium mining, Cancer Alley in Louisiana, which some call Death Alley, and polluted water in Flint, Michigan, among innumerable other examples. Sacrifice zones means that people are being sacrificed, and if you cannot have sacrifice zones, then you cannot have racism and capitalism. If you have nowhere to put pollution, you cannot pollute.[49]

Yet, we continue to find ways to justify the unjustifiable, especially when conservative commentators turn to blame, claiming that people of color and the poor need to be self-reliant, or, basically, that they get what they deserve. This is a twisted version of Emersonian self-reliance, which focused on inner listening and acting out of what we learn, true to inner genius rather than conforming to outward society. But it is also a gross misreading of history, in which we are supposed to lift ourselves out of deeply rooted structural inequalities, in economics, in education, in everything, without structural help. We are called to be responsible for ourselves, but we are also called to be responsible to each other. No one deserves to live in Cancer Alley, and Marjorie Richard and others should not have to fight so hard for big oil to do the right thing; and Dakota Sioux did not deserve to have a pipeline built through their land, threatening their water supply, after it was rejected by white neighbors, and no one in Flint deserves lead poisoned water.

And this brings us back to the skewed master-slave communicative system, in which oppression is maintained by backgrounding, radical exclusion, incorporation, instrumentalism, and homogenization or stereotyping. But when you have a society built on intergenerational wealth inequality, and unequal access to the power to change things, that is what you are going to get. But a movement grows out of intersectional environmentalism, and thus communication that perceives shared suffering and shared goals despite not being able to share interiors and achieve complete understanding, which is the kind of community building that Peters celebrates. Leah Thomas, in her article "Why Every Environmentalist Should Be Anti-Racist," writes that Eric Garner's desperate "I can't breathe" before dying at the hands of the police may be interpreted more broadly to include people of color and the poor who live with poor air quality and are disproportionately stricken with asthma. The same system of oppression undermines planetary health and the health of the disenfranchised, and thus forces must be joined, and the good fight must be fought.[50]

The above should already be clear, but is increasingly clear as coronavirus disproportionately affects the black community, who are more likely to be essential workers, and more likely to already have health issues due to living in unhealthy environments, including from the stress that goes along with it. That is something that they most certainly don't deserve. And no one deserves to live in a climate-compromised world, but that is where we are already heading, and it will be much worse for those who already live in compromised environments. Hopkins writes, "If Black lives mattered, then *all* lives would matter," and that "We will never stop climate change without ending white supremacy." He admits this sounds daunting, given how entrenched racism is, but he is convinced that that is the only way forward.[51]

But just like individual crisis leads to growth, our collective lostness is pushing us to search and evolve. There is the argument, somewhat New Agey, yet supported by the positive possibilities of the developmental holarchy, that individual and collective crisis is leading to a great awakening, taking the baseline awareness of a significant portion of humanity to a higher level—a level of soulful connections as we become increasingly aware of ourselves as world citizens immersed within the larger intelligence of the world soul, or the bounce rather than the crash.

In *Dark Night, Early Dawn*, Christopher M. Bache, a professor of religion, ruminates that our sense of separateness, or "species-ego," has been breaking down for some time, due to factors like increased ecumenism; social movements like the fight for women's rights and equality, as the damage done by the patriarchal-ego was challenged and deconstructed; the questioning of the Newtonian-Cartesian world view and the birth of quantum theory; confronting death via nuclear annihilation; the emergence of a global economy; the explosive rate of change due to new technologies, which, in relation to the evolutionary timeline, is unprecedented; and barrier-breaking global communications systems.[52]

Yes, global communications, the internet and social media, may be driving the breakdown of the isolated ego, and isolation generally. Perhaps my students, or the millennial generation, intuitively know the potential of the global mind and thinking together in a grand classroom. If so, this potential needs focus from deep reading and time in nature. Otherwise, the grandness degrades into students, and citizens, isolated in their rooms sending selfies, watching porn, and distracted from the actual world, leading to more, rather than less, separateness. Of course, now we also have the wild card, brought on by human expansion into wild places, of pandemics and quarantining, which remind, like Treadwell not treading well with grizzlies and Plumwood not reading the danger signs when entering crocodile territory, that some borders should not be crossed.

Still, for Bache, all these factors, along with ecocrisis, are pushing us to have a "dark night of the species-soul," as techno-industrial capitalist structures are not sustainable. Something has got to give. Our individual dark nights, while painful and often lengthy, lead to transformation, to an awakening of a new sense of identity and new awareness; our collective dark night may do the same.[53]

Bache cites Elgin's prescient 1992 *Awakening Earth*, where he writes that the early decades of the twenty-first century will be "superheated." Elgin meant this in terms of the intense pressure put upon humanity from factors like the ones articulated by Bache—to which we must add the wealth gap and other adversity trends—to change societal structures and habits.[54] Well, the early decades of the twenty-first century are indeed superheated, literally, from climate disruption, which is intensifying still more the need for transformation.

We are feeling the heat, internally and externally.

For Bache, the internal goes deep into the collective unconscious, where our suffering but also new realizations are held, as if stirred in a caldron, ready to bubble up into the lives of those prepared for learning from confronting external challenges to psyche and planet, which, of course, is similar to Sheldrake's arguments for morphogenetic fields. Bache takes this dialogic exchange between the individual and the collective further into New Age territory, although stopping short of claiming we can create reality, focusing instead on individual transformation as a catalyst for collective evolution, or, as Jung put it: "Does the individual know that he (or she) is that makeweight that tips the scales?"[55]

This is the enlightened guru argument, in which the consciousness of one shines a path for the many, except that the many may already be actualizing a more enlightened awareness. Whether this awareness also streams into the collective unconscious, making new realizations available to everyone, but especially those ready to listen to and be re-formed by the fire of love, remains open to debate. From the perspective of a great awakening, however, a leap in consciousness has been gestating on multiple levels for some time—conscious, unconscious, collective unconscious, as well as transcendent dimensions—as running through them all is Spirit unfolding or eros, whether or not we have been aware.

Better to be aware, or consider the possibility. The Integral stages beckon, characterized by gathering the best aspects of our pre-rational mythic past, wisdom traditions that point toward spiritual insights that, while still evolving, have universal resonance, and our more rationality-focused modern and postmodern present, most productively expressed as logos marked by world-centric care and concern. Of course, there is also the argument that crisis is pushing us to evolve gradually, in fits and starts, with some meandering yet learning in steps, as well as a third argument: We are being pushed to awaken, but after the collapse of civilization as we know it. Only a whopper of a crisis will wake us up, and climate disruption will oblige.

Great awakening, steps, collapse, or something else. We explore these possibilities in my classes, but, most importantly, I share with students that there is no shortage of world citizens participating in the great work of our time. Whether they would like join is the core question of our combustible, crisis ridden era, as well as my teaching and liberal arts pedagogy generally. Humans have crossed a line, but a growing spiritual awareness of the unfolding universe and earth story may well lead to a flourishing Anthropocene. Thomas Berry argued that we have moved from "suicide to homicide to genocide to ecocide to geocide"—a depressing assessment to consider—but he also stated we are fortunate to live at such a crucial moment, charged with consciously celebrating life and life force and playing a creative role in the evolutionary epic.[56] Such creativity demands rational and imaginative and spiritual inquiry, as wicked problems must be met with wicked, multidimensional solutions.

If you have kids, like I do, who will inherit the results of today's responses, you have probably read Dav Pilkey's *Captain Underpants* book series, featuring two 4th graders, George and Harold, in their school adventures, whether with Principal Krupp, who they can turn into the diaper-clad Captain Underpants by snapping their fingers—you need to read it for yourself—or their teacher, Ms. Tara Ribble, who makes her students miserable by always following up good news, like "you all get extra recess," with bad news, "but you'll need to use that time to do extra homework." Her students respond with a rousing "Yea!" after every bit of good news, and then a disappointed "Aw, man!" after the bad. Ms. Ribble plays the game all day long, and so can we when confronting the daily influx of eco-social happenings on the planet.

An earth-focused coronavirus recovery could create 400 million jobs, says one report. Yea! Over 5,600 fossil fuel companies have received at least 3 billion in coronavirus relief aid. Aw, man! A Washington D.C. judge ordered the Dakota Access Pipeline to shut down pending a risk assessment by the Army Corps of Engineers, a significant activist victory. Yea! The virus and economic downturn and crisis in leadership have emboldened far right extremists who want their own revolution. Aw, man! Regenerative ocean farming of seaweed and shell fish may be coming soon, which, along with bringing food will also soak up carbon. Yea! An over-100° heat wave in Siberia thawed its tundra and caused a rash of fires. Aw, man! And on and on and on it goes.

In *Captain Underpants*, mean-spirited Principle Krupp turns into the free-spirited Captain when the boys need him to save the day, which he does with a resounding "Tra-la-laaa," but, of course, there is no frolicking superhero coming to our rescue; but, as Paul Hawken has wisely stated: "If you look at the science about what is happening on earth and aren't pessimistic, you don't understand the data. But if you look at the people working to restore this earth and the lives of the poor, and you aren't optimistic, you haven't got a pulse."[57] The constant influx of good and bad news can dizzy our heads and knot our hearts, but if we are looking for hope, and faith, we may find it in Spirit-in-action, expressed by those with a pulse working for change, but also in the inevitable growing pains of the pulse-less that lead to growth. We may also find it by finally discarding a reductionistic anthropocentric view of communication and embracing the wider vision of an ecology of communication.

We most certainly face tough choices as we communicate our way into the future, with some new technology choices making my head spin. I had the good fortune of hearing Karen Bakker, Professor and Canada Research Chair at the University of British Columbia, speak on "The Future of Environmental Communication in an Era of Digital Disruption." Bakker reviewed numerous examples of the "interspecies internet" that supposedly allows interspecies communication via digitized environments, with individual animals, insects, and plants, as well as whole ecosystems, hooked up and monitored via sensors, satellites, drones, AI, and virtual reality. The interspecies internet, not to be mistaken with the non-technological wood wide web, is like a planetary macroscope revealing hidden worlds, the result being "smart earth" and "smart ocean" capabilities that produce huge amounts and continual flow of real-time data. Bakker, while acknowledging criticisms, stated that the "tsunami of data" enables us to see and hear what we would not otherwise be able to see and hear, providing scientists and policy-makers with information that may be used to monitor differing crises and respond through conservation efforts.[58]

Much of the research and gizmos, which are mostly available now, are funded by big corporations like IBM, Microsoft, and Hewlett-Packard, as well as NASA and the U.S. Pentagon. Are you ready for weaponized bugs? It's possible, along with conservation possibilities like monitoring poaching and keeping animals in parks and humans out, or honeybees wearing no-see-um backpacks to detect colony collapse; however, if you have watched the techno-futurist series *Black Mirror*, you have already witnessed bees as mechanized weapons. Or consider cyber-enhanced working dogs, already here, replete with cameras and sensors that read the pooch's heart rate and emotions, providing, of course, more data, which may be used in search and rescue missions—anyone who watches Marvel movies know that this one reminds of Tony Stark as Iron Man. Other positive uses include citizen science apps, in which the public can participate in receiving data, but also collecting it; earthquake warnings providing a couple minutes of preparation time; and carbon footprint analyses as you move through the day, nudging behavior in sustainable directions, like biking instead of driving.[59]

But these examples are small potatoes, or GMO-enhanced small potatoes, because we are talking about social media for the planet, or as the planet—it's dizzying—like Wildbook, the equivalent of Facebook for animals, used to barcode, monitor, and protect zebras and whale sharks, and digitized Lake George, which, via an array of sensors, discloses over 400 million depth measurements, and Fitbit for the ocean, generating over 200 data types, including salinity, temperature, and acidification. There are also robotic fish with cameras used to monitor the Great Barrier Reef, which can be controlled via a video game style console, solar-powered Seaswarm, which collects data along the surface, floater systems that act like garbage trucks cleaning up ocean plastic, and underwater robots that kill invasive species by zapping or poisoning with a lethal injection. There are also ocean Pingers that send out acoustic alarms via frequencies audible to whales, dolphins, and porpoises to aid them in avoiding entanglement in fishing nets. Hmm, what could go wrong? In the case of bioacoustics and fishing nets, my first thought was that the same technology could be used to find and kill whales, and my second was to question the practice of using fishing trawlers and mile-long nets at all, which destroy spawning grounds, making it impossible for fish stocks to regenerate.[60]

But wait, there's more, or bigger, better, and more.

Virtual reality in the form of fish avatars tagged with sensors that take us into the world of live fish, or the Magicverse, in which Google glasses or Halolens enhance the physical environment by adding multiple layers of digital information, hacking our senses and immersing us in animal habitat to provide the experience of being nonhuman—goodbye to the imaginative play of the Council of All Beings, hello to wearing a helmet and walking around the virtual woods, or actual woods, but with species coming alive via technology instead of the practice of mythic-animistic communication. The argument, which is being experimentally tested, is that virtual experiences stimulate our emotions and may cause us to care, eliciting the call to responsibility.[61] But does it last, like the wilderness effect or spiritual experience, or is it a fleeting high that may be a catalyst for knowing more, and seeking more, but fades without a practice? And does it keep us inside, or staring at screens even when outside; and if we habituate to the virtual, what becomes of the actual world in comparison? What about heightening perception through Kaza's conversations with trees, or Kimmerer's grammar of animism and aesthetics, or Abram's bodily empathy and sensuous receptivity and reciprocity, or Gagliano's practice of emptiness and dialogue with plants?

I have a friend who had a life-changing nondual spiritual experience while snorkeling, due to feeling deeply immersed in the underwater world of fish. She fully entered the beauty of their world and then when she re-entered our world, and saw a platter of cooked fish, she realized that it would be the ultimate hypocrisy to eat them. That began a journey of vegetarianism and veganism and working for animal rights. Does virtual reality have such potential? I doubt it, but perhaps, if used as tool it has some merit. Or, perhaps it is another gizmo that ultimately leads to a treadmill of abuse, like agriculture and pesticides and bioaccumulation and resistance. Again, pesticides have situational use, with targeted spraying and organic pesticides as fitting choices—my local and beloved family-run orchard uses such methods to protect their berries and apples—but we must remember the dangers of indiscriminate use and that biocides are often a better descriptive term. I am not interested in being a technophobe, and I admit that walking the woods, experiencing, as best as human technology can approach, what it is like to be dragonfly, would be stimulating, although I am not fond of the idea of not seeing the actual world while wearing a silly helmet, and remain interested in the needed work of deconstructing technology as savior.

But there is still more. Bio-hybrid robots or symbionts, like mothbots and slugbots, which are part bug, part machine; in other words, cyborgs. Cyborg organisms are here and are the next wave, which, once again, can be used for good or ill, or well-being and information gathering or weaponizing and distraction. Bakker mostly catalogued the new and emerging digital toys, but also mentioned justifiable concerns, and the long literary history of mad scientist machinations gone awry, with Frankenstein likely being the most well-known. But symbionts also remind of Haraway's cyborg manifesto, which celebrates the inevitability of the human-nature-technology nexus. Bakker also spoke of inevitability, given that the tech is already here and more is coming, but also asked the question we must ask: Are we communing with nature or anthropocentric versions of nature (I'd say the latter) or the same question that emerges for all claims of interspecies communication: are we encountering others or just our own projections? The difference here, though, is that virtual reality is mediated, and the loss of physicality while you are having a physical reaction is not a minor loss.[62]

Bakker ended her talk by turning to the example of Karl von Frisch, the bee master who spent every day monitoring honey bees, eventually discovering the waggle dance, which bees use to communicate the location of nectar. With new digital tools, his years of study could have been reduced immeasurably.[63] But that raises the question of measuring at all, or data collection, which, again, is basic science and much needed, but, in most of these examples, represents the familiar bias of "if we only had more data, then all would be well," or panglossian data as savior. Do we need to know more via more data or to know better? This may be a false dichotomy, but what we most need to know is that collecting and cataloguing, while providing baseline knowledge and stability, as Bateson argued, needs to be balanced with the flexibility to think new thoughts. And new thoughts lead to an inevitability far more important than new technologies: ethics inspired by the pattern that connects. After all, the genome project, within the context of capitalist ownership, was turned into corporations patenting the building blocks of life. Also, while Frisch could have collected data much more quickly, his non-mediated focused observations mirrored Zen "just sitting" and the converse practiced by Kaza, which are their own rewards and also reward with information and insights not found in any other way. And so, we are left with the question of whether we should embrace the interspecies internet, or, which parts are fitting and which parts are unfitting.

I can see the benefit of collecting more data, depending on what we do with it, and using innovative machines to clean up ocean plastic, and even taking a virtual reality trip to the Great Barrier Reef, viewing, and feeling, your way around the intricate and diverse seascape only to watch it die before your eyes, approximating, but not nearly matching, Leopold's witnessing the death of a fierce, green fire in the eyes of the wolf; but what makes my head spin the most is the notion that all this improves our ability to listen. Bioacoustics allows us to hear what we cannot hear, whether whale song or the clicking of plants, and that is both wonder and gift, yet Bakker stated that all these gizmos allow us to not only communicate about the environment but communicate with it. But are we dialoguing or is this more monologue, or enhanced monologue that discloses information, just like I-It science using instrumentation discloses much needed information? We have moved from the microscope to the macroscope, but they are of the same mediated techno-kind. Another example was a robo-lobster that could scuttle across the ocean floor collecting data, and it was mentioned that such robotics are a form of biomimicry, given the precise replication of lobster movements; but it is mere replication, or mechanical cloning, not learning that comes from deep listening. All these gizmos may lead us to be more connected in one way, but more disconnected in others—conceal as we reveal—with our awareness more heightened but also less heightened. We bring little to these interactions other than mediated technology and species are forced to respond.

But some choices, at least for me, are not that tough; although, again, practicing an ecology of communication does not necessarily lead to agreement, only that we are focusing our energy on the right questions. There is the debate, for example, between the EcoModernists, who have a manifesto that basically argues that the best response to ecocrisis is more growth, more progress, and more technology, including investment in nuclear power, given such energy does not emit CO_2, and the Degrowth folks, who also have a statement but espouse the opposite view, focusing on sufficiency, downsizing, and eco-social justice. This debate was played out in the Michael Moore produced film *Planet of the Humans*, which created quite a stir because of its dated misinformation on solar and wind power, and the contention that environmental leaders have been co-opted by big business. The misinformation nearly ruins the film, but the message of degrowth is salient and the path we should take, leading us to need far less energy while we also focus on energy efficiency. We cannot return to our hunter-gatherer past, but climate emergency is pushing us to mimic hunter-gatherer lifeways as much as possible, and listen to Thoreau's call to "simplify, simplify, simplify." Unfortunately, the film, while making some good points, has become a case study of miscommunication and the failure to integrate differing views.[64]

It is also clear to me, by practicing an ecology of communication, that we need to turn to democratic ecosocialism, which would include plenty of free market structures. While we must rid ourselves of the myths of bigger, better, and more, the right kind of bigger government is needed to tackle big and wicked problems. Moral or conscious capitalism, while having merit on smaller scales, is not the larger fix we need. On the other hand, Finland, a smaller and less diverse country, has embraced capitalism reined in by government, and effectively so.[65] Sometimes this debate feels like semantics and splitting hairs, but other times it does not, as capitalism by definition treats nature as capital, or a standing reserve of resources, and egregious, immoral behavior abounds, including disaster capitalism, and pandemic capitalism, in which billionaires make bank while others suffer.[66] Major change to economic life is obviously essential in crisis times, especially the insanity of externalities, in which big oil and other big corporate institutions do not pay for the damage they cause, foisting the cost onto unsuspecting tax-paying publics. Or, as Kimmerer puts it, the Indigenous idea of land as a commonly held gift was replaced with the notion of private property, and "the battle between land as sacred home and land as capital stained the ground red."[67] After all, the ultimate gift event is the earth's bounty, which, given reciprocal relations, makes demands on us; for Kimmerer, the earth gives and gives and thus asks us for a return: the call to gratitude, to pay attention, to recognize the personhood of all beings, to reciprocity, and to change.[68]

We must also take the long view while acting in the short term. The paleoclimatologist Curt Stager argues that based on CO_2 cycles in history, we can make general predictions for the future. 350 ppm is good guide for where we want to be, and in 50,000 to 150,000 years we will likely go way below that, approaching an uninhabitable ice age. In other words, we are wisely concerned now about going too far over, but what about too far under in the deep future? This takes us into strange territory, especially in regard to the ethic of intergenerational responsibility and thinking with the seventh generation in mind. If our concern is with our generation through to the seventh generation, then we should do everything we can to stop further warming the planet. But for 50,000 to 150,000 years in the future, that may be a different story. Of course, if humans don't get through this crisis, we will not even reach the next. One reasoned response, then, is to keep fossil fuels in the ground, not only for our sake, but so that those in the far off future can burn it to intentionally warm the planet.[69]

To Stager, we have crossed a line in the Anthropocene, but from the perspective of a future ice age, that is a good thing; we will need to manage the planet, to the degree that we can, to survive. But that raises the question: what does sane, or fitting, management look like? We have made so many foolish attempts, and are considering more like geoengineering; mucking around nearly always leads to a treadmill of abuse marked by unsuspected new problems. In all instances, in all responses, we must practice logos, and an ecology of communication, but Stager's analysis makes it clear that life on this planet for humans is tenuous. Conditions have created us, and conditions change on this Gaian self-organizing home hurtling through space, and thus recognizing the absurd is also a rational response as we live in sympathy with wider intelligence but also our own intelligence.

There is no end of authors weighing in on possible fitting responses to ecocrisis, and thus there is no end of solutions. In *The Nature of Design*, the leading environmental educator David Orr argues that everything will need to be redesigned with the earth in mind, and Benyus's biomimcry supplies many of the means to do so.[70] In *Mobilizing the Green Imagination*, Anthony Weston, professor of philosophy and environmental studies, argues for welcoming rising waters, turning New Orleans into a city of canals like Venice, among other surprising ideas that embrace a changed world.[71] And in *What We Think About When We Try Not to Think About Global Warming*, Per Espen Stoknes, a psychologist, economist, and entrepreneur, integrating science with the humanities, asks that we re-imagine climate as living air, as it is not an object but "a grand subject, to which we are subjected." We exist within the air like fish immersed in water, and yet, because it is usually invisible to our eyes, except when experiencing floating clouds or sunsets or wind, we take it for granted, spewing the byproducts of industrial life into what gives us life.[72]

The air is historically and rightly correlated with breath and spirit, and for Stoknes it overflows with exchange of information: information of beauty and pattern and voices that should make us think twice before despoiling it and information of disruption and chaos and symptoms that convey messages of what happens when we do.[73] But, most often, we fail to listen. Thoreau offers many practices, but the most compelling, the most needed today, is the act, and art, of deep listening: we must attend to the voice of Spirit, to the voices of nature, and to inner voice.

We can begin to listen, and some are listening, and engaging the key questions of our time with seriousness, but also an unburdened, light heart. The world has changed, deeply—Bill McKibben's 2011 book on warming, entitled *Eaarth*, argues that we have changed the planet to the point that it needs to be renamed—yet sustainability solutions will come as we embrace purpose but also play despite dire predictions; or rather, purpose emerges from identifying with the play of Spirit, with humans destined to be creative and have some fun, expressing our full capacities by thinking and acting our way to soulful, sustainable living. Or rather, after you go to jail, pick some berries with your friends, reconnecting with the joy that fuels compassionate activism, and yes, some *conscious* rage. Burning yourself out from just jail and just rage makes one ineffective in the collective fight for justice; we are likely to bring forth unfit responses, not to mention what it does to the individual soul.

We must also embrace the dialogic interplay among the work of Schrag, Wilber, Shepard, and Bateson, while remaining mindful of Peters' existential grounding in what often seems like the impossibility of communication.

Schrag has provided a constructive postmodern vision, deconstructing appeals to Truth and certainty but not leaving us bereft of ethical response. Transversal logos, in which we traverse the past for insights, embrace the present responsibility of discernment, and are willing to rethink in the future, or the continual practice of participation and distanciation, provides a baseline for communicative action. Of all the selves explored in this project—the ecological self, the relational self, the cybernetic self—his responding self is the most compelling. We express many selves, but always with the ethical call to respond. And thus what is fitting and unfitting is the ultimate question for humans living on a finite planet, but also for our particular culture wars age, with Magic-Mythic, Rational, and Pluralistic citizens duking it out in the public sphere, which is increasingly the social media sphere. Some sanity is much needed, and transversal logos and the fitting response provides guidance, as does the appeal to the transcendent gift event, in which giving is receiving and receiving unfolds into future giving. All of which aids rational communication and our ability to more fully express a fitting responsiveness.

Wilber has also provided a constructive postmodern vision, although his meta-theory constructs more than Schrag, placing rational communication within the context of insights disclosed via spiritual experiences. In many ways, Wilber's take on religion mirrors Dogen's fourteenth-century *Mountains and Rivers* sutra, in which nature spirits, or some higher Being, or God, are central to premodern tribal and traditional societies, but then God disappears from the making-sense-of-reality equation during modern and deconstructive postmodern times, or God is dead and thus humans are in charge, living in accordance with the myths of linear progress, unlimited growth, and technology as savior. However, then another shift occurs and God reappears, or rather, over the course of human history we have gone from God is everywhere to God is nowhere to God is everywhere again.[74] But just like mountains are mountains again but with the difference of perceiving from a higher nondual level, God reappears and is interpreted at Integral and Super Integral, with the dualism of dualisms overcome and thus transcendent and immanent dimensions harmonized. Out of that harmony, that sympathy with intelligence, comes spiritual communication as spiritual experience providing insights, but also spiritual communication as the integration of all levels of the developmental holarchy, at least as ideal, and thus a fitting responsiveness.

Shepard's mythic-animistic vision acts as the counter-balance, the so much needed grounding on this actual earth. While Wilber argues that we must transcend and include our way up the developmental holarchy, Shepard makes sure we include and is adamant about what must be included if we are to keep the world running properly: the embrace of a dialogic life-world filled with messages, which are only heard if we practice attentiveness to the aliveness, reciprocity, and gifts of nonhuman others and specific places, and tell stories and participate in rituals that allow us to learn from predator-prey converse. Our senses and sensuous imagination must not be overcome, whether from worship of transcendent sky gods, the move to agriculture, or further "progress" to techno-industrial structures. Despite the downsides of hunter-gatherer lifeways, which Wilber happily articulated, community living within specific places holds lessons for us today, perhaps the ultimate lessons, as modernist myths must be re-imagined within the context of degrowth marked by sufficiency, vital needs, and appropriate technology. While we need to grow up, we must not grow away from our rootedness within this diverse earth, and thus we must not grow away from the practice of mythic-animistic communication, which is essential if we are to respond to ecocrisis.

Bateson's vision of the pattern that connects is marked by the good kind of transcending. The Cartesian I-It world view, while leading to so much scientific discovery, and so much new technology, has also led to so much destruction. Thus, it must be overcome and replaced with a different epistemology: living within the larger context of mind, and thus thinking via rigor and imagination and perceiving via the grace of aesthetic recognition and empathy. We do not need religious experience to approach the sacred, but we do need receptivity to beauty, which allows us to live in sympathy with order and randomness and practice I-Thou science, discovering that relationship is the basis of the biosphere and the core of who we are. Ecocrisis reflects a crisis of perception, and we have a snowball's chance in hell of surviving unless we recognize that we live within a larger communicative system expressed as the continual exchange of information. For Bateson, the most fitting response is to acknowledge this wonder, and open ourselves to this wonder, by practicing aesthetic communication.

Peters, on the other hand, reminds us of the dream of communication, of the mistake of reaching too high and expecting too much, thinking that communication is best described via Socratic eros, in which mutual understanding is easily reached, or worse, telepathy, instead of acknowledging the inevitability of miscommunication and the hard work of community-building and decision-making. All true, yet also partial, because when community-building and decision-making occur within the context of anthropocentric bias, it is inevitably unfitting. This bias is not up to the task or makes matters worse. Acknowledging the dream of communication, as necessary as it is, must be balanced by what Thomas Berry calls the dream of the earth, inspired by Native American spirituality, in which the larger earth community is characterized by a communion of subjects rather than a collection of objects.[75] Peters is not unresponsive to this understanding, which is worth making as mutual as possible. In his latest book, *The Marvelous Clouds*, he proposes a new theory of media that transcends and includes the human, in which nature, in all its forms, speaks to us.[76]

Emerson wrote, "The fact that I am here certainly shows me that the soul had a need of an organ here. Shall I not assume the post?"[77] And Thoreau surmises, as only he can: "Every man must once more learn the points of the compass as often as he wakes, whether from sleep or from any abstraction. In fact, not till we are lost do we begin to realize where we are and the infinite extent of our relations."[78] These quotations may elicit hope—out of our lostness may come new awakenings and soulful responsibilities—but hope or not, this we know: millions of decisions have gotten us here, now, to an age of ecocrisis. And this I hope we come to know: we respond more fittingly, and with more responsibility, when we practice an ecology of communication.

Whitman, in his preface, wrote "This is what you shall do," a bold, or perhaps arrogant, exhortation, which he knowingly undercuts when he further advises to "dismiss whatever insults your soul" and "re-examine all that you have been told," including from books. In humility, and honesty, I will attempt my own far-less poetic list, as so much depends on what we *consciously* bring to our encounters with others versus what we unconsciously bring via interpellation and our habit bodies.

This is what you shall bring: the desire to move up the developmental holarchy and to move others up, knowing that the higher we go, the wider our awareness and more discerning our responses, the more effortless effort, the more we are called to responsibility; the discipline of transversal rationality, participating in and distancing ourselves from traditions with the knowledge that there are no final answers, no one and done final fitting response; the artful ability of discerning the unfit, recognizing the pathology of monologic forms of knowing and communication, as well as the pitfalls of differing forms of communication when they are poorly interpreted and expressed; and the much-needed practice of logos and skillful engagement within the public sphere where impactful decisions are made, while being mindful of the inevitability of miscommunication and that some dualisms, some divides, some contradictions, resist integration, and thus we must work to keep the conversation going. Ultimately, we must bring a fitting responsiveness, as best we can, listening within and without to the voice and voices of nature; and a commitment to play, not just work, reflecting the dialogic interplay of rational, spiritual, mythic-animistic, and aesthetic communication, integrating or prioritizing based on the needs of particular situations. We listen, we care, and we act, engendering the *possibility* of responses that are more fitting.

And so, if I may be bold like Whitman: the practice of an ecology of communication is no less than the self-correcting steering wheel of evolution. Our communicative action acts on us, and then we respond, and respond, ad infinitum, creating new worlds.

NOTES

1. Climatechange12.com, *Disruption*, Sept. 7 2014, youtube video, 52:27, https://www.youtube.com/watch?v=uWPj6CxtsGo.
2. See Walt Whitman, *Leaves of Grass*, in *Poetry and Prose*, edited by Justin Kaplan, (New York: Library of America, 1982).
3. Walls, *Henry David Thoreau: A Life*, 394–395.
4. Whitman, *Leaves of Grass, in Poetry and Prose.*
5. Ibid.
6. Naomi Oreskes and Erik M. Conway, *Merchants of Doubt: How a Handful of Scientists Obscured the Truth on Issues from Tobacco Smoke to Global Warming*, (New York: Bloomsbury, 2010), 216.
7. Ibid., 220–221.
8. Ibid., 216–217.
9. Ibid., 217–218.
10. Ibid., 219.
11. Ibid., 224–226.
12. Ibid.
13. Ibid., 228–229.
14. See Elizabeth D. Blum, *Love Canal Revisited: Race, Class, and Gender in Environmental Activism* (Lawrence: University of Kansas Press, 2008).
15. Goldmanprize.org, "The Goldman Environmental Prize 2004," https://www.goldmanprize.org/recipient/margie-richard/.
16. Sandra Steingraber, *Living Downstream: An Ecologist's Personal Investigation of Cancer and the Environment* (Cambridge, MA: Da Capo Press, 2010).
17. Steingraber.com, http://steingraber.com/category/civil-disobedience/.

18. Oreskes and Conway, *Merchants of Doubt*, 229.
19. Ibid., 236.
20. Ibid., 236–239.
21. https://www.nicolasnierenberg.com/.
22. Justin Gillis, "Naomi Oreskes: A Lighting Rod in a Changing Climate," *The New York Times*, June 15, 2015, https://www.nytimes.com/2015/06/16/science/naomi-oreskes-a-lightning-rod-in-a-changing-climate.html.
23. Ibid.
24. See *Disruption.*
25. Gillis, "Naomi Oreskes."
26. Neil Postman, *Amusing Ourselves to Death: Public Discourse in the Age of Show Business*, (New York: Penguin, 1985).
27. James Hoggan, *I'm Right and You're an Idiot: The Toxic State of Public Discourse and How to Clean It Up* (Canada: New Society Publishers, 2016).
28. Andrew Postman, "My Dad Predicted Trump in 1985: It's Not Orwell, He Warned, It's *Brave New World*," *The Guardian*, Feb. 2, 2017, https://www.theguardian.com/media/2017/feb/02/amusing-ourselves-to-death-neil-postman-trump-orwell-huxley.
29. Martin Heidegger, *Discourse on Thinking*, translated by John M. Anderson and E. Hans Freund (New York: Harper and Row, 1966).
30. Yale Divinity School, "YDS Climate Panel Pushes Past Denial and Despair," March 14, 2019, https://divinity.yale.edu/news/yds-climate-panel-pushes-past-denial-and-despair.
31. Ava Kaufman, "Bruno Latour, the Post-Truth Philosopher, Mounts a Defense of Science," *The New York Times*, October 25, 2018, https://www.nytimes.com/2018/10/25/magazine/bruno-latour-post-truth-philosopher-science.html.
32. Thoreau, *"A Natural History in Massachusetts*, in *The Portable Thoreau*, 56.
33. Vine Deloria, Jr., "If You Think About It, You Will See that It is True," in *Spirit and Reason: The Vine Deloria Jr. Reader* (Colorado: Fulcrum Pub., 1999), 42-60.
34. Henry David Thoreau, *Cape Cod* (New York: Penguin Books, 1987).
35. Naomi Oreskes and Erik M. Conway, *The Collapse of Western Civilization: A View from the Future* (New York: Columbia University Press, 2014).
36. Gillis, "Naomi Oreskes."
37. Roger E. Meiners, Pierre Desrochers, and Andrew P. Moriss, *Silent Spring at 50: The False Crises of Rachel Carson* (Washington, DC: The Cato Institute, 2012).
38. Chad Montrie, *The Myth of Silent Spring: Rethinking the Origins of American Environmentalism* (Oakland, CA: University of California Press, 2018).
39. Forbes.com, "BP Goes for Public Relations Makeover to Get Beyond Gulf Spill, February 7, 2012, https://www.forbes.com/sites/greatspeculations/2012/02/07/bp-goes-for-public-relations-makeover-to-get-beyond-gulf-spill/#2a50f5513fab.
40. Michelle Goldberg, "John Lewis Believed America Would Survive Trump," *The New York Times*, July 30, 2020, https://www.nytimes.com/2020/07/30/opinion/john-lewis-legacy.html.
41. Peters, *Speaking into the Air*, 245.

42. Elgin, *Promise Ahead*, 97.

43. Jem Bendell, "Deep Adaptation: A Map for Navigating Climate Tragedy," Jem Bendell,com, July 2028, revised July 2020, https://jembendell.com/2019/05/15/deep-adaptation-versions/.

44. Iflas.com, "International Scholars Warning on Societal Disruption and Collapse," https://iflas.blogspot.com/2020/12/international-scholars-warning-on.html

45. Walls, *Henry David Thoreau: A Life*, 441–443.

46. Ibid., 488–489.

47. Sandra Laville, "Human-made Materials Now Outweigh Earth's Entire Biomass," *The Guardian*, December 9, 2020, https://www.theguardian.com/environment/2020/dec/09/human-made-materials-now-outweigh-earths-entire-biomass-study

48. David Zirin, *Welcome to the Terrordome: The Pain, Politics, and Promise of Sports* (Chicago, IL: Haymarket Books, 2007), 16-22.

49. Hop Hopkins, "Racism Is Killing the Planet," *Sierra*, June 8, 2020, https://www.sierraclub.org/sierra/racism-killing-planet.

50. Leah Thomas, "Why Every Environmentalist Should Be Anti-Racist," *Vogue*, June 8, 2020, https://www.vogue.com/article/why-every-environmentalist-should-be-anti-racist.

51. Hopkins, "Racism is Killing the Planet."

52. Christopher Bache, *Dark Night, Early Dawn: Steps to a Deep Ecology of Mind* (New York: SUNY Press, 2000), 213-256.

53. Ibid.

54. Ibid,

55. Quoted in Bache, 244.

56. Thomas Berry, *The Great Work: Our Way into the Future* (New York: Bell Tower, 1999), 74.

57. Paul Hawken, "Commencement Address to the Class of 2009, University of Portland, May 3, 2009," paulhawken.com, www.paulhawken.com/multimedia/UofP_Commencement.pdf.

58. Karen Bakker, "The Future of Environmental Communication in an Era of Digital Disruption," Presentation, International Environmental Communication Association (IECA) Conference, University of British Columbia, Canada, June 2018.

59. Ibid.

60. Ibid.

61. Ibid.

62. Ibid.

63. Ibid.

64. Filmsforaction.org, "Films for Action's Statement on Planet of the Humans," April 25, 2020, https://www.filmsforaction.org/articles/films-for-actions-statement-on-planet-of-the-humans/.

65. Anu Partanen and Trevor Corson, "Finland Is a Capitalist Paradise," *The New York Times*, December 7, 2019, https://www.nytimes.com/2019/12/07/opinion/sunday/finland-socialism-capitalism.html.

66. See Joel Kovel, *The Enemy of Nature* (New York: Zed Books, 2002).

67. Robin Wall Kimmerer, "Greed Does Not Have to Define Our Relationship to the Land," Lithub.com, June 4, 2020, https://lithub.com/robin-wall-kimmerer-greed-does-not-have-to-define-our-relationship-to-land/.

68. Robin Wall Kimmerer, "Returning the Gift," Center for Humans and Nature, Centerforhumansandnature.org, May 2014, https://www.humansandnature.org/returning-the-gift.

69. Curt Stager, *Deep Future: The Next 100,000 Years of Life on Earth* (New York: St. Martin's Press, 2011), 1-12.

70. David Orr, *The Nature of Design: Ecology, Culture, and Human Intention* (Oxford: Oxford University Press, 2004).

71. Anthony Weston, *Mobilizing the Green Imagination* (Canada: New Society Publishers, 2012).

72. Per Espen Stoknes, *What We Think About When We Try Not to Think About Global Warming* (White River Junction, VT: Chelsea Green, 2015), 203.

73. Ibid.

74. Wilber, *The Religion of Tomorrow*, 420–421.

75. Thomas Berry, *The Dream of the Earth* (California: Sierra Club Books, 1988).

76. John Durham Peters, *The Marvelous Clouds: Towards a Philosophy of Elemental Media* (Chicago: University of Chicago Press, 2015).

77. Emerson, "Spiritual Laws," *The Portable Emerson*, 206.

78. Thoreau, *The Heart of Thoreau's Journals* (New York: Dover Pub., 1961).

Works Cited

Abram, David. *The Spell of the Sensuous: Perception and Language in a More-Than Human World*. New York: Pantheon, 1996.

———. "Perceptual Implications of Gaia." *Dharma Gaia*. Edited by Allan Hunt Badiner, 75–92. Berkeley, CA: Parallax Press, 1990.

Abraham-Hicks, https://www.abraham-hicks.com/.

Abraham-Hicks Publications. "Abraham: How Source Sees Natural Disasters." April 13, 2012. YouTube video, 10:10. https://www.youtube.com/watch?v=aupLxCokSZI.

Andrews, Barry. *Transcendentalism and the Cultivation of the Soul*. Amherst, MA: The University of Massachusetts Press, 2017.

Althusser, Louis. *Lenin and Philosophy and Other Essays*. Translated by Ben Brewster. New York: Monthly Review Press, 1971.

Aristotle. *Ethics*. Translated by J. A. K. Thomson. London: Penguin Classics, 1953.

AtKisson, Alan. *Believing Casandra: An Optimist Looks at a Pessimist's World*. White River Junction, VT: Chelsea Green, 1999.

Bache, Christopher. *Dark Night, Early Dawn: Steps to a Deep Ecology of Mind*. New York: SUNY Press, 2000.

Bakker, Karen. "The Future of Environmental Communication in an Era of Digital Disruption." Presentation, International Environmental Communication Association (IECA) Conference, University of British Columbia, Canada, June 2018.

Barfield, Owen. *Saving the Appearances*. London: Faber and Faber, 1988.

Barlow, Connie. *Green Space, Green Time: The Way of Science*. New York: Springer Verlag, 1997.

Bateson, Gregory. *Steps to an Ecology of Mind*. New York: Ballantine, 1972.

———. *Mind and Nature: A Necessary Unity*. New York: Bantam, 1979.

———. *A Sacred Unity: Further Steps to an Ecology of Mind*. Edited by Rodney E. Donaldson. New York: HarperCollins, 1991.

Bateson, Gregory and Mary Catherine Bateson. *Angels Fear: Toward an Epistemology of the Sacred*. New York: Bantam Books, 1987.
Bateson, Mary Catherine. "Foreword," in Gregory Bateson, *Steps to an Ecology of Mind*. Chicago: University of Chicago Press, 2000.
Beitman, Bernard D. "Jung's Scarab as a Psychotherapeutic Technique." *Psychology Today*. August 12, 2015. https://www.psychologytoday.com/us/blog/connecting coincidence/201508/jung-s-scarab-psychotherapeutic-technique.
Bendell, Jem. "Deep Adaptation: A Map for Navigating Climate Tragedy." Jem Bendell.com. Revised July 2020. https://jembendell.com/2019/05/15/deep-adaptation-versions/.
Benyus, Janine. *Biomimicry: Innovation Inspired by Nature*. New York: HarperCollins, 1997.
Berkes, Fikret. *Sacred Ecology*. New York: Routledge, 2012.
Berman, Morris. *The Reenchantment of the World*. New York: Bantam, 1984.
Berry, Thomas. *The Dream of the Earth*. California: Sierra Club Books, 1988.
———. *The Great Work*. New York: Bell Tower, 1999.
Berry, Thomas and Brian Swimme. *The Universe Story: From the Primordial Flaring Forth to the Ecozoic Era--A Celebration of the Unfolding of the Cosmos*. San Francisco: HarperCollins, 1992.
Berry, Wendell. *What are People For?* San Francisco: North Point Press, 1990.
———. *Home Economics*. San Francisco: North Point Press, 1987.
———. "Whose Head is the Farmer Using? Whose Head is Using the Farmer?" In *Meeting the Expectations of the Land*. Edited by Wes Jackson, Wendell Berry, and Bruce Colman. San Francisco: North Point Press, 1984.
Blum, Elizabeth D. *Love Canal Revisited: Race, Class, and Gender in Environmental Activism*. Lawrence, Kansas: University of Kansas Press, 2008.
Boone, J. Allen. *Kinship with All Life*. San Francisco: HarperCollins, 1954.
Botkin, Daniel B. *No Man's Garden: Thoreau and the New Vision for Civilization and Nature*. Washington, DC: Island Press, 2001.
Bridgman, Richard. *Dark Thoreau*. Lincoln, NE: University of Nebraska Press, 1982.
Buber, Martin. *I and Thou*. Edited by Walter Kaufmann. New York: Scribner's, 1970.
———. *Pointing the Way*. New York: Harper & Row, 1963.
Buell, Lawrence, ed. *The American Transcendentalists: Essential Writings*. New York: Random House, 2006.
Carbaugh, Donal. "Response Essay: Environmental Voices Including Dialogue with Nature, Within and Beyond Language," in *Voice and Environmental Communication,* edited by Jennifer Peeples and Stephen Depoe, 241–256. England: Palgrave MacMillan, 2014.
———. "Quoting 'The Environment': Touchstones on Earth." *Environmental Communication: A Journal of Nature and Culture*, *1*(1) (May 2007): 64–73.
———. "Just Listen: 'Listening' and Landscape Among the Blackfeet." *Cultures in Conversation*. New York: Routledge, 2005.
Carney, Scott. *What Doesn't Kill Us*. New York: Penguin/Rodale, 2017.
Charlton, Noel. *Understanding Gregory Bateson: Mind, Beauty, and the Sacred Earth*. Albany: SUNY Press, 2008.

Climatechange12.com. *Disruption*. Sept. 7 2014. YouTube video, 52:27. https://www.youtube.com/watch?v=uWPj6CxtsGo.

Cramer, Jeffrey S. *Solid Seasons: The Friendship of Henry David Thoreau and Ralph Waldo Emerson*. Berkeley, CA: Counterpoint, 2019.

Deep Green Resistance. https://deepgreenresistance.org/en/.

Deetz, Stanley A. *Democracy in an Age of Corporate Colonization: Developments in Communication and the Politics of Everyday Life*. New York: SUNY Press, 1992.

Deloria, Vine. Jr. *Spirit and Reason: The Vine Deloria, Jr. Reader*. Colorado: Fulcrum Pub., 1999.

Devall, Bill and George Sessions. *Deep Ecology*. Utah: Gibbs-Smith, 1985.

Diamond, Jared. *Collapse: How Societies Choose to Fail or Succeed*. New York: Penguin, 2005.

Dossey, Larry. *One Mind: How our Individual Mind is Part of a Greater Consciousness and Why It Matters*. Colorado: Hay House, 2013.

Dunsmore, Roger. *Earth's Mind: Essays in Native Literature*. Albuquerque, NM: University of New Mexico Press, 1997.

Ehrlich, Paul R. and Anne H. Ehrlich. *Betrayal of Science and Reason: How Anti-Environmental Rhetoric Threatens Our Future*. Washington D.C.: Island Press, 1996.

Eisenstein, Charles. *Climate: A New Story*. Berkeley, CA: North Atlantic Books, 2018.

Elgin, Duane. *Promise Ahead: A Vision of Hope and Action for Humanity's Future*. New York: Quill, 2001.

Emerson, Ralph Waldo. *The Portable Emerson*. Edited by Carl Bode & Malcolm Cowley. New York: Penguin Books, 1981.

———. "Nature" in *Nature/Walking*. Edited by John Elder. Boston: Beacon Press, 1991.

Esbjorn-Hargens, Sean and Michael E. Zimmerman. *Integral Ecology: Uniting Multiple Perspectives on the Natural World*. Boston: Shambhala/Integral Books, 2009.

Ferguson, Marilyn. *The Aquarian Conspiracy: Personal and Social Transformation in Our Time*. New York: Penguin/Tarcher, 2007.

Feuerstein, Georg. *Holy Madness: The Shock Tactics and Radical Teachings of Crazy-Wise Adepts, Holy Fools, and Rascal Gurus*. New York: Penguin Books, 1992.

Filmsforaction.org. "Films for Action's Statement on Planet of the Humans." April 25, 2020. https://www.filmsforaction.org/articles/films-for-actions-statement-on-planet-of-the-humans/.

Fisher, Andy. *Radical Ecopsychology: Psychology in the Service of Life*. New York: SUNY Press, 2002.

Forbes.com. "BP Goes for Public Relations Makeover to Get Beyond Gulf Spill. February 7, 2012. https://www.forbes.com/sites/greatspeculations/2012/02/07/bp-goes-for-public relations-makeover-to-get-beyond-gulf-spill/#2a50f5513fab.

Fox, Margalit. "Jose Arguelles, New Ager Focused on Time, Dies at 72." *New York Times*, April 2, 2011. https://www.nytimes.com/2011/04/03/us/03arguelles.html.

Fox, Matthew and Rupert Sheldrake. *The Physics of Angels: Exploring the Realm Where Science and Spirit Meet*. New York: HarperCollins, 1996.

Frazier, Vidya. *Ascension: Embracing the Transformation*. Sarasota, FL: First Edition Design Publishing, 2015.

Friskics, Scott. "Dialogic Relations with Nature," *Environmental Ethics*, 23, Issue 4 (Winter 2001): 392–393.

Fuller, Margaret. "Recollection of Mystical Experiences," in *The American Transcendentalists: Essential Writings*, edited by Lawrence Buell. New York: Modern Library, 2006.

———. "The Great Lawsuit," in Lawrence Buell, ed., *The American Transcendentalists: Essential Writings*. New York: Modern Library, 2006.

Fun and Budget with Tinesha Davis. "Abraham Hicks' First Talk on the Virus." March 2020. YouTube video, 13:05. https://www.youtube.com/watch?v=wUYPa5H3Pnk.

Furlong, Monica. *Zen Effects: The Life of Alan Watts*. Woodstock, VT: Skylight Paths Publishing, 2001.

Gagliano, Monica. *Thus Spoke the Plant: A Remarkable Journey of Groundbreaking Scientific Discoveries and Personal Encounters with Plants*. Berkeley, CA: North Atlantic Books, 2018.

Gagliano, Monica, John C. Ryan, & Patricia Viera, *The Language of Plants: Science, Philosophy, Literature*. Minnesota: University of Minnesota Press, 2017.

Gandhi, M. K. *Non-Violent Resistance (Satyagraha)*. New York: Dover Press, 2001.

Gebser, Jean. *The Ever-Present Origin*. Translated by Noel Barstad with Algis Mickunas. Athens, Ohio: University of Ohio Press, 1985.

Gillis, Justin. "Naomi Oreskes: A Lighting Rod in a Changing Climate." *The New York Times*. June 15, 2015. https://www.nytimes.com/2015/06/16/science/naomi-oreskes-a-lightning rod-in-a-changing-climate.html.

Gleick, James. *Chaos: The Making of a New Science*. New York: Penguin Books, 1987.

Goldberg, Michelle. "John Lewis Believed America Would Survive Trump." *The New York Times*. July 30, 2020. https://www.nytimes.com/2020/07/30/opinion/john-lewis legacy.html.

Goldmanprize.org, "The Goldman Environmental Prize 2004," https://www.goldmanprize.org/recipient/margie-richard/.

Goodall, Jane and Marc Bekoff, *The Ten Trusts: What We Must Do to Care for the Animals We Love*. San Francisco: Harper San Francisco, 2003.

Greenway, Robert. "The Wilderness Effect and Ecopsychology." *Ecopsychology: Restoring the Earth, Healing the Mind,* edited by Theodore Roszak, Mary E. Gomes, and Allen D. Kanner, 122–135. San Francisco: Sierra Club Books, 1995.

Greenway, Robert. "Healing by the Wilderness Experience." *Wild Ideas*, edited by David Rothenberg, 182–193. Minneapolis: Univ. of Minnesota Press, 1995.

Griffin, Susan. *Woman and Nature*. New York: Harper & Row, 1978.

———. *The Eros of Everyday Life*. New York: Doubleday, 1995.

Hadot, Pierre. *Philosophy as a Way of Life*. Oxford: Blackwell Publishing, 1995.

———. *What is Ancient Philosophy?* Cambridge, MA: Harvard University Press, 2004.

Hall, Matthew. *Plants as Persons: A Philosophical Botany*. New York: SUNY Press, 2011.

Hanegraaff, Wouter J. *New Age Religion and Western Culture: Esotericism in the Mirror of Secular Thought*. New York: SUNY Press, 1998.

Haraway, Donna J. *Simians, Cyborgs, and Women: The Reinvention of Nature*. New York: Routledge, 1991.

———. *When Species Meet*. Minnesota: The University of Minnesota Press, 2008.

Harding, Walter. *The Days of Henry Thoreau: A Biography*. Princeton, NJ: Princeton University Press, 1992 (1962).

Harries-Jones, Peter. *A Recursive Vision: Ecological Understanding and Gregory Bateson*. Toronto: University of Toronto, 1995.

Haskell, David George. *The Songs of Trees: Stories from Nature's Great Connectors*. New York: Penguin Books, 2017.

Hawken, Paul. *Blessed Unrest: How the Largest Social Movement in History is Restoring Grace, Justice, and Beauty to the World*. New York: Penguin Books, 2008.

———. "Commencement Address to the Class of 2009, University of Portland, May 3, 2009." paulhawken.com, www.paulhawken.com/multimedia/UofP_Commencement.pdf.

Hawken, Paul, Amory Lovins & L. Hunter Lovins. *Natural Capitalism*. Boston: Back Bay Books, 2000.

Heidegger, Martin. *Discourse on Thinking*. Translated by John M. Anderson and E. Hans Freund. New York: Harper and Row, 1966.

———. *The Question Concerning Technology*. New York: Harper & Row, 1977.

Hedges, Chris. "Onward Christian Fascists." *Truthdig*, December 30, 2019. https://www.truthdig.com/articles/onward-christian-fascists/.

Herzog, Werner, dir. *Grizzly Man*. 2005; U.S.: Discover Docs/Lion's Gate, DVD.

Hick, John. *The Fifth Dimension: An Exploration of the Spiritual Realm*. London: OneWorld Publications, 2013 (1999).

Hixson, Lex. *Coming Home*. New York: Larsen Publications, 1995.

Hogan, Linda, Deena Metzger, and Brenda Peterson, eds. *Intimate Nature: The Bond Between Women and Animals*. New York: Ballantine Publishing Group, 1998.

Hoggan, James. *I'm Right and You're an Idiot: The Toxic State of Public Discourse and How to Clean It Up*. BC, Canada: New Society Publishers, 2016.

Homestead, William. "The Language that All Things Speak." In *Voice and Environmental Communication*, edited by Jennifer Peeples and Steve Depoe. New York: Palgrave MacMillan, 2014.

Hopkins, Hop. "Racism Is Killing the Planet." *Sierra*. June 8, 2020. https://www.sierraclub.org/sierra/racism-killing-planet.

Hunt, Tam. "Could Consciousness All Come Down to the Way Things Vibrate?" *The Conversation*. November, 2018. https://theconversation.com/could-consciousness-all come-down-to-the-way-things-vibrate-103070.

Initiative for Leadership and Sustainability. "International Scholars Warning on Societal Disruption and Collapse." https://iflas.blogspot.com/2020/12/international scholars warning-on.html.

Integral World, http://www.integralworld.net/.

IPBES. https://ipbes.net/global-assessment.

James, William. *The Varieties of Religious Experience*. New York: Penguin, 1982.

Jensen, Derrick. *Listening to the Land: Conversations about Nature, Culture, and Eros*. San Francisco: Sierra Club Books, 1995.

———. *A Language Older Than Words*. New York: Context Books, 2000.

———. *The Myth of Human Supremacy*. New York: Seven Stories Press, 2016.

Jung, C. G. *Memories, Dreams, Reflections*. New York: Vintage Books, 1989 (1961).

Kane, Sean. *The Wisdom of the Mythtellers*. Ontario: Broadview Press, 1994.

Kaufman, Ava. "Bruno Latour, the Post-Truth Philosopher, Mounts a Defense of Science." *The New York Times*. October 25, 2018. https://www.nytimes.com/2018/10/25/magazine/bruno-latour-post-truth-philosopher science.html.

Kaza, Stephanie. *The Attentive Heart: Conversations with Trees*. New York: Ballantine, 1993.

Keller, Evelyn Fox. *A Feeling for the Organism*. New York: W. H. Freeman, 1983.

Kellert, Stephen and Edward O. Wilson, eds. *The Biophilia Hypothesis*. Wash. DC: Island Press, 1993.

Kimmerer, Robin Wall. *Gathering Moss: A Natural and Cultural History of Mosses*. Corvallis: Oregon State University Press, 2003.

———. *Braiding Sweetgrass: Indigenous Wisdom, Scientific Knowledge, and the Teachings of Plants*. Minnesota: Milkweed Editions, 2013.

———. "Returning the Gift." Center for Humans and Nature, Centerforhumansand-nature.org. May 2014. https://www.humansandnature.org/returning-the-gift.

———. "Greed Does Not Have to Define Our Relationship to the Land." Lithub.com. June 4, 2020. https://lithub.com/robin-wall-kimmerer-greed-does-not-have-to-define-our-relationship to-land/.

Klimo, Jon. *Channeling: Investigations on Receiving Information from Paranormal Sources*. Los Angeles, CA: Jeremy P. Tarcher Inc., 1987.

Kovel, Joel. *The Enemy of Nature*. New York: Zed Books, 2002.

Kripal, Jeffrey J. *The Flip: Epiphanies of Mind and the Future of Knowledge*. New York: Bellevue Literary Press, 2019.

Krohn. Elizabeth G. and Jeffrey J. Kripal. *Changed in a Flash: One Woman's Near-Death Experience and Why a Scholar Thinks it Empowers Us All*. Berkeley, CA: North Atlantic Books, 2018.

Laville, Sandra. "Human-made Materials Now Outweigh Earth's Entire Biomass." *The Guardian*. December 9, 2020. https://www.theguardian.com/environment/2020/dec/09/human-made-materials-now outweigh-earths-entire-biomass-study

Lebeaux, Richard. *Young Man Thoreau*. Amherst, MA: University of Massachusetts Press, 1977.

Lehrer, Jonah. "How the City Hurts Your Brain," *Boston Globe*, Boston.com, January 2, 2009. http://archive.boston.com/bostonglobe/ideas/articles/2009/01/04/how_the_city_hurts_yo r_brain/.

Lemire, Elise. *Black Walden: Slavery and Its Aftermath in Concord, Massachusetts*. Philadelphia, PA: University of Pennsylvania Press, 2009.

Leopold, Aldo. *A Sand County Almanac*. New York: Ballantine Books, 1970.

Lesser, Elizabeth. *The New American Spirituality: A Seeker's Guide*. New York: Random House, 1999

Levi-Strauss, Claude. *Myth and Meaning*. New York: Schocken Books, 1979.

Lynne McTaggart. "Lynne McTaggart: A Live Intention to Limit the Spread of Coronavirus." March 19, 2020. YouTube video, 24:07. https://www.youtube.com/watch?v=aTDJ15t5PPo.

Macy, Joanna. "The Ecological Self." *Sacred Interconnections*. Edited by David Ray Griffin. Albany: SUNY Press, 1990.

Marder, Matthew. "To Hear Plants Speak." In *The Language of Plants: Science, Philosophy, Literature*, edited by Monica Gagliano, John C. Ryan, and Patricia Vieira. Minneapolis, MN: University of Minnesota Press, 2017.

Martin, Calvin Luther. *In the Spirit of the Earth*. Baltimore: John Hopkins University Press, 1992.

MattKahn.org. https://mattkahn.org/.

Matt Kahn All For Love. "The Pleidian Prophecy." June 2013. YouTube video, 1:03:44. https://www.youtube.com/watch?v=-5WElscYSFw&t=785s.

———. "Thy Will Be Done," March 2020, YouTube video, 1:37:55. https://www.youtube.com/watch?v=Ct2vWHz-DM0&t=3757s.

McBay, Aric, Lierre Keith, and Derrick Jensen. *Deep Green Resistance*. New York: Seven Stories Press, 2011.

McKibben, Bill. *The End of Nature*. New York: Anchor Books, 1999.

———. *The Comforting Whirlwind: God, Job, and the Scale of Creation*. Cambridge, MA: Cowley Publications, 2005.

———. *Earth*. New York: Henry Holt & Co., 2010.

———. "Let's Agree Not to Kill Each Other." *The New York Times*, October 20, 2018. https://www.nytimes.com/2018/10/20/opinion/sunday/lets-agree-not-to-kill-one another.html

McTaggart, Lynne. *The Intention Experiment: Using Your Thoughts to Change Your Life and the World*. New York: Simon & Schuster, 2007.

Meiners, Roger E., Pierre Desrochers, and Andrew P. Moriss, *Silent Spring at 50: The False Crises of Rachel Carson*. Washington, DC: The Cato Institute, 2012.

Merleau-Ponty, Maurice. *The Primacy of Perception*. Illinois: Northwestern University Press, 1964.

Millennium Assessment Report. https://www.millenniumassessment.org/documents/document.356.aspx.pdf.

Montrie, Chad. *The People's History of Environmentalism in the United States*. London and New York: Continuum International Publishing Group, 2011.

———. *The Myth of Silent Spring: Rethinking the Origins of American Environmentalism*. Oakland, CA: University of California Press, 2018.

Mooney, Chris. "30 Years Ago Scientists Warned Congress on Global Warming. What They Said Sounds Eerily Familiar." *Washington Post*, July 11, 2016. https://www.washingtonpost.com/news/energy-environment/wp/2016/06/11/30-years

ago-scientists-warned-congress-on-global-warming-what-they-said-sounds-eerily familiar/.

Mueller, Martin Lee. *Being Salmon, Being Human: Encountering the Wild in Us and Us in the Wild*. Vermont: Chelsea Green, 2017.

Naess, Arne. "Self-Realization." In *Deep Ecology for the 21st Century*. Edited by George Sessions. Boston: Shambhala, 1996.

NASA, https://climate.nasa.gov/scientific-consensus/.

Nierenberg, Nicholas. https://www.nicolasnierenberg.com/.

NOAA, noaa.gov. August 2020, https://www.esrl.noaa.gov/gmd/ccgg/trends/data .html.

Omega Institute, "Statement Regarding John of God Allegations," eomega.org: https://www.eomega.org/about-omega/press center/statement-from-omega-regarding john-of-god-allegations.

Oppenheimer, Mark. "The Queen of the New Age," *New York Times*, May 4, 2008. https://www.nytimes.com/2008/05/04/magazine/04Hay-t.html.

Oreskes, Naomi and Erik M. Conway. *Merchants of Doubt: How a Handful of Scientists Obscured the Truth on Issues from Tobacco Smoke to Global Warming*. New York: Bloomsbury, 2010.

———. *The Collapse of Western Civilization: A View from the Future*. New York: Columbia University Press, 2014.

Oreskes, Naomi and Geofrey Supran. "What Exxon Mobil Didn't Say About Climate Change." *The New York Times*, August 22, 2017. https://www.nytimes.com/2017 /08/22/opinion/exxon-climate-change-.html.

Orr, David. *The Nature of Design*. Oxford: Oxford University Press, 2004.

Otto, Rudolf. *The Idea of the Holy*. London: Oxford, 1923.

Oxford Academic, November 5, 2019, https://academic.oup.com/bioscience/article/7 0/1/8/5610806.

Panek, Richard. *The 4 Percent Universe: Dark Matter, Dark Energy, and the Race to Discover the Rest of Reality*. New York: Mariner Books, 2011.

Partanen, Anu and Trevor Corson. "Finland Is a Capitalist Paradise." *The New York Times*. December 7, 2019. https://www.nytimes.com/2019/12/07/opinion/sunday/ finland socialism-capitalism.html.

Peters, John Durham. *Speaking into the Air: A History of the Idea of Communication*. Chicago, IL.: The University of Chicago Press, 1999.

———. *The Marvelous Clouds: Towards a Philosophy of Elemental Media*. Chicago, IL.: University of Chicago Press, 2015.

Plec, Emily. "Perspectives on Human-Animal Communication: An Introduction." In *Perspectives on Human-Animal Communication: Internatural Communication*, edited by Emily Plec, 1-11. New York: Routledge, 2013.

Plumwood, Val. *Feminism and the Mastery of Nature*. New York: Routledge, 1993.

Pollan, Michael. *The Omnivore's Dilemma: A Natural History of Four Meals*. New York: Penguin, 2006.

Postman, Andrew. "My Dad Predicted Trump in 1985: It's Not Orwell, He Warned, It's *Brave New World*." *The Guardian*, Feb. 2, 2017. https://www.theguardian.com/me dia/2017/feb/02/amusing-ourselves-to-death-neil postman-trump-orwell-huxley.

Postman, Neil. *Amusing Ourselves to Death: Public Discourse in the Age of Show Business*. New York: Penguin, 1985.

Premack, Laura. "John of God: My Encounter with Brazil's Accused Faith Healer." *The Conversation*, January 4, 2019. https://theconversation.com/john-of-god.

Project Drawdown. https://www.drawdown.org/.

Radin, Dean. *Supernormal: Science, Yoga, and the Evidence for Extraordinary Psychic Abilities*. New York: Random House/Crown Pub., 2013.

Ramsey, Ramsey Eric. *The Long Path to Nearness: A Contribution to a Corporeal Philosophy of Communication and the Groundwork of an Ethics of Relief*. New York: Humanity Books, 1998.

———. "Procreation in a Beautiful Medium: Eroticizing the Vectors of Communicative Praxis," in *Experiences Between Philosophy and Communication,* edited by Ramsey Eric Ramsey & David James Miller (New York: SUNY Press, 2003), 194–195.

———. "Communication and Eschatology." *Communication Theory*, 7(4), (2006): 343–361.

Ramsey, R. Eric & David James Miller, eds. *Experiences Between Philosophy and Communication*. New York: SUNY Press, 2003.

Reich, Wilhelm. "Character and Society." Translated by Theodore P. Wolfe. *Journal of Orgonomy* 8(2) (1974): 116–129.

Richardson, Robert D. Jr. *Henry Thoreau: A Life of the Mind*. Berkeley, CA: University of California Press, 1986.

Riley, Christopher. "The Dolphin Who Loved Me: The NASA Funded Project That Went Wrong." *The Guardian*, June 8, 2014. https://www.theguardian.com/environment/2014/jun/08/the-dolphin-who-loved-me.

Riley, Tess. "Just 100 Companies Responsible for 71% of Global Emissions, Study Says." *The Guardian*, July 10, 2017. https://www.theguardian.com/sustainable-business/2017/jul/10/100-fossil-fuel-companies-investors-responsible-71-global-emissions-cdp-study-climate-change.

Roads, Michael J. *Talking with Nature/Journey into Nature: A Michael Roads Reader*. California: HJ Kramer, 2003.

Robinson, David M. *Natural life: Thoreau's Worldly Transcendentalism*. Ithaca, NY: Cornell University Press, 2004.

Roszak, Theodore. *The Voice of the Earth*. New York: Simon & Schuster, 1992.

Sancho, Miguel and Kimberly Brown, "Arizona Sweat Lodge: The Inside Story of James Ray's Fatal Retreat," abcnews.com: https://abcnews.go.com/Primetime/james-arthur-ray arizona-sweat-lodge/story?id=11016900.

Schrag, Calvin O. *Communicative Praxis and the Space of Subjectivity*. Bloomington: Indiana University Press, 1986.

———. *The Resources of Rationality*. Bloomington: Indiana University Press, 1992..

———. *The Self After Postmodernity*. New Haven: Yale University Press, 1997.

———. *God as Otherwise Than Being: Toward a Semantics of the Gift*. Evanston, IL: Northwestern University Press, 2002.

Schulz, Kathryn. "Pond Scum: Henry David Thoreau's Moral Myopia." *The New Yorker*, 91(32), Oct. 19, 2015. https://www.newyorker.com/magazine/2015/10/19/pond scum.

Science Daily. "Using Thoreau, Scientists Measure the Impact of Climate Change on Wildflowers," March 14, 2019. https://www.sciencedaily.com/releases/2019/03/190314192645.htm.

Seed, John, Joanna Macy, Pat Flemming, & Arne Naess. *Thinking Like a Mountain: Towards a Council of All Beings*. Philadelphia: New Society Publishers, 1988.

SF Environment. "Celebrating Black Environmentalists During Black History Mont https://sfenvironment.org/article/celebrating-black-environmentalists-duringblack historymonth#:~:text=%20Celebrating%20Black%20Environmentalists%20during%20 ack%20History%20Month,his%20career%20working%20to%20solve%20social...%20 ore%20.

Sheldrake, Rupert. *Dogs That Know When Their Owners Are Coming Home*. New York: Three Rivers Press, 1999.

———. *The Sense of Being Stared At: And Other Aspects of the Extended Mind*. New York: Crown Publishers, 2003.

———. *Science Set Free: 10 Paths to New Discovery*. New York: Crown Pub., 2012.

Sheldrake, Rupert & Matthew Fox. *The Physics of Angels*. San Francisco: Harper San Francisco, 1996.

Shepard, Paul. *Nature and Madness*. Athens, GA: University of Georgia Press, 1998 (1982).

———. "Interview: The Unreturning Arrow." *Talking on the Water*. Edited by Jonathan White. San Francisco: Sierra Club Books, 1994.

———. *The Only World We've Got: A Paul Shepard Reader*, edited by Paul Shepard. San Francisco, Sierra Club Books, 1996.

———. *Coming Home to the Pleistocene*. Washington, DC: Island Press, 1997.

———. "A Post-Historic Primitivism." *Limited Wants, Unlimited Means*. Edited by John Gowdy. Washington, DC: Island Press, 1998.

Silko, Leslie Marmon. *Ceremony*. New York: Penguin Books, 1977.

Simmons, Nancy Craig. "Speaking for Nature: Thoreau and the 'Problem' of Nature Writing." In *Thoreau's Sense of Place: Essays in American Environmental Writing*. Edited by Richard J. Schneider. Iowa City: University of Iowa Press, 2000.

Slater, Lauren. "Monkey Love." *The Boston Globe*. Boston.com news, March 21, 2004. http://archive.boston.com/news/globe/ideas/articles/2004/03/21/monkey_love/.

Smith, Harmon. *My Friend, My Friend: The Story of Thoreau's Relationship with Emerson*. Amherst, MA: University of Massachusetts Press, 2001.

Snow, C. P. *The Two Cultures and A Second Look*. Cambridge: Cambridge University Press, 1959.

Spangler, David & William Irwin Thompson. *Reimagination of the World: A Critique of the New Age, Science, and Popular Culture*. Sante Fe: Bear & Co., 1991.

Stager, Curt. *Deep Future: The Next 100,000 Years of Life on Earth*. New York: St. Martin's Press, 2011.

Steingraber, Sandra. *Living Downstream: An Ecologist's Personal Investigation of Cancer and the Environment*. Cambridge, MA: Da Capo Press, 2010.

———. Steingraber.com. http://steingraber.com/category/civil-disobedience/.

Stoknes, Per Espen. *What We Think About When We Try Not to Think About Global Warming*. White River Junction, VT: Chelsea Green, 2015.

Strogatz, Steven H. *Sync: How Order Emerges from Chaos in the Universe, Nature, and Daily Life*. New York: Hyperion, 2003.

Sullivan, Robert. *The Thoreau You Don't Know: What the Prophet of Environmentalism Really Meant*. New York: HarperCollins, 2009.

Sutcliffe, Steven J. *Children of the New Age: A History of Spiritual Practices*. New York: Routledge, 2003.

Taylor, Dorceta. "Interview: How Green Groups Became So White." Yale Environment 360. https://e360.yale.edu/features/how-green-groups-became-so-white-and-what-to-do-about it.

TED. Suzanne Simard. "How Trees Talk to Each Other." Aug. 30 2016. YouTube video, 18:25. https://www.youtube.com/watch?v=nH2BLg43QSQ.

Theharmonicconvergence2020.org: https://www.theharmonicconvergence2020.org/.

Thomas, Leah. "Why Every Environmentalist Should Be Anti-Racist." *Vogue*. June 8, 2020. https://www.vogue.com/article/why-every-environmentalist-should-be-anti-racist.

Thoreau, Henry David. *The Heart of Thoreau's Journals*. Edited by Odell Shepard. New York: Dover Pub., 1961

———. *The Portable Thoreau*. Edited by Carl Bode. New York: Penguin Books, 1975.

———. *Cape Cod*. New York: Penguin Books, 1987.

———. "Walking" in *Nature/Walking*. Edited by John Elder. Boston: Beacon Press, 1991.

———. *Material Faith: Thoreau On Science*. Edited by Laura Dassow Walls. New York: Houghton-Mifflin, 1999.

———. *The Maine Woods*. Edited by Jeffrey S. Cramer. New Haven, CT: Yale University Press, 2009.

———. *The Portable Thoreau*. Edited by Jeffrey S. Cramer. New York: Penguin Books, 2012.

———. "Slavery in Massachusetts," American Transcendentalism Web. https://archive.vcu.edu/english/engweb/transcendentalism/authors/thoreau/slaver mass.hml.

Tom Tom Foundation. "Panel discussion: Is there life after death, at the University of Virginia with scholars from the Department of Perceptual Studies." May 30, 2018, 1:41, YouTube video. https://www.youtube.com/watch?v=4RGizqsLumo&t=11s.

Trungpa, Chogyam. *Cutting Through Spiritual Materialism*. Boston: Shambhala, 1987.

Tompkins, Peter and Christopher Bird. *The Secret Life of Plants*. New York: Harper & Row, 1973.

Turner, Frederick. *Beyond Geography: The Western Spirit Against the Wilderness*. New Brunswick: Rutgers University Press, 1992 (1980).

Underhill, Evelyn. *Mysticism: The Preeminent Study on the Nature and Development of Spiritual Consciousness.* New York: Penguin/Image Classics, 1990 (1911).

Union of Concerned Scientists. August 2020. https://www.ucsusa.org/resources/1992-world scientists-warning-humanity.

Vatican. http://www.vatican.va/content/francesco/en/encyclicals/documents/papa francesco_20150524_enciclica-laudato-si.html.

Vox. "Lake Erie Now has Legal Rights, Just Like You." April 29, 2019. YouTube video, 8:08. https://www.vox.com/future-perfect/2019/2/26/18241904/lake-erie-legal-rights personhood-nature-environment-toledo-ohio.

Vulliamy, Ed. "Seattle Fears Green Rage." *The Guardian*, November 27, 1999. https://www.theguardian.com/world/1999/nov/28/wto.theobserver.

Waldman, Scott and Benjamin Hulac, "This is when the GOP turned away from climate policy," *E&E News*, December 5, 2018. https://www.eenews.net/stories/1060108785.

Waldman, Scott and Nijna Heikkenen, "As Climate Scientists Speak Out, Sexist Attacks Are on the Rise." *E&E News*, August 22, 2018. https://www.scientificamerican.com/article/as climate-scientists-speak-out-sexist-attacks-are-on-the-rise/.

Walls, Laura Dassow. *Seeing New Worlds, Henry David Thoreau and Nineteenth Century Natural Science.* Madison, WI: The University of Wisconsin Press, 1995.

———. "Introduction" in *Material Faith: Thoreau On Science*, edited by Laura Dassow Walls. New York: Houghton-Mifflin, 1999.

———. *Henry David Thoreau: A Life.* Chicago: The University of Chicago Press, 2017.

Walsh, Neale Donald. *Conversations with God: An Uncommon Dialogue, Book 1.* New York: J. P. Putnam & Sons, 1995.

Walsh, Roger. *The Spirit of Shamanism.* New York: Tarcher/Putnam, 1990.

Walter, Sandra. Sandrawalter.com, retrieved January 2020. https://sandrawalter.com/.

Watkins, Alan and Ken Wilber. *Wicked and Wise: How to Solve the World's Toughest Problems.* Great Britain: Urbane Publications, 2016.

Welwood, John. "Human Nature, Buddha Nature: An Interview by Tina Fosella." *Tricycle*, spring 2011. https://tricycle.org/magazine/human-nature-buddha-nature/.

Weston, Anthony. *Mobilizing the Green Imagination.* BC, Canada: New Society Publishers, 2012.

Whitman, Walt. *Poetry and Prose.* Edited by Justin Kaplan. New York: Library of America, 1982.

Wilber, Ken. *The Eye of Spirit: An Integral Vision for a World Gone Slightly Mad.* Boston: Shambhala, 1997.

———. *No Boundary: Eastern and Western Approaches to Spiritual Growth.* Boston, MA: Shambhala, 1979.

———. *Grace and Grit: Spirituality and Healing in the Life and Death of Treya Killam Wilber.* Boston: Shambhala, 1993.

———. *Sex, Ecology, Spirituality: The Spirit of Evolution.* Boston: Shambhala Publications, 1995.

———. *A Brief History of Everything.* Boston: Shambhala, 1996.

———. *The Atman Project*. Wheaton: Quest Books, 1996 (1980).
———. *The Marriage of Sense and Soul*. New York: Random House, 1998.
———. *One Taste*. Boston: Shambhala, 1999.
———. *A Theory of Everything*. Boston: Shambhala, 2000.
———. *The Religion of Tomorrow: A Vision of the Future of the Great Traditions*. Boston: Shambhala Publications, 2017.
———. *Trump and a Post-Truth World*. Boston: Shambhala Publications, 2017.
Wilber, Ken, ed. *Quantum Questions*. Boston: Shambhala Publications, 1984.
Wilder, Carol. "The Palo Alto Group: Difficulties and Directions of the Interactional View for Human Communication Research." *Human Communication Research*, 5(2) (Winter, 1979): 171–185.
Wilshire, Bruce. *Wild Hunger: The Primal Roots of Modern Addiction*. Maryland: Rowman & Littlefield, 1998.
Wohllenben, Peter. *The Hidden Life of Trees: What They Feel, How They Communicate*. Vancouver: Greystone Books, 2015.
Yale Climate Communication. https://climatecommunication.yale.edu/publications/the-francis effect/.
Yale Divinity School. "YDS Climate Panel Pushes Past Denial and Despair." March 14, 2019. https://divinity.yale.edu/news/yds-climate-panel-pushes-past-denial-anddespair.
Zimmerman, Michael. *Contesting Earth's Future: Radical Ecology and Postmodernity*. Berkeley: The University of California Press, 1994.
———. "A Transpersonal Diagnosis of the Ecological Crisis." *Ken Wilber in Dialogue*. Edited by Donald Rothenberg and Sean Kelly. Wheaton: Quest Books, 1998.
Zirin, *Welcome to the Terrordome: The Pain, Politics, and Promise of Sports*. Chicago, IL: Haymarket Books, 2007.

Index

About the Author

William Homestead is an associate professor in the Communication Studies Department at New England College and has had a long association with the Ometeca Institute, a nonprofit devoted to the integration of the sciences and the humanities. His work with Ometeca, along with his interdisciplinary degrees (MA in communication studies, MS in environmental studies, MFA in creative writing), study with a spiritual teacher, and hiking experiences, provided much of the insight and inspiration for writing *An Ecology of Communication*. He is also a member of the International Environmental Communication Association (IECA) and The Thoreau Society and has been teaching his course, *The Voice of Nature*, for many years. He lives in Vermont and spends much time walking in the woods with his dog, Snoopy, who was named by his three children.

www.ingramcontent.com/pod-product-compliance
Ingram Content Group UK Ltd.
Pitfield, Milton Keynes, MK11 3LW, UK
UKHW012118150926
475560UK00002B/172

9 781793 618146